INTEGRATIVE BIOPHYSICS

Integrative Biophysics

Biophotonics

Edited by

Fritz-Albert Popp
*International Institute of Biophysics,
Neuss, Germany*

and

Lev Beloussov
*Faculty of Biology,
Moscow State University,
Moscow, Russia*

KLUWER ACADEMIC PUBLISHERS
DORDRECHT / BOSTON / LONDON

A C.I.P. Catalogue record for this book is available from the Library of Congress.

ISBN 1-4020-1139-3

Published by Kluwer Academic Publishers,
P.O. Box 17, 3300 AA Dordrecht, The Netherlands.

Sold and distributed in North, Central and South America
by Kluwer Academic Publishers,
101 Philip Drive, Norwell, MA 02061, U.S.A.

In all other countries, sold and distributed
by Kluwer Academic Publishers,
P.O. Box 322, 3300 AH Dordrecht, The Netherlands.

Cover Image: Arrangement of Gurwitsch's experiment with onion roots

Printed on acid-free paper

All Rights Reserved
© 2003 Kluwer Academic Publishers
No part of this work may be reproduced, stored in a retrieval system, or transmitted
in any form or by any means, electronic, mechanical, photocopying, microfilming, recording
or otherwise, without written permission from the Publisher, with the exception
of any material supplied specifically for the purpose of being entered
and executed on a computer system, for exclusive use by the purchaser of the work.

Printed in the Netherlands.

TABLE OF CONTENTS

Introduction to integrative biophysics
M. Bischof 1

Bio-electromagnetism
G. J. Hyland 117

Developmental biology and physics of today
L. V. Beloussov 149

Cellular and molecular aspects of integrative biophysics
R. van Wijk 179

Physical basis and applications of delayed luminescence
F. Musumeci 203

Biological effects of electromagnetic fields on living cells
J.-J. Chang 231

Detectors for the quantized electromagnetic field
J. Swain 261

Detection of photon emission from biological systems
X. Shen 287

*Photon emission from perturbed and dying organisms
– the concept of photon cycling in biological systems*
J. Slawinski 307

*Mitogenetic radiation, biophotons and non-linear
oxidative processes in aqueous media*
V. Voeikov 331

Biophotons: Ultraweak photons in cells
H. J. Niggli, L. A. Applegate 361

*Biophotons - background, experimental results,
theoretical approach and applications*
F.-A. Popp 387

The physical basis of life
R. P. Bajpai 439

Definition of consciousness. Impossible and unnecessary?
M. Lipkind 467

Chapter 1

INTRODUCTION TO INTEGRATIVE BIOPHYSICS

Marco Bischof
International Institute of Biophysics,
 Kapellener Str., 41472 Neuss, and
Future Science & Medicine,
Gotlandstr.7, D-10439 Berlin, Germany.
e-mail: mb@marcobischof.com

1. INTRODUCTION

If you look through a number of contemporary textbooks and introductions of biophysics, in order to find out what exactly biophysics is and what the field covers, you will be quite bewildered. Even the titles of the books with their diversity of names for the field, which include, besides biophysics itself, *"Medical Physics", "Medical and Biological Physics", "Physical Biology", "Physical Bases of Medicine and Biology", and "Molecular Biology"*, already show the existence of different interpretations and tendencies.

Looking at the tables of content, we find that some books have the application of physical devices and measurements to physiological problems as the only subject of biophysics. Some others are structured according to the different categories of physical phenomena, i.e. the mechanical, thermal, acoustical, electromagnetic and nuclear, gaseous, fluid etc. aspects of living systems, or according to the different functional systems in the organism. In not a few instances, life processes such as transport processes, chemical reaction kinetics, the acid-base-balance, and diffusion processes are treated without reference to the organizational level concerned. In general - with a few notable exceptions - the notion of a hierarchy of subsystems within the biological system, and any kind of overall view of the organism as a whole, is completely absent or relegated to the introduction. Rarely we find a chapter on theoretical biophysics aimed at a synthesis of all these partial aspects of the biophysical treatment of organisms.

Most of the very recent textbooks are dominated by, if not exclusively devoted to, investigations of molecular structures and processes, proteins, nucleic acids, genetic mechanisms, and membrane processes. One of the major modern textbooks, *"Biophysics"* by W.Hoppe et al. (1982), has a clear emphasis on the molecular level (macromolecules, enzymes, molecular interactions, energy transfer etc.). Much less space is devoted to the subcellular and cellular levels (organelles, cell architecture, membranes, neurobiophysics), and even less to the level of organs, organ systems and whole organisms (in subjects such as parts of biomechanics, cybernetics) or to whole populations and the interaction of organisms with the environment (environmental biophysics). Physical methods of investigation take very extended space ($1/6$ of the 950 pages of the volume), while, typically, the treatment of the biological effects of electromagnetic fields is restricted to ionizing radiation and that of electrical fields and currents in the organism to membrane potentials, electroreception in fish, and, surprisingly, the control of differentiation and growth by ionic currents.

We find that biophysics covers a very disparate field of investigations; no two textbooks agree on its scope and range, and no one seems to cover all the relevant questions. A notable exception is Otto Glasser's *"Medical Physics"* of 1944 (with two more volumes in 1950 and 1960), which to my knowledge still remains one of the most complete treatments of the full range of biophysical investigations, not surpassed in its scope by anything later; it also includes physical therapies, mitogenetic radiation (now called biophoton emission), the biological effects of light and electromagnetic fields, and a chapter on bioelectric fields by H.S.Burr. The statement of J.R.Loofbourow made in 1940 that there is "no clear agreement, even among biophysicists, as to what the term biophysics means" (Loofbourow, 1940) basically is still valid. At the beginning of the 2nd Russian Biophysical Congress held in August 2000 in Moscow, the luminaries of Russian biophysics once again discussed the question "What is biophysics ?". As it soon became clear that nobody was able to give a strict definition of biophysics, someone in exasperation and half jokingly offered the following one: "Biophysics is the research published in the journals of biophysics". In other words, the situation has not much changed since the 1940's, except for the fact that in the last decades biophysics has so much become identified with molecular biology that today there may not be many biophysicists who are aware that there are other legitimate kinds of biophysics. Just recently, the journal *Nature* reported in an editorial about the heavily funded effort of the total genetic mapping, sometimes called by the military epithet of a "Manhattan project of the life sciences", as the "new physics-biology agenda" of US science (N.N., 1999). This not-so-new agenda actually goes back to the 1930's, when the pioneers of molecular biology first conceived their program of total technological control over life (Kay, 1993).

However, not all biophysicists feel comfortable with the usurpation of biophysics by the reductionistic, molecular-genetic approach to biology, and the number of those who develop and support an entirely different kind of "new biophysics" is steadily increasing. A look into the history of biology and biophysics shows that there always have been alternative traditions to the approach now dominating these fields, giving support to the concept of "integrative biophysics" that is proposed here.

I would like to point out the significance of the recent parallel trend towards an "integrative medicine" to which contribute, besides biomedicine, many other fields of knowledge and healing practice, such as social and environmental medicine, psychosomatic medicine and consciousness research, naturopathic medicine, homeopathy and non-Western medical systems, transpersonal psychotherapy, methods of "body work", yoga, meditation, biometeorology and geomedicine, chronobiology, and not least, biophysical models and methods (Bischof, 2000b; Milburn, 1994, 2001). In a similar way, many different aspects of the study of life are contributing to the emerging field of integrative biophysics.

2. THE CONCEPT OF INTEGRATIVE BIOPHYSICS

Since there is no agreement as to what constitutes biophysics, we may feel free to attempt a redefinition of the field, based on several recent trends converging towards such a concept. As we have already pointed out in our first attempt at such a redefinition (Bischof, 2000a), traditional biophysics has up to now been based on classical physics and equilibrium thermodynamics, and thus mainly needs a redefinition in terms of the revolution brought about by the last few decades of quantum-mechanical experiments and interpretations, and of non-equilibrium thermodynamics. It indeed will entail a revolution based on physical concepts – however, not of the kind alluded to in the *Nature* editorial. While the physical view will be fundamental, it will not be that of classical physics, and the goal will not be the reduction of biology to physics, but an understanding of the physics of the living, and physics must not replace, but support profound biological understanding. As many authors have predicted, the new biophysics may even lead to the recognition that the study of life can yield insights into basic physical laws more fundamental than those obtained from the investigation of nonliving matter, and thus may become a new field of fundamental research in physics. In contrast to molecular biology, the new biophysics should be more than just an empirically based bioengineering technology; it will need epistemological and philosophical foundations. Its goal should be to develop an adequate theory of life, and it should balance the mastery of life with the understanding of life.

As a complement to the onesidedness of the molecular approach, the new biophysics will focus on holistic aspects of organisms, and will attempt to provide a vision able to synthesize the wealth of molecular details accumulated by molecular biologists. The most significant feature of quantum theory for biology is its intrinsic fundamental holism. As Primas points out, the atomistic-molecular view of matter and the reductionistic-mechanist philosophy have no longer any scientific foundation, according to the actual understanding of quantum theory (Primas, 1990, 1999). The description of reality by isolated, context-independent, elementary systems such as quarks, electrons, atoms, or molecules is only permissible under certain specific experimental conditions, and these entities cannot in any way be considered as "fundamental building stones" of reality. Besides the molecular one, there are other, fundamentally different descriptions, complementary to the molecular one, which are quantum-theoretically equivalent and equally well founded. Quantum theory is much richer in possibilities than is admitted in the worldview of molecular biology.

For quantum mechanics, the scientific theory most widely recognized as fundamental and best confirmed by experiment, material reality forms an unbroken whole that has no parts. These holistic properties of reality are precisely defined mathematically by the Einstein-Podolsky-Rosen (EPR) correlations (Einstein, Podolsky, & Rosen, 1935; Bell, 1987). In quantum mechanics, it is never possible to describe the whole by the description of parts and their interrelations. This holistic view of quantum theory, although the phenomena on which it is based are not yet completely understood theoretically, cannot be rejected anymore because the strange EPR quantum correlations of non-interacting and spatially separated systems have been amply demonstrated in many experiments (Aspect et al., 1982; Aspect & Grangier, 1986; Selleri, 1988; Duncan & Kleinpoppen, 1988; Hagley et al., 1997). Therefore the world-view of classical physics, atomism and mechanistic reductionism, definitely cannot anymore be the basis of our worldview, and of biophysics. Quantum mechanics has established the primacy of the unseparable whole. For this reason, the basis of the new biophysics must be the insight into the fundamental interconnectedness *within* the organism as well as *between* organisms, and that of the organism *with the environment*. The new biophysics therefore will be an *integral*, i.e., holistic biophysics.

For the same reason it must also be an *integrative*, i.e., inter- or transdisciplinary, discipline, and has to truly integrate biological, biochemical and medical expertise into its physical models, but should also connect to knowledge from fields such as geophysics, biometeorology, heliobiology etc. The determination of biophysical parameters of the organism, like that of molecular processes in biochemistry or molecular biology or the procedures of any other specialized approach, has no value in itself, but only in the context of the whole range of biological knowledge. Excellence in biophysics will always be based on a good grasp of biological and medical base knowledge – and a "feeling for the organism" (Fox-Keller, 1983) - , and starting from there it will then be possible to transcend the onesided anatomical, molecular-biological approach and to expand its scope by using biophysical concepts and methods. The necessity and justification of an integrative approach results also from quantum theory, which teaches that science from now on has to get accustomed to the simultaneous use of different complementary viewpoints and methods, and cannot anymore in good conscience be practiced from a single approach only.

The new biophysics will be based on quantum theory, and not classical mechanics – therefore it may also be called "quantum biology" –, and also, instead of equilibrium thermodynamics, it must refer to non-equilibrium thermodynamics. Organisms clearly are open systems far from equilibrium. Other central concepts and features of the new biophysics will be coherence, macroscopic quantum states, long-range interactions, non-linearity, self-organization and self-regulation, communication networks, field models, interconnectedness, non-locality, and the inclusion of consciousness.

I postulate that field thinking and field models will have to be one of the central elements of the new biophysics, as a complement to the molecular view, as a means to synthesize the wealth of its details, and as an instrument to adequately model thinglessness, interconnectedness and non-locality – therefore bioelectromagnetics will play a central role in the new biophysics. However, recent experiments have shown that the existence of hitherto unknown, non-electromagnetic fields in and between organisms cannot be excluded. Of course, the field aspect of the organism has to be seen in close connection, and constant interaction, with the solid aspect. Attention should also be paid to the field aspect of biochemical processes, for example in collective processes, reactivity and molecular recognition.

I suggest that the existence of a pre-physical, unobservable domain of potentiality in quantum theory, which forms the basis of the fundamental interconnectedness and wholeness of reality and from which arise the patterns of the material world, may provide a new model for understanding the holistic features of organisms, such as morphogenesis and regeneration, and thus provide a foundation for holistic biophysics – therefore I propose that the usefulness of the theories of the physical vacuum for understanding the phenomena of life is investigated - one important aspect of their usefulness may be as a link between the domain of biology and consciousness.

I postulate that the new biophysics needs to extend its interdisciplinarity even beyond natural science. Consciousness cannot be excluded anymore from biophysics, although the difficulties of such an extension should not be underestimated. There is now enough evidence showing that consciousness is a causal factor in biology and not just an inconsequential epiphenomenon. Starting from the analysis of the phenomenology and the experimental evidence for mind-body interaction, field models and vacuum theories may provide the necessary tools for bridging the mind-body gap. Integrative biophysics is also in a good position to investigate the physical bases of the subjective field experiences of interpersonal fields, field-like states of consciousness and streaming feelings in the body experienced in various altered states of consciousness, occurring, e.g., in the practice of yoga, meditation, Chinese medicine, body therapies, shamanism, spiritual healing etc. Among other things, it may be interesting to investigate the correlation of normal and extraordinary states of consciousness to the biophoton emission of the person during these states, as well as with EEG and bioelectric parameters, for instance. However, in order to attain the goal of holistic understanding it is also necessary to acknowledge the limits of the scientific approach, and value the goal of understanding highly enough to include non-observables into our models, if this supports understanding.

As to the methods used, integrative biophysics will mainly use biophotonic and other bioelectromagnetic techniques, but any other methods suitable for the investigation of holistic functions and features of living organisms, such as self-regulation in growth and healing, morphogenesis etc., will be used as well. Concerning the applications, the development of non-invasive methods to assess the functional state of whole and intact organisms, and that of the organism's subsystems in the context of the whole organism, is its main interest. Measurements of the organism's own fields are well suited to make internal regulation processes accessible at the surface of the body.

3. HISTORY OF THE CONCEPT

Biophysics is a relatively young field of science. Systematic scientific investigations by means of physical instruments, and the explanation of their results by physical and mathematical concepts, have been carried out since about 1840, and as a separate discipline it has established itself only since about the 1920's, if not the 1940's. However, physical approaches to the understanding of life certainly have existed before that time, and so have holistic or integral/integrated biophysical approaches and practices which can be taken as exemplary antecedents of an integrated biophysics.

3.1 Exemplary Pioneers

First I would like to summarize the work of some early pioneers whose work may be considered as inspiring exemplars of integrative biophysics, namely Jagadis Chunder Bose, Alexander L. Chizhevsky, Vladimir I.Vernadsky, Ludwig von Bertalanffy, Walter Beier, and Solco W. Tromp, and to describe two exemplary institutional efforts to develop a new understanding of life based on physical concepts. After this, I will give a historical overview over the development in this field, mainly in the 20th century, and will try at the same time to delineate the main topics of the integrative approach to life.

3.1.1 Jagadis Chunder Bose

The first Indian scientist to receive international recognition, Jagadis Chunder Bose (1858-1937), was a biophysicist before biophysics existed as such (Bose, 1902, 1907, 1915; Geddes, 1930; Susskind, 1970; Dasgupta, 1999). Educated in London and Cambridge in the 1880's, he became a brilliant inventor of instruments, which he used to perform delicate experiments first in physics, then in plant physiology. His research career started in 1894 with work on the electromagnetic waves discovered in 1888 by Heinrich Hertz, mainly investigations of polarization, refraction, reflection and other optical properties. He improved on the "coherer" invented by Eduard Branly in 1890 and perfected by Oliver Lodge, a receiver and detector of electric waves, and was one of the first to produce waves of very short wavelengths (microwaves in modern language). Besides the microwave generator, he also designed microwave lenses, wave guides, electromagnetic horns, and an artificial retina, and first showed that semiconductor rectifiers could detect radio waves. Although generally not mentioned as such in histories of the field, Bose was also one of the pioneers of radio telegraphy; he made experiments proving the feasibility of radio transmission at about the same time as Marconi and Lodge, but never actually tried to transmit any signals.

In connection with his coherer work he discovered in 1899 that the molecular structure of some metals changed under the influence of electric waves, either by increasing or decreasing their electrical resistance; some of them even showed automatic recovery from this change. Another effect was the complete loss of sensitivity to the electric waves which he called "fatigue". He christened this property of materials "electric touch" and ascribed it to a structural modification of surface materials. This work which makes him a pioneer of solid-state physics, already contained the first hints of the subsequent venture into the investigation of certain analogies between the behaviour of inorganic and organic matter. This research occupied Bose from 1900 onwards, and also marked his crossing-over from physics to biology. The concept of fatigue, probably created in analogy to the well-known mechanical fatigue of metals, subsequently carried him into ascribing a common responsiveness to the inorganic and organic worlds, and attributing life-like properties to inanimate matter, thus blurring the boundary between the living and the nonliving. After having proposed this hypothesis in 1900 in a paper read to the International Physical Congress in Paris, he tested it in in a series of experiments conducted in 1900-1902 which showed that not only elementary metals, but also metallic compounds and non-metals such as chlorides, bromides, oxides and sulphides, exhibit similar effects. He also compared the behaviour of these inorganic substances under stimulation to the response of animal muscle, and they showed very similar reaction curves. Bose concluded that metals, when subjected to a succession of electrical vibrations, like muscle undergo the protracted state of fatigue physiologists call "tetanus". In a similar way as drugs and stimulants modify the response of living matter, inorganic matter also could be "poisoned" and then be "revived" by an antidote. The response curves of living and inanimate matter were so similar that one could not distinguish between them.

Bose concluded that there was a continuity between the living and the nonliving. In line with an ancient tradition in physiology, he regarded response (Albrecht von Haller's "irritability"), especially electric response, as the criterion of life. As inorganic matter was also responsive, there must be a continuity between living and nonliving. However, while he regarded electrical response as a sufficient condition for life, physiologists of his time like Augustus Waller made it clear this was only a necessary condition. This made Bose a vitalist in the eyes of many colleagues, but he himself rejected and attacked vitalism and certainly was not a vitalist, but also not content to be an ordinary mechanist. His idea was rather that "the responsive processes seen in life are foreshadowed in non-life" (Bose, 1920), and he tended to ascribe life to all matter. His provocating theses caused a heated response by physiologists denying that metal responds in the way that living matter does, and rejecting that responsiveness of ordinary plants could be compared to that of "sensitive plants" like mimosa, the main experimental subject used by Bose, and that of animal muscle.

In the time of 1903-1913, Bose became a plant physiologist which he remained for the rest of his life. As he had pioneered India's research in modern physics in 1895, he now became his country's pioneer in modern plant physiology. He set out to systematically investigate the effect of stimuli on plants by exact methods of measurement and registration. He used mechanical, electrical, chemical and thermal stimuli to elicit mechanical, chemical and electrical responses, and studied the velocity of transmission of the excitatory wave, the electrical response and the electrophysiology of plants in general, and the growth and movement of water in plants, among other things (Bose, 1907, 1915, 1923). His conclusion was that all plants are excitable, at least to electrical stimulation - not only the sensitive ones that possess motile organs. In contrast to the prevailing view that the transmission of excitation was due to mere mechanical movement of water in the plant, Bose was convinced that the stimulus was transmitted by protoplasmic change, as in animal tissue; for him, living organisms were not just mechanical stimulus-response devices, but had a kind of memory. As a crucial test for this hypothesis he regarded the experiment whether the polar effects, found by the German physiologist Eduard Pflüger in animal tissue, could also be found in plants. According to Bose, this would be inexplicable by the hypothesis of water movement. From the positive outcome he concluded that a basic property of protoplasm common to plants and animals must underlie the similarity of response to stimuli.

When in 1913 the Royal Society published the first time one of Bose's biological writings in their *"Philosophical Transactions"*, this signalled the formal acceptance of the Indian scientist as a plant physiologist by the English biological establishment, and initiated the period of his success. In 1917, he was knighted by King George V. and got his own research institute in Calcutta, the Bose Research Institute that still exists. In 1920, he was elected as a fellow of the Royal Society. However, the rest of his life was almost exclusively devoted to defending his theses and was characterized by constant controversies with his fellow scientists which separated the biologists into "Bosephiles" and "Bosephobes" (Pierce, 1927). His peers generally admired the originality of the instruments he invented and were overwhelmed by the ingenuity and wealth of his experiments. But most of them were exasperated or even outraged by the interpretation he gave his results.

Today, the judgment on his scientific merits should not be as harsh at it was then. As to his work on plants, the 1970 edition of the *"Encyclopedia Britannica"* wrote that "his outstanding work was so much in advance of his time that the precise evaluation of it was controversial" (Vol.3, 1970). Concerning his ideas, a contributor to the *Dictionary of Scientific Biography* (Susskind, 1970) wrote that "today when biophysics is a generally recognized discipline..., [Bose's ideas] seem less controversial and may even be taken as foreshadowing Norbert Wiener's cybernetics".

Bose's work was highly interdisciplinary and in his complex personality he managed to bridge many contradictory tendencies in a way that we recognize today as conducive to scientific creativity. He embodied a highly creative synthesis between his contemplative and metaphysical heritage as an Indian native, and the experimental/mechanistic side of his Western scientific training. For Bose, a friend of the poet Rabindranath Tagore and the mystic Swami Vivekananda, scientist and poet or mystic were not separate and contradictory callings; he regarded scientific and poetical understanding as each incomplete without the other. Also, he held that the scientist, like the poet and the mystic, should strive to see "beyond the expressed" and take the reality of the unmanifest behind the appearances into account.

3.1.2 Alexander L. Chizhevsky

The Russian biophysicist Alexander L. Chizhevsky (1897-1964, sometimes also quoted as "Tchijevsky") was a pioneer in the study of the effects of physical factors of atmospheric and cosmic origin on life processes; he is recognized as the founder of heliobiology and of the investigation of the effects of air ions on life in general and on human health and behavior (Chizhevsky, 1930, 1936, 1940, 1968, 1973; Sigel, 1975; Yagodinskii, 1987). Referring to Claude Bernard, he spoke of the "cosmic milieu" in which the earth existed, and of the "atmospheric milieu" which included atmospheric electricity. From 1920 to 1940, he established in extensive statistical studies the correlation of the sunspot cycle with many phenomena in the biosphere; he showed that the physical fields of the earth should be included among the basic causes affecting the condition of the biosphere. In 1935 he discovered the metachromasia of bacteria, now known as the Chizhevsky-Velkhover effect, which makes it possible to predict solar emissions influencing man on earth as well as in space. From 1919 to 1930, Chizhevsky was the first to experimentally determine the opposite physiological effect of negative and positive air ions on living organisms, the pathological effect of air with reduced ion concentration, and the stimulating effect of negative air ions. In 1930-36 he developed and propagated the use of artificial aerionization in medicine, agriculture, and stock raising. In the 1930's, Chizhevsky also investigated the electrodynamic processes in flowing blood (Chizhevsky, 1936). He showed that the organism's colloids, of which, among other things, the blood is composed, precipitate (ions with electric charge opposite to the charge of colloids) or stabilize (same charge) under the influence of air ions. He demonstrated not only that negative ions shifted the pH to the alkaline side, but also that they stimulated the mitogenetic radiation of blood, while positive ions inhibited it. In 1939, Chizhevsky was elected (in absentia) as president of the International Congress on Biological Physics and Space Biology in New York, in recognition of the fundamental contributions to the biophysics of his time.

3.1.3 Vladimir I. Vernadsky

Although not a biophysicist in the narrow sense, the Russian geologist, mineralogist and one of the founders of geochemistry and biogeochemistry, Vladimir I.Vernadsky (1863-1945), one of the leading Russian intellectual figures of the 20th century, must be considered as another one of the forefathers of integrative biophysics (Vernadsky, 1954-60, 1998; Bailes, 1990). Working mainly in the 1920's, 1930's and 1940's, Vernadsky has proposed a teleological concept of life which can be considered as the basis of ecology and system theory. He showed the function of the distribution and migration of chemical elements in the earth's crust and of their concentration in mineral deposits, and demonstrated the essential significance of "living substance" for these geological processes. From his school originated many new directions of the geological and biological sciences, of ecology, biophysics and bioelectromagnetics. From 1914 onwards, he took up the concept of the "biosphere" from Eduard Suess and gave it a precise quantitative and qualitative meaning; with this he became one of the founders of ecology and provided the foundation of James Lovelock's "Gaia theory". He pointed out how much all life was embedded in planetary and cosmic connections and how central the influence of the sun was on all life, which was inconceivable without the interaction with the electromagnetic radiation field surrounding it. Vernadsky also created the concept of the "noosphere" which may be known from the work of Teilhard de Chardin; he saw it as a new dimension of the biosphere characterized by the evolutionary partaking of man in the natural processes.

Some of the key tenets of the pupil of Pasteur and Pierre Curie can be summarized as follows: The biosphere is a part of every living organism, which on the one hand is inseparably connected to it, on the other hand is an autonomous organism. Everything in nature is intricately connected to everything else. Life is tied in to sensitive regulation processes by means of the surrounding electromagnetic fields, including sunlight. The distribution of energy in the life processes and their deviation from the state of thermodynamic equilibrium constitute an inner space-time structure in the living organisms which is different from linear euclidian geometry. Life is characterized by asymmetries (symmetry breakings) which can be demonstrated all the way down to the molecular level (e.g., in the optical activity of substances).

Vernadsky sought to develop a holistic, interdisciplinary science in which also intuition and speculation had their legitimate places. He considered the entire planet as a whole and the life on it as a part of the geological cycles. In his famous work *"The Biosphere"* (1986) he wrote in 1926: "The living organism of the biosphere has to be studied empirically as a special body not completely reducible to known physical-chemical systems". It was not sufficient to describe the phenomena of life by mere material and energetic properties; future scientists would extend the concept of living matter with additional factors besides matter and energy. Vernadsky himself has identified one of the most essential of these new factors as information, long before cybernetics and information theory developed this concept.

3.1.4 Ludwig von Bertalanffy

One of the most important pioneers of integrative biophysics may well be the Austrian-Canadian biologist Bertalanffy (1901-1972), whose 100th birthday was celebrated in 2001 (see //www.bertalanffy.org). When he started his life as a scientist in Vienna, biology was involved in the famous mechanism-vitalism controversy (Bischof, 1995b). As a biologist with philosophical leanings, he became already in the early 1920's "puzzled about the obvious lacunae in the research and theory of biology, and dissatisfied with the then prevalent reductionist-mechanist approach in science", which "appeared to neglect or actively deny just what is essential in the phenomena of life". As a consequence, he advocated an "organismic conception in biology which emphasizes consideration of the organism as a whole or system, and sees the main objective of the biological sciences in the discovery of the principles of organization at its various levels" (Bertalanffy, 1968). His goal as a scientist was to overcome specialization and reductionism by view on the whole, the "system", and an interdisciplinary and integral approach which investigates the subject in its entirety with all its relations to neighbouring disciplines.

This view also resulted from his disputes with the members of the "Vienna circle" of logical empiricists (Moritz Schlick, Rudolf Carnap, Otto Neurath, Karl Popper). He had obtained his Ph.D. in philosophy under their leader, Moritz Schlick, with a thesis on Gustav Theodor Fechner, and had participated in their discussions in the 1930's, but was never able to accept their variant of positivism. Contrary to their views, he was convinced that absolute objectivity in science was not possible and that interest in human values was fundamental for a scientist.

Since the late 1920's, Bertalanffy strived to arrive at a theoretical biology (Bertalanffy, 1927a, 1932) and, at the same time, he also started to think about the significance of the revolutionary changes in physics for biology (Bertalanffy, 1927b). His book on *"Theoretical Biology"* (Bertalanffy, 1948), first published in 1932, with a second volume in 1942, is still one of the classic and fundamental treatments of holistic biology. Its theoretical penetration of biological principles, processes and structures has prepared the way for a modern physical conception of the organism from a holistic viewpoint and for this reason – even if some details certainly are not up-to-date anymore – is still one of the fundamental texts of integrative biophysics.

In an effort to concretize the organismic program and his theory of life, and based on his experimental work on metabolism and growth, Bertalanffy advanced in 1940 his groundbreaking theory of the organism as an open system (Bertalanffy, 1940). In its basic outlines already put forth in 1926, the theory sprang from the key question that had always concerned his research: How comes, and how is it compatible with the natural laws, that organisms spontaneously develop into higher degrees of order, how do the phenomena of "self-organization" emerge without recourse to any mystical "life force"? Like the organismic school of developmental biologists in England and the USA (Bischof, 1998a), Bertalanffy was opposed to any narrow mechanistic-reductionistic approach, but equally to any kind of vitalistic principle which would situate organisms outside of the laws of physics and chemistry; any explanation had to be in line with the laws of science.

His solution came from the insight that organisms are not isolated but open to their environment; Bertalanffy transposed the already known physical concept of "open systems" to biology and demonstrated the fundamental significance of this insight. This allowed him to construct the first ever theoretical biology in the strict scientific sense, and signified his entry into biophysics. He recognized that a biophysical model of the organism required an extension of conventional physical theory by way of a generalization of kinetic principles and thermodynamic theory. At this time, it was generally assumed that the state of physical systems was that of a closed system. Statistical mechanics had given a solid foundation to the Second Law of Thermodynamics, the law of entropy; as a consequence, it was clear that isolated systems inevitably must reach a state of complete energetic equalization, thus maximal disorder and maximal entropy. From this it was deduced that every system had to gravitate towards disorder and equalization.

In 1932 Bertalanffy showed that a true equilibrium can only occur in closed systems, while in open systems, a special type of disequilibrium is the predominant and characteristic feature. He postulated that living organisms are in fact such open systems, where the tendency to increase the entropy which stemmed from the organism's own entropy production, could be compensated by an influx of negative entropy (free energy), which made entropy reduction possible. As a consequence, the organism must exist in a state fundamentally different from the equilibrium of thermodynamics, which is characterized by maximal entropy and lack of dynamic. For this state, Bertalanffy introduced the notion of "Fliessgleichgewicht" (equilibrium of flux), now usually called "steady state", which is really not an equilibrium, but a state of stationary non-equilibrium. Open systems are in continuous interaction and exchange with the environment; they continuously exchange their material constituents while keeping their structural integrity. With this concept, not only the reconciliation of biology with thermodynamics was made possible, but it also provided the physical foundation for important biological concepts such as organization, equifinality, hierarchical structure, homeostasis, and regulation. Bertalanffy considered Driesch's teleological concepts of self-regulation and equifinality, for instance, as fundamental properties of open systems. "Every healing process, every dynamic restitution following a disturbance, every return to the normal state after a clinical intervention is the re-establishment of an equifinal steady state". He equated the "equifinal self-preservation of the organism" with the "natural healing power" spoken of by physicians in the tradition of Hippocratic medicine (Bertalanffy, 1975). With the concept of the organism as an open system, Bertalanffy also triggered the further development of the thermodynamics of irreversible systems, which in the 1960's especially was brought forward by the work of Ilya Prigogine (see below).

As a consequence of the concept of the organism as an open system, Bertalanffy then developed his "General System Theory" first announced in 1945 (Bertalanffy, 1945, 1968). In it, the holistic view he had developed on the example of biological systems was extended to other disciplines, such as philosophy, psychology, chemistry, ecology, economics, history, and politics. General System Theory is an attempt to identify general laws of the behaviour and properties of systems of whatever nature, which cannot be predicted from the properties of their components. With his system theory, Bertalanffy wanted to overcome the mechanistic conception of the sciences and to establish a humanistic alternative on the basis of scientific laws. For this reason, he always remained very critical towards the many mechanistic variants of system theory, such as systems engineering, operations research, and even Norbert Wiener's cybernetics. With his systems approach, Bertalanffy also prepared the introduction of the concept of complexity into biology, and he already recognized the role of nonlinearity which can lead to chaotic, unpredictable dynamics. He was also a pioneer of the mathematical penetration of systemic thinking, and with his use of mathematical methods in biology he became a pioneer of modelling and mathematical biophysics, as later developed by Nicolas Rashevsky and Robert Rosen, among others. Rashevsky and Rosen promoted in the late 1950's, 60's and 70's the use of relational and similarity considerations and optimization principles in biophysics (Rashevsky, 1960, 1962; Rosen, 1958-59, 1967, 1970, 1972).

3.1.5 Walter Beier

In this tradition stood also the work of Walter Beier and his school at the Institute of Biophysics of the University of Jena (GDR) in the 1960's and 70's. Applying the findings of the school of Bertalanffy to biophysics, Beier has developed a systemic, comprehensive approach always aiming at the wholeness of the system considered. In this approach, a system is not viewed as a conglomerate of parts, and it does not investigate isolated phenomena, but rather the wide field of interactions. Biophysics is understood as a complex, interdisciplinary branch of science, born as a result of the striving for synthesis of the sciences separated by the classical demarcation lines, and of the development of new disciplines sited at the border between classical disciplines. "The nature and the object of contemporary biophysics follow from the striving for comprehensive knowledge on the problems of life, and therefore are intimately touching the original meaning of physics as the natural science per se, even though its development has first turned towards nonliving nature. The efforts at the conceptual integration of biology and physics contain one of the most exciting problematics of our century" (Beier, 1965). While including the investigation at the molecular and cellular levels of organization, Beier stresses that those at the systemic level should also be included. He also emphasizes the central importance of theoretical studies in biophysics. He bases his theoretical approach on the work of Timofeeff-Ressovsky, Bertalanffy, and Rashevsky, who in his view are responsible for the introduction of mathematical-theoretical considerations into biophysics. However, Beier states that theoretical biophysics is still far from the goal of creating an "axiomatics of biology". His approach can be used as one of the starting points for the development of modern integrative biophysics.

3.1.6 Solco W. Tromp

The Dutch geologist and micropalaeontologist Solco W. Tromp (1909-) became in the late 1950's and 1960's one of the leading biometorologists and founder of the International Society of Biometeorology, one of the research organizations that may serve as a model for the organization of work in the field of integrative biophysics. Before that time, he published a unique and unconventional book containing what may be called an early concept of integrative biophysics (Tromp, 1949). In *"Psychical Physics"*, Tromp attempts to give an explanation of dowsing, radiesthesia, hypnosis and other altered states of consciousness, certain distant influences between organisms and the homing instinct of animals, by an analysis of the influence of external electromagnetic fields on psychic and physiological phenomena in living organisms. In the first two chapters, the voluminous book comprises a comprehensive review of "electromagnetic fields in and around living organisms" which includes bioluminescence, "cell radiation" and bioelectricity, with an extensive discussion of Gurwitsch's "mitogenetic radiation". Next he treats the susceptibility of organisms to electric, magnetic and electromagnetic fields and discusses the physico-chemical nature of the cell and its functions, the properties of colloids and liquid crystals, the high sensitivity of the components of the cells of living systems and of the systems as wholes, its limits and conditions, and finally the geological, geophysical, atmospheric and cosmic factors influencing life on earth, including biometorology. In chapter III Tromp looks at "divining and kindred phenomena" in detail, including various theories trying to explain them, his own experiments that he performed in the physical and physiological laboratories of Leiden University, and those of other researchers, and also unexplained phenomena such as mesmerism, hypnotism and the homing sense of animals. In the last chapter finally he proposes his concept of "psychical physics". In Tromp's view the problems of dowsing and radiesthesia belong to the wider field of "para-psychical problems", into which he includes homeopathy and other unexplained phenomena, and which are connected to the "web of electromagnetic forces which seem to regulate all living process on earth". He is convinced that these problems are of a more general significance and "ask for more extensive physical research". They are so extremely complicated that it is impossible for individual scientists to tackle them alone; a group effort is called for. Their solution is "not only of academic importance, but might throw a completely new light on the medical, physical, chemical, and biological sciences, and its influence will without doubt be considerable". The study of this large field should be taken out of the hands of charlatans and be studied by scientists of good repute. It "has been given by the author the name of psychical physics", but "it could also have been called biophysics", only that of course "existing biophysical laboratories do not concentrate on the use of physical methods in the study of psychical phenomena. The vitalists strongly object to the name of psychical physics, as they are convinced that psychical phenomena are ruled by (...) laws fundamentally different from the physical laws". But "such a distinction is not based on facts; psychic phenomena can be approached by using physical and physicochemical methods".

In a farsighted and daring paper that Tromp published in 1961, these ideas are further extended (Tromp, 1961). We find that in order to understand the place of fields of research such as psychophysics, biophysics and biometeorology in the order of the sciences, he came to classify them into the category of "border sciences". These are defined by him as "those branches of science which interconnect the fringes of well-established basic sciences" such that "new independent sciences" are created. The border sciences "also comprise these types of fundamental research which penetrate into completely unknown realms of human knowledge, until recently considered the domain of vague, unrealistic quasi-scientists and unfortunately often the hunting-ground of unscientific charlatans". According to Tromp, "the greatest achievements in the future of science will result from the interaction of existing and recently developed branches of science, an interaction that creates 'border sciences' ". Tromp differentiates between "established border sciences" and "non-established border sciences" which "may in not too far distant future, develop into new independent branches of border sciences". The former comprise, e.g., geology, psychophysics – which includes "physical neurology" and "physical embryology"– geo-ecology (which concerns the interaction of environment and living organisms, including bioclimatology and biometorology), "biorhythmics", and finally, cybernetics. Among the non-established border sciences he mentions astronautics and "supersensorics", concerning "certain phenomena of perception based on presently not explainable physiological and physical mechanisms" in man and animals, like telepathy, clairvoyance, hypnosis, altered states of consciousness, yogic phenoemena, and the homing instinct of animals. Therefore, many of the scientific disciplines which below are described as contributions or branches of integrated biophysics, would be classified by Tromp as border sciences.

However, Tromp makes clear that for him border sciences are of high significance for the future of mankind. As very complex branches of human knowledge, they can only be developed by the common interdisciplinary effort of many experts in various scientific disciplines, cooperating in a truly universal spirit. Besides interdisciplinarity and international cooperation, they require great mental flexibility – requirements that may not be had without a drastic change in our educational systems, such as the introduction of an early teaching of the interconnectedness of the specific sciences. Tromp also has some good remarks about the reasons for the great resistance of orthodox science frequently encountered by the border sciences. The first reason is that "Scientists are often less unbiased and less open-minded than one would expect". He quotes Alexis Carrel, the great physiologist and Nobel laureate, who wrote in his book *"Man, the Unknown"* (1935): "Our mind has a natural tendency to reject the things that do not fit into the frame of the scientific and philosophical beliefs of our time. After all, scientists are only men. They are saturated with the prejudices of their environment and of their epoch. They willingly believe that facts that cannot be explained by current theories do not exist. Evident facts having an unorthodox appearance are suppressed. By reason of these difficulties the inventory of the things which could lead to a better understanding of the human being has been left incomplete". A second reason for the resistance to the border sciences is the interest shown by charlatans and "cranks" in these subjects which makes a scientist reluctant to enter this field of research. Thirdly, the acceptance of some of the phenomena, e.g., precognition, requires a fundamental revision of basic concepts of the natural sciences, physics in particular, "overthrowing several fundamental pillars of knowledge and creating a feeling of mental unrest and instability which most scientists cannot endure". Finally, Tromp suggests the establishment of a "Universal Academy of Border Sciences" which may be started with the faculty in which work is already most advanced, and then could be extended to the other border sciences.

3.1.7 The Cambridge Biotheoretical Gathering

There are also two group efforts that may serve as inspirational models for the development of integrative biophysics: the *"Biotheoretical Gathering"* of the Cambridge Group of Theoretical Biology active in the 1930's (Waddington, 1968-72; Abir-Am, 1987) and the Paris *"Institut de la Vie"* founded in 1969 and active until the early 1990's.

The first group was formed in 1932 by several speakers of the session on "The historical and contemporary interrelationships between the physical and the biological sciences" at the 2nd International Congress for the History of Science that took place in London in the Summer of 1931 (Abir-Am, 1987), and existed till the outbreak of World War II. The founders of the group, theoretical biologist Joseph H. Woodger, biochemist and embryologist Joseph Needham, experimental embryologist Conrad H. Waddington, mathematician and philosopher of science Dorothy M.Wrinch, and crystallographer John D. Bernal, were non-conformist scientists striving for change in both science and society and were mainly concerned with recasting the historical relationships between the physical and the biological sciences. Most of them subscribed to various forms of holistic views of the organism and considered the concept of biological fields as highly relevant to the understanding of morphogenesis (Bischof, 1998). They viewed biology as the "science of the future" and their common goal was the development of a theoretical biology. According to Abir-Am the group's activities may be divided into three phases. In the early 1930's, they developed a "common theoretical discourse on the epistemological relationship between the biological and the physical sciences", stimulated by the example of D'Arcy Wentworth Thompson, Alfred N. Whitehead and Ludwig von Bertalanffy. In the mid-1930's, the group formulated a "transdisciplinary research program, integrating problems and experimental resources from specific branches in the disciplines of biology, physics, mathematics, and philosophy". Named "mathematico-physico-chemical morphology", the research program "embodied a new rationale for the unity of science and encouraged theoretical and experimental syntheses across the institutionalized boundaries of the above disciplines". In the late 1930's, the group tried to establish a research institute designed to validate this transdisciplinary program and for this purpose negotiated with various science policy agents such as the University of Cambridge, the Rockefeller Foundation, and the Royal Society. For some time, the Rockefeller Foundation was interested in supporting the establishment of such an institute. They funded several smaller research projects of the group, but Warren Weaver, director of its Natural Science Division, eventually terminated support for the project, which ended the activities of the group and the Cambridge career of Needham and Waddington. The main reasons were the renown of the group members as political and scientific radicals and institutional resistance to the transdisciplinarity of the scientific program and to the pluralistic and participatory model of collaboration and organization practiced in the group and proposed for the institute. In retrospective, the transdiciplinary research programs the Cambridge Group of Theoretical Biology had articulated without doubt were one of the main contributions to the emergence of molecular biology; however, their conception of a new type of biology, based on the fusion of morphology with biophysics, may well have led to an alternative conception of molecular biology, had it not failed to get funded and institutionalized.

3.1.8 The Institut de la Vie

Thirty years later, another multidisciplinary effort to "mobilize and conjugate the resources of contemporary science for the fundamental understanding of life" was initiated by the Parisian *"Institut de la Vie"* under the direction of Maurice Marois, professor of medicine at the Sorbonne. It was, however, Herbert Fröhlich at whose suggestion this series of conferences was created (Fröhlich & Hyland, 1995). Fröhlich was introduced to the *Institut* – which up to this moment had been more sociologically oriented - some years after its foundation, when he accompanied his wife Fanchon to the mountain village of Alpbach (Tyrol) where she attended one of the famous Alpbach conferences on science and life, while he only wanted to do some mountain-climbing. At a lunch hosted by Marois, the founder of the Institut asked him what could be done to bridge the "gap between physics and biology". Fröhlich, then president of the International Union of Pure and Applied Physics (IUPAP), suggested that an international conference should be convened to discuss possible cross-fertilizations. An organizing committee was set up – comprising Pierre Auger, Alfred Fessard, Herbert Fröhlich, Pierre Grassé, A.Lichnerowicz, Ilya Prigogine, Leon Rosenfeld and Marois – and at the end of June 1967 the *"1st International Conference on Theoretical Physics and Biology"* took place in the Palais des Congrès at Versailles, jointly sponsored by IUPAP and the French Minister of Scientific Research and Atomic and Spatial Questions, Maurice Schumann. This was the first of a series of conferences of the Institut on this topic (Marois, 1969, 1971, 1973, 1975, 1997), held every two years, except when the lack of funds did not allow it, and attracting many prominent scientists (including several Nobel prize winners) from both the physical and biological sciences, among them Albert Szent-Györgyi, Francis Crick, Jacques Monod, François Jacob, Walter M. Elsasser, Albert Dalcq, Franz Halberg, Eugene Wigner, Henry Margenau, K.Mendelssohn, R.S.Mulliken, Severo Ochoa, Lars Onsager, Manfred Eigen, Hermann Haken, Isidor Rabi, S.L.Sobolev, Etienne Wolff, and Laszlo Tisza.

Here again the attempt was made to develop a theoretical biology, conceived as "a special form of biology which would be the pendant of theoretical physics and which may throw light on those parts of the domain of life that have remained obscure" (Marois, 1969). As the Danish physicist Leon Rosenfeld wrote in the same volume, they were not the usual kind of scientific conferences, but were designed as "experiments in establishing contact between two branches of science, theoretical physics and biology, which so far have developed independently of each other, but which are now converging towards a common ground". Pierre Auger, President of the scientific commission of the UNESCO, emphasized that it was "necessary in a time of specialization to break the barriers between specialities and even between disciplines, and to return to a unitary concept of science as that of antiquity which only made way under the pressure of the immense masses of new and uncorrelated knowledge", in order to find "the grand principles and general laws which cross the entire field of the sciences, be they exact or natural". He predicted that theoretical physics and biology would henceforth enter into a definitive and close union. However, one of the principal postulates of the Institut de la Vie was that it was not sufficient to use classical physics for explaining biological mechanisms; it was necessary to treat living systems by recourse to quantum physics which was considered more adequate. One of the most significant contributions of the Institut de la Vie was probably that Herbert Fröhlich found in these meetings the adequate resonance and expertise to develop the ideas on coherence in biological systems that have proved so essential for the development of quantum biophysics. At the first conference of the Institut de la Vie he gave the famous lecture on *"Quantum mechanical concepts in biology"* (Fröhlich, 1969) in which he introduced the concepts of long-range phase correlations and coherence. In his introduction to the second conference, he criticized the assumption that all important biological principles will be ultimately expressed in terms of molecular properties. He pointed out that "some of the most important properties of materials imply concepts which have no meaning in terms of single particles or of their pair interactions. Order in physical systems is one of these concepts". Ordering by long-range phase correlations, as found in connection with the properties of superconductors and superfluids and also present in lasers, was a feature not explainable by the properties of single molecules that Fröhlich suggested was also essential in biological systems.

3.2 The Founders of Biophysics

For a better understanding of the development of biophysics I would now like to return to its origins. The onset of the scientific study of life was also the beginning of biophysics. The founders of biophysics were a group of German scientists in the mid-19th century who first did systematic physiological studies by using physical methods. Although certain isolated problems in physiology had been treated before 1840 in a physical manner, by such men as Ernst Heinrich Weber, Richard von Volkmann, Hugo von Ziemssen and Johannes Müller, the so-called "Berlin school of physiologists", including Hermann von Helmholtz, Emil DuBois-Reymond, Ernst von Brücke, and Carl Ludwig, were the first to conduct a "broadly planned, systematic investigation of an extensive field of physiological phenomena in accordance with the most rigourous physical methods" and assert as a general law that "a vital phenomenon can only be regarded as explained if it has been proven that it appears as a result of the material components of living organisms interacting according to the laws which those same components follow in their interaction outside of living systems", writes Adolf Fick, pupil of Carl Ludwig and author of the first textbook of biophysics, in his *Collected Writings* (1904). In 1847, this group had set up a program which seems to anticipate that of modern molecular biophysics (Cranefield, 1957). This program is reflected in the following statements by DuBois-Reymond, taken from Cranefield. In 1841 he wrote "I am gradually returning to Dutrochet's view 'the more one advances in the knowledge of physiology, the more one will have reasons for ceasing to believe that the phenomena of life are essentially different from physical phenomena", and in 1842: "Brücke and I have sworn to make prevail the truth that in the organism no other forces are effective than the purely physical-chemical". In 1848, in the preface to his famous *"Untersuchungen über thierische Electricität"*, DuBois-Reymond asserted that "it cannot fail that...physiology...will entirely dissolve into organic physics and chemistry".

In 1856 Adolf Fick, strong adherent of all the goals of the Berlin school, published the first textbook of biophysics (Fick, 1856). But the program of the Berlin school of reducing physiology to physics and chemistry was quite premature; 19th century physics and chemistry were not sufficient to provide molecular explanations or any other complete physical models for living systems. As Cranefield has shown, the Berlin school has failed to carry out its program of 1847 and has mainly lived on through its physiological approach. It was very successful in establishing experimental methods - which they were, however, by no means the first to introduce. Their antivitalistic stance, the assertion that the explanation of life processes must be sought in the ordinary laws of inorganic nature, had its most significant importance in the implication of an intelligible and accessible causality, making life processes amenable to experimental investigation. The only biophysical fields developed by it that continued to be of importance in classical physiology are the mechanical and thermodynamical investigation of muscular contraction and the electrical study of the nerve impulse. General physiology, however, till World War I followed a quite different course than the one outlined by the 1847 program. It is only in the program of the molecular-biological variety of biophysics predominating since about 1960 that this program seems to be carried out. The Berlin school also set the agenda for the creation of "scientific medicine" through its influence on the reform of medical research and education in the USA brought about by the famous "Flexner report" in 1910. The leaders of the reform which led to the exclusive acceptance of experimental science as the basis of medicine and the marginalization of homeopathy, electrotherapy and other alternative therapies and schools of thought by 1930, were a group of American pupils of Ludwig, among them W.H.Welch, H.P.Bowditch, F.P.Mall, and J.T.Abel (Flexner & Flexner, 1941).

However, the usual characterization of the German founders of biophysics at the beginning of the 19th century by historians of science as staunch reductionists and opponents of vitalistic and idealist ideas in biology has to be corrected, as Culotta has shown (Culotta, 1974). In reality they were not in such a sharp opposition to the romantic *Naturphilosophie* spirit of the time, and nearer to the antireductionistic approach of Claude Bernard, than is generally assumed. This is especially true if we consider men such as Ernst Heinrich Weber and Gustav Theodor Fechner who constitute a subgroup of the Berlin school that has also been called the "Goettingen school of physiology" (Gallagher, 1984). They even recognized consciousness as the "ultimate problem of biology" and saw the necessity of including it into biology, but at the same time realized that science was not yet prepared to do so (Culotta, 1974). However, it was the French school of physiology founded by Claude Bernard, who really set the course of general physiology till World War I, and can also be considered as the ancestors of a more holistic school of thought in biophysics.

3.3 The Physiology of Regulation, the Fluid Matrix of the Body and Physical Chemistry

3.3.1 Claude Bernard: Inner Environment and "Terrain"

Bernard's concept of the constancy of the "milieu interne" (internal environment), first formulated in 1859 and fully developed in 1865, was central to his experimental work and theoretical outlook and had a considerable influence on later developments in physiology (Bernard, 1859, 1865; Holmes, 1963; Olmsted, 1967). He maintained that "protective functions" (regulative mechanisms) existed in the organism that maintained the internal environment in order to preserve its "vital activity". The "dislocation and perturbation of these functions" first caused pathological changes in the fluid "terrain" (soil) in which cells and organs were embedded, and finally may lead to sickness and death. Bernard also emphasized the interconnection of the organism with the environment; internal regulation was based on full communication with and active response to the changing environment. The maintenance of the constancy of the inner environment had allowed the development of higher organisms and the independence of life from its oceanic origins. Bernard first identified the inner environment only with blood plasma, then extended it to the lymph, and finally emphasized that it consisted of the totality of the body fluids.

Bernard's concept of the "terrain", closely related to the "milieu intérieur", is one of the most seminal and fruitful contributions of early physiology to biophysics and also had a great influence on holistic medicine, especially in France. As Charles Richet, the famous French immunologist, remarked in 1927 in the preface to a book on the "terrain" by his pupil Héricourt (Héricourt, 1927), it originated in the observation that the same pathological factors often produced a great variation in response in different individuals, therefore the physiology – the tissues, immune system, and body fluids - must be, like the psychological personality, very differentiated and vary from one individual to the next. In 1936 the French biochemist Wladislaw Kopaczwewski writes in a paper on the same subject that the terrain actively counterbalances the pathogenic influences (either from external or internal origin) to a certain degree and acts a barrier to noxious agents (Kopaczwewski, 1936). In the period of 1870 to 1920, the success of Pasteur's and Koch's bacteriological work caused a tendency to overvalue external pathogenic factors and to underestimate the role of internal factors, i.e., the terrain and the active regulatory role of the organism. Pasteur also recognized the role of the terrain, but saw it merely as a passive, static and unchanging stage for the activities of the microorganisms, comparable to the culture media he used in his laboratory.

3.3.2 Physical Chemistry and the Homeostasis of the Fluid Matrix

With the concept of the terrain the founder of experimental physiology not only became one of the fathers of the research in physiological regulation, but also triggered a renaissance of humoral physiology based on the introduction of physical chemistry – mainly the concept of dilute ionic solutions, developed by Jacobus H. van't Hoff and Svante Arrhenius - into physiology. The regulation of the internal environment of the body fluids by the control of hormones and electrolytes characterized the work of pioneers like René Quinton, Léon Frédérick, Joseph Barcroft, Ernest H. Starling, Lawrence J. Henderson and Walter B. Cannon.

Starling, pioneer of hormone research, who with W.M.Bayliss in 1902 first isolated a hormone (secretin) and in 1905 introduced the notion of "hormone", was a pioneer in the investigation of peripheral circulation, the role of osmotic forces and the fluid exchange between capillaries, intercellular tissues and the lymph system. He showed the existence of an equilibrium between the blood pressure in the vascular system and the osmotic forces in the peripheral capillaries and the interstitial tissue (now known as "Starling equilibrium"). He was convinced that these processes were an expression of the "wisdom of the body".

Henderson, influenced by the hippocratic idea of a healing force of nature, studied the regulation of acid-base-balance by the various chemical buffer systems, mainly in the blood. In *"Fitness of the Environment"* (1914) he developed the idea, derived from his physiological work, that the physical and chemical properties of the inorganic environment, especially those of water, carbohydrate compounds and carbonic acid, were fit for the development of organic life. This correspondence could not be by chance. In this fitness of the environment for life the unique properties of water played a prominent role.

Cannon in 1929 introduced the the concept of "homeostasis" as a mechanism stabilizing physiological variables within a certain range of variation; this took place mainly in the *"fluid matrix of the body"*. He also introduced the concept of stress, defined as the straining of the homeostatic mechanisms by the breaching of critical levels of the physiological variables; this could in his view be caused by emotions, fear hunger, rage or other emergency situations. In his view, bodily reactions always were protective, purposive, appropriate and "wise" (Cannon, 1929, 1932).

This line of research was later continued by Hans Selye (1907-1982) who in 1936 started a series of experiments to study the nonspecific, systemic responses of the organism to a wide range of stressful challenges like physical exertion, exposure to radiation, heat and cold, traumatic wounds and loss of blood, disease, injury, infection, danger and fear. He had already earlier on made the observation that many different acute and chronic diseases were all accompanied by the feeling of exhaustion and the same group of symptoms, and wanted to understand what this general "state of being sick" was. In the theory of the "General Adaptation Syndrome" (GAS) which Selye formulated as a result of his experiments (Selye, 1936, 1946, 1950; Weiner, 1992), he identified this general state of being sick as a result of an unspecific, systemic and uniform reaction of the organism in which adrenal corticosteroid hormones play a central role and which is characterized by three stages: (1) The "alarm phase" in which the stress is experienced as a hardship, and the organism reacts with restlessness and tenseness, loss of appetite and sexual drive, increased cell-membrane poermeability, and loss of temperature and tension, or even depression and shock, depending on the severity of the stress. (2) The "stage of resistance", where the first reaction to the stress is counteracted by neural and hormonal activation, the organism becomes used to the stress, and the appetite is restored. (3) The "stage of exhaustion" in which the organism cannot stand the stress anymore and gives up, the manifestations of the alarm reaction return and changes in arterial blood vessels occur. Selye's concept of GAS was used to explain the pathogenesis of many diseases, named by Selye "diseases of adaptation", and his concept of stress found wide application in medicine and physiology, but even more in the social and behavioural sciences (Weiner, 1992).

3.3.3 The French School of Biophysics

In the tradition of Bernard, the French biophysics school of Fred Vlès and Wladislaw Kopaczewski continued the humoral-physiologic approach with physico-chemical studies of acid-base-balance and reduction-oxidation potentials and their influence on the proteins and colloids of the "terrain" of the organism (Vlès, 1927, 1929, 1936 a,b; Vlès et al., 1931, 1936c; Reiss, 1926, 1940, 1943/43; Reiss et al., 1942/43; Kopaczewski 1921-22, 1923, 1926, 1930-38, 1933; Garnier, 1959; Benezech, 1962). Vlès, who like the other members of this school held that water was the basis of biology because its properties determined many of the properties of living substance, also was one of the first to recognize and study the biological significance of water structure (Vlès, 1936 d,e). Vlès and his pupils perfected the pH and redox measurement of blood and other body fluids established by Michaelis (1914, 1933) and Clark (1926, 1928) and later taken up and popularized by Louis-Claude Vincent (1954, 1956 a,b), Ernst Ziegler (1960), and Werner Kollath (1968). (For recent biophysical work based on Vincent's approach see Országh, 1990, 1992, and Fougerousse, 1992; a contemporary textbook on electrolytes and acid-base balance in the body fluids is Eccles, 1993). An outstanding early example of the clinical application of the study of tissue pH is Delore's work on its role in pulmonary tuberculosis (Delore, 1926).

3.3.4 A New Concept of the Cell

As usual, the new approach of physical chemistry has not only led to many advances in biological understanding, but also has caused a number of misconceptions that required correction. One of the most fundamental of these probably is the assumption that the cell is in essence a membrane-covered aqueous solution of free proteins, ions, and other small and large molecules. The "membrane-pump theory" of the cell - first proposed in 1877 by the plant physiologist Wilhelm Pfeffer, who also discovered osmosis, established in 1902 by Helmholtz' pupil Julius Bernstein and later refined by Frederick Donnan (1911), Alan L.Hodgkin (1949), Andrew F.Huxley and Kenneth S.Cole - which is based on this assumption, is still one of the paradigmata of contemporary biology. According to the membrane theory, the difference in ionic concentrations between the cytoplasm and the extracellular medium − mainly in sodium, potassium and chloride ions − which also serves to explain the biopotentials − is due to the semipermeability of the membrane (older version of the theory), or to the "active transport" by ATP-driven ionic pumps.

However, since the 1950's a number of alternative conceptions have developed, which trace their origin to early observations of the gelatinous (colloid) properties of living matter, and of the anomalous behavior of water associated with certain biomolecules. Like Abraham Trembley, C.F.Wolff and other biologists before him, Felix Dujardin in 1835 recognized that entity as a "living jelly" that since Jan Evangelista Purkinye (1839) is known as "protoplasm". After the introduction of the concept of the "colloid" by Thomas Graham in 1861, many biologists and physiologists have conceived of the protoplasm as a hydrated polyphasic colloidal system (Fischer & Moore, 1907; Fischer, 1910; Fischer & Hooker, 1933; Schade, 1909, 1921a; Kopaczewski, 1923, 1926a, 1930-38; Lichtwitz, Liesegang & Spiro, 1935; Fischer & Suer, 1951). In 1908, Benjamin Moore and Herbert Roaf, and in the 1920's, Ernst & Scheffer (1928) suggested that the potassium ions in cells, the most abundant intracellular ions, existed in a bound and unionized form. Most prominently, Heinrich Schade already in 1908 recognized the cell as a polyphasic colloid (Schade, 1909) and wrote in his pioneering book on *"The Physical Chemistry in Internal Medicine"* (1921) that "The body of the cell is neither uniformly solid nor uniformly fluid; each cell rather forms a (...) 'microheterogenous system', a mixture of gel-like and sol-like masses and true solutes in a common medium, the colloid nature giving the characteristic feature to the whole as well as to its parts". Martin H. Fischer postulated in 1910 that living cells, together with their intercellular substance, form a water-binding colloid that cannot be compared to a suspension or solution of molecules in water, but rather is a "water-free" hydrated system in which the water is not free, but bound to its colloids, mainly proteins in the form of alkaline and acidic proteinates with a high hydratation potential (Fischer, 1910, 1951). In 1938, Fischer and Suer suggested that protoplasm was the "union of protein, salt and water in a giant molecule" (Fischer & Suer, 1938).

In the view of Schade, Fischer and many other biologists of the time, the water in the cell and in the extracellular spaces was not free but existed in the form of "bound water", also called "Schwellungswasser" (swelling or imbibition water) (Newton & Gortner, 1922; Gortner, 1938). The phenomenon of the strong water affinity of certain biomolecules was first discovered in 1848 by Carl Ludwig of the "Berlin school of physiology" and again observed in 1917 by J.R.Katz. However, in the 1930's Nobel laureate Archibald Vivian Hill and others seemed to show that bound water and bound potassium ions did not exist, and thus the view of the living cell as a membrane-enclosed solution of free ions and free water was saved for a while (Hill, 1930; Hill & Kupalov, 1930). In addition, colloid chemistry in general lost its explaining power in the 1940's. Although in 1938 we find a declaration for the polyphasic colloid character of the cell by the eminent neurophysiologist Charles Sherrington, the more consequential proposition of two major alternatives to the membrane theory of the cell had to wait till the early 1950's; it initiated a controversy that even today has not come to its decisive end.

In 1951 and 1952, Afanasy S.Troshin, student of the eminent Russian cell physiologist Dimitri H.Nassonov, presented his pioneering studies on cell permeability, which were based on his teacher's work and provided the foundation for his "sorption theory" of the cell (Troshin, 1956, 1958, 1966; Nassonov, 1959). Like the other variants of the "phase theories" or "coacervate theories" proposed as alternatives to the membrane theory, this was based on the interaction of macromolecules and ions with cell water and the protein-lipid membrane. Troshin pointed out that in contrast to the low-protein extracellular medium, the interaction of electrolytes with proteins in the cytoplasm is so strong that ionic mobility and therefore electrogenic activity of ions is strongly reduced; the cytoplasm forms a "complex coacervate" in which proteins, ions and most of the water are strongly bound. Cytoplasm and extracellular medium therefore form two distinct phases. A coacervate (a concept introduced into colloid science by Bungenberg de Jong in 1929) is formed when hydrophilic proteins, together with water, form droplets and amoeboid masses that are distinctly demarcated from the ambient fluid. It encloses 80-90% water and remains hydrophilic. Coacervation depends on a strong enough dipole moment and electric charge of the proteins and on the electrolyte content of the medium. Through coacervation the protoplasma obtains the property of liquid-crystallinity (Recently the topic of liquid-crystallinity, first discovered in 1888 by F.Reinitzer and O.Lehmann, as a property of living systems has been again discussed by several authors – see Mishra, 1975, Ho, 1996). Troshin's work was followed up in Germany by Jakob Segal and his group (Segal, 1958; Segal et al., 1983).

The most momentous of these alternative cell theories was the "Association-Induction Theory" put forward at approximately the same time by the Chinese-American cell physiologist Gilbert N. Ling (Ling, 1960, 1962, 1970, 1984, 1992, 2001). It was then extensively tested in the 1960's and 1970's, most notably by Freeman Cope (1970), Carlton Hazlewood (Beall et al., 1967) and Raymond Damadian by means of the newly developed Nuclear Magnetic Resonance (NMR) technology. The subsequent invention of Magnetic Resonance Imaging (MRI) by Damadian actually was a result of Damadian's testing of the Ling theory of cell water, when he discovered a difference in NMR signals from water in cancer cells to those from normal cells (Damadian, 1971). The insight that any theory attempting to explain the unequal distribution of ions between the inside and outside of cells, the dynamic aspects of solute and water exchange and the bioelectric phenomena of cells requires knowledge of the physical state of water in the cells, has stimulated research on the role of cell water and and its supramolecular structure in the following decades, especially after these concepts first reached a wider scientific audience and recognition in 1965 at a conference on the biological role of water structure in the New York Academy of Sciences (Whipple, 1965). The phase theory of the cell assumes the existence of aqueous compartments (phases) in the cell with distinct metabolic activities, and the exclusion of solutes (e.g., sodium), to some extent, from the cell by the changed solvent properties, and also predicts that changes in the physiological state of the organism can alter the physical properties of cellular water (see, e.g., Cope, 1969, 1970, 1973, 1975, 1976, 1977; Drost-Hansen & Clegg, 1979; Clegg, 1981, 1983, 1984, 1987, 1992; Hazlewood, 1971, 1979, 1991; Tigyi et al., 1991; Wiggins, 1971, 1972, 1990).

Of course, the concept of a highly structured cell is also strongly supported by the recent discoveries of the subtleties of cellular architecture. In addition to the discoveries of the first cell organelles (Golgi bodies, mitochondria) in the late 19th century, the application of electron microscopy, beginning with the work of the Belgian-American cytologist Albert Claude in 1945, brought the discovery of the endoplasmatic reticulum, of microtubuli, microfilaments, intermediate filaments, and finally the microtrabecular lattice (Porter, 1981, 1984). Today, it is clear that we have to view the cytoskeleton as a highly structured three-dimensional network of these elements which is very dynamic and labile. Its elaborate structure is very easily disassembled under the influence of small perturbations, and thus oscillates between a gel state in which the structure is up and visible and a sol state in which it is dissolved. Recent molecular-biological research has also shown that the structures of the cytoskeleton pass through the cellular and nuclear membranes and thus establish a continuity between extracellular matrix, cytomatrix and the nucleus and its contents (Pienta & Coffey, 1991).

The discovery of the importance of water structure for the understanding of cell and organism is also significant in connection with the evidence we will describe later, for a role of the structure of cellular water in the reception of environmental information, including ambient electromagnetic fields, by the humoral system formed of body water and extracellular matrix.

3.3.5 Synthesis of Regulative Mechanisms

In 1918, the German pathologist Friedrich Kraus (1858-1936) made a first attempt to synthesize the findings of physical chemistry, the physiology of neural and humoral regulation and colloid science from the standpoint of constitutional medicine. He emphasized the functional unity of the various regulative mechanisms of the endocrine, the vegetative-nervous and the physico-chemical systems, and combined them in his concept of the "vegetative system" which is the functional unity of the organism's colloids, the body fluids with their electrolytes, and the vegetative nervous system. In the "moisture theory of life" he proposed in his book *"Allgemeine und Spezielle Pathologie der Person"* (Kraus, 1918-26), a reformulation of humoral physiology, he viewed the organism as an oscillatory system of ionotropic colloids whose state constantly changes between the liquid "sol" and the solid "gel" state, and whose basic mechanisms are mechanical (hydraulic) and electronic (but with conductors of second order). According to Kraus, the main work of the organism consists of swelling and de-swelling; the transition from the wet to the less wet state performing the work. Kraus saw the stream of water through the body, driven by the dynamics of protoplasm, as one of the major features of the living state, and he considered the degree and the distribution of water saturation and the turgor of the tissues as a measure of life. One of the important elements of the "vegetative system" was thus the "vegetative streaming" of the body fluids in the vascular system, the lymph system, the interstitial spaces (connective tissue) and the cell. However, "the colloidal biosystem is not only driven and regulated by water, but also by specific salt effects respectively by electrical interface potentials", and thus the antagonism of electrolyte cations and anions was important. He considered the electrolyte content of tissue as an essential element of constitution and the distribution and shifting of ions in the organism as an important mechanism of life. The changes in the distribution of electrolytes cause shifts of body water, nervous excitation and changes of the state of colloids. In the tradition of colloid science, surface tension and interface potentials which are seen by Kraus as an expression of the "vegetative system", are considered as essential factors in all phenomena of living substance – "they determine the form of the fluid system and the distribution of solutes" (for more recent work on surface forces in biology, see Ninham, 1982).

Kraus' early effort for a synthesis was continued in the work of later pioneers such as Schade, Ricker, Eppinger, Hauss and Pischinger. They introduced important new elements for such a synthesis missing in his work, mainly the fundamental insight on the central function of the extracellular matrix.

3.3.6 The Extracellular Matrix

While for a long time the investigation of the terrain followed Claude Bernard in almost exclusively emphasizing the role of the body fluids (especially blood), from the first decades of the 20th century onwards some physiologists and pathologists also started to pay attention to the connective tissue. Now usually called extracellular matrix, this is strangely enough one of the most fundamental systems of our organism that at the same time has been largely neglected by physiology and medicine since the time of Virchow.

One of the earliest and most groundbreaking pioneers in this field was the already mentioned founder of molecular pathology, Heinrich Schade (1876-1935) (Schade, 1912a & b, 1913, 1921 a & b, 1922, 1923a,b,c – on Schade see also Hadjamu, 1974). Schade may well be called the founder of modern connective tissue research; he was the first physician in the 20th century to recognize the "organ function of connective tissue" and to establish the nature of the extracellular matrix as an active physiological system in its own right. He criticized its characterization by Rudolf Virchow as "connective tissue", as it fulfilled far more than just connecting and supporting functions. To this contempt of connective tissue by the founder of "cellular pathology", not surprising because of the predominance of cellular mass over the portion of extracellular tissue and by reason of its poor visibility under the earlier staining methods, Schade opposed his colloid-chemical concept. Starting with physicochemical researches on the biological functions of connective tissue in 1911, he experimentally demonstrated some of the fundamental systemic functions of connective tissue, its colloidal nature, the significance of pH for connective and other solid tissues, and of hydrogen ions for pathology in general. He summarized his findings in the groundbreaking work *"Physical Chemistry in Internal Medicine"* (1921). Thus he established connective tissue as a central issue in pathology and thereby laid the ground for the later work by Gustav Ricker, Hans Eppinger, jun., Heinrich Thiele, Werner H.Hauss, Alfred Pischinger, Otto Bergsmann, and Hartmut Heine, among others.

The work of these researchers was concerned with identifying and more precisely characterizing Claude Bernard's "terrain" which is the "locus" of the functional disturbances which predate pathological evidence, as Gustav von Bergmann had shown in 1932 (Bergmann, 1932). Bergmann (1878-1955), prominent professor of internal medicine, had made clear that the morphological changes of pathological evidence must be considered as the end result of a development of disturbed functions that may have taken years. He therefore demanded more attention to the "latent and larvated early forms of illness", which should be recognized and reacted to already before the occurrence of objective morphological signs, by constitutional thinking, clinical tests of functions and the abolishment of artificial demarcations between "functional disease " and "organic illness". Already in 1912, Felix Buttersack (1865-1950) had localized the latent illnesses in the connective tissue and had introduced the notion of "ground tissue" to discriminate between supporting tissue and soft connective tissue (Buttersack, 1912).

Ricker (1870-1948), a pathologist, was a protagonist of "neural pathology" and maintained that the neural regulation of body liquids and vascular permeability was a key process in physiology. He postulated in his "relational pathology" (Ricker, 1905, 1924), against Virchow, that there is no independent activity of cells; he held that the network of relations between nervous system, blood, liquids of the tissues and tissue itself has to be viewed as a single system. He was convinced that tissue alterations of other than nervous etiology do not occur without forcible destruction. The tissue as a living innervate could not but react neurally, because the complete reception of stimuli and the reaction to it was mediated by the nervous system that was especially made for the processing of stimuli. In 1924 Ricker proposed a "law of stages" according to which different neural stimulations cause different changes of flow in the capillaries and consequently changes in tissue. Weak stimuli (first stage) on the vascular innervation excite the dilatators. In the scecond stage (medium stimulation) the constrictors are excited which slows down the capillary and venous flows. Stronger stimulation causes ischemia and anemia. In the third stage (strongest stimulation) the excitability of the constrictors is suppressed, but also that of the dilatators which remain excitable for a longer time, resulting in a "prestatic state" (very slow blood flow in the phase preceding stasis) and "red stasis" (cessation of blood flow).

Eppinger's "permeability pathology" (1949) is also a neural pathology, based on the work of Hess (1948). The permeability changes Eppinger describes are regulated by the vegetative nervous system; vasodilatation is subject to the vagus. Processes like inflammation, serous and more progressive vascular permeability that allows even red blood cells (hemorrhagic permeability) and white blood cells (purulent inflammation) to pass, therefore are vagus excitations. Vasoconstriction which causes a diminishment of permeability and hyperplastic states, is subject to the sympathicus.

The colloid scientist Thiele (1902-1990), a pupil of Nernst, Haber and Einstein and one of the pioneers of histophysics, demonstrated in work spanning the time from the early 1940's to the late 1960's that the interaction of complex colloidal gels with ions is the basic structuring mechanism in organic structures, especially in connective tissue (Thiele, 1947, 1954,1967; Thiele & Schacht, 1957).

The cardiologist Hauss (born 1907) worked with his group on the biochemistry of connective tissue and on the role of the "non-specific mesenchyme reaction" (NMR) in arteriosclerosis, infarction, myocardial necrosis and rheumatic illnesses (Hauss & Losse, 1960; Hauss & Junge-Hülsing, 1961; Hauss et al., 1968). Hauss demonstrated that the NMR is the basis reaction of the organism in its response to any kind of stimuli; the metabolism of the mesenchyme cells that produce the ground substance and the structural fibers of connective tissue, reacts with high sensitivity to all kinds of stimuli to which the organism is subjected, including infections, toxins, proteins and other foreign substances, lack of oxygen, mechanical stimuli, the influence of the weather, noise, radiation, and emotional stimuli (see below). This reaction was called non-specific because it was triggered reproducibly not by a specific factor, but by many different non-specific factors and events. Hauss and collaborators also demonstrated that the NMR was a pathogenetic principle of universal significance. Many different diseases with multifactorial etiology, caused by various noxious agents, including everyday conditions like working and eating habits, muscular, emotional or mental stress, and banal infections, originate in this reaction which is characterized by the fact that the process induced may easily smoulder for years and even decades without obvious objective or even subjective signs. They also showed that it is possible to achieve a physiological "conversion" (Umstimmung), i.e., increase the tolerance of the connective tissue system and lower the degree of the mesenchyme reaction, by training it with a combination of stresses, starting with low doses and increasing the dose during 2-4 weeks.

The Austrian histologist and embryologist Pischinger (1899-in 1980's), one of the founders of tissue histochemistry, has worked in connective tissue research since 1948 and has developed a "theory of ground regulation" (Pischinger, 1991; Bischof, 1992; van Wijk & Linnemans, 1993; Heine, 1997), according to which the "ground regulation system" (GRS) of the extracellular matrix (ECM) is the most fundamental and oldest regulatory and information system in the organism, older than the nervous and hormone systems. It is the basis of all general and nonspecific immune responses and the carrier of all systemic functions, and also triggers the specific regulations. It controls the physicochemical and bioelectrical situation as well as the bioenergetic processes, such as the water, oxygen, electrolyte and heat metabolism, the acid-base balance, and the redox potential. In this task its interaction with water and water structures plays a key role.

The anatomical substrate of ground regulation is the soft connective tissue (the "ground substance") which fills all the extracellular spaces of the organism and reaches every cell. It consists of a network of highly polymerized glycoproteins in which the connective tissue cells (fibrocytes, mast cells and macrophages) and the structural glycoproteins (collagen, elastin, finbronectin, and laminin) are embedded. As was already recognized by Carl B. Reichert in 1845 and by Georg von Rindfleisch in 1869, proved by electron microscopy in 1949 by Eppinger and further elucidated in 1961 by Hauss, blood capillaries and vegetative endfibers nowhere in the body directly reach the organ cells, but end blindly in the ECM. This makes the ECM a "transit space": every exchange of fluids or other materials between the vascular system and the parenchyme cells (organs) is mediated by the ECM, and also the nerves secrete their active compounds (adrenaline, acetylcholine) into the intercellular fluid. It is the connective tissue cells that take care of the complex interaction between capillaries, vegetative fibers and stationary cells, by secreting prostaglandins, interleukins, interferons, proteases, and protease inhibitors. By the same mechanism, they trigger the phases of the immune reaction and regulate the permeability of the capillaries. They are also responsible for the synthesis and decomposition of the ground substance and the structural glycoproteins, and thus are able to regulate the sponge-like structure of the ECM and the width of its pores, on which the passage of water and substances through the transit space depends.

The structural changes in the ECM are mainly due to changes in the structure of the proteoglycan-water complexes of the ground substance, which are the most essential element of the fundamental homeostasis brought about by the GRS. Because of their strong electronegativity, proteoglycans and glycosaminoglycans (GAG) possess the strongest water-binding and ion-exchanging ability of all tissue substances. Based on this, they guarantee osmotic, electrolytic, and ionic equilibrium in the ground substance and establish the electrostatic tonus of the tissue. They also scavenge the surplus electrons and protons in the extracellular space in the form of oxygen and hydroxyl radicals produced in all enzymatically controlled chemical reactions. In this way they homeostatically regulate the redox potential of the organism and counteract the inflammatory tendencies of the ground tissue.

This makes the GRS a highly complex and interconnected humoral regulatory system with innumerable feedback cycles and high redundancy, able to influence even the central nervous and hormonal systems by way of the capillaries and vegetative nerves. The regulatory capacity of the ECM is highly significant for the disease process; in all acute and chronic dieseases and in cancer, regulative disturbances and structural modifications of the ECM have been demonstrated, which also show in changed redox potentials. The electrostatic ground tonus established by the negative electrical charge of the proteoglycans reacts on every change in the ECM with fluctuations of the potential, that become manifest in redox measurements of the body fluids (method Vincent), or those of electric skin potential and resistance. Therefore, these methods can be used to assess the ground regulation.

By reason of the redox properties of ECM, every change of electrical potential in it represents an information; by means of such informations the GRS also exerts an influence on connective tissue and organ cells. This can happen by way of the potential fluctuations of the glycocalyx in the cell membrane, which are able, if strong enough, by depolarizing cell membranes of muscles and nerve cells or by activating transmitters like cAMP, to transmit information into the cell interior and to influence the structure of cell architecture and of cell water or cytoplasmic enzymes. By way of the recently discovered connections between cytoskeleton and nucleus, an information transfer to the nucleus is also possible (Pienta & Coffey, 1991; Oschman, 1996). The proteogylycan-water complexes of the ECM are a very fast information system; because of their nature as dissipative systems, their structural changes can spread rapidly over the ECM and be utilized as information by the cells. This phase change is due to the high lability of the liquid-crystalline water-polysaccharide complex and therefore can be triggered by minimal stimuli without any energy expenditure worth mentioning.

The important work of Pischinger, based on that of the earlier pioneers, was continued more recently by Bergsmann (1994a,b, 1998) and Heine (Heine, 1997; Heine & Anastasiadis, 1992), among others. The biophysics of the ground system was first studied by the Austrian physician F.Kracmar (Kracmar, 1971) and is more extensively treated in Barrett (1976), Heine (1997), Bergsmann (1994a,b, 1998) and Oschman (2000).

3.3.7 Bioelectronics and the Solid-State Model

Another important contribution to integrative biophysics comes from solid-state physics, or condensed matter physics. Its emergence as an autonomous discipline was connected with a strong opposition against reductionism in physics, as it had mainly developed in particle physics (Jordi, 1998). Among the prominent spokesmen of this movement were Eugene Wigner, Michael Polanyi, and Philip W. Anderson. Solid-state physics is characterized by the application of quantum mechanics to macroscopic phenomena, the tendency to work with realistic and phenomenological models of macroscopic systems, and the notion that the principles organizing macroscopic phenomena cannot be completely accounted for by the laws of atomic and molecular behaviour. As a heritage of its origin in wartime research, solid-state physics also was multidisciplinary from its onset and govened by the conviction that different scientific disciplines have to stimulate each other by sharing methods and conceptual tools. This is the reason why this branch of physics became one of the important contributors to integrative biophysics. One of the developments in the field consequential for integrative biophysics was the development of the theory of superconductivity by Fritz and Heinz London in the late 1920's and subsequently by Vitaly Ginzburg and Lev Landau in the 1950's.

If already the colloid-chemical and phase concepts of the cell described above can be considered as steps towards a shift from the "ideal-gas" model of biochemistry and physical chemistry to the "ordered-crystal" model of solid-state science, the work of the great biochemist Albert Szent-Györgyi, based on a novel solid state approach, has led to another, related field of biophysical relevance which can be seen to continue the tradition of humoral-physiologic thinking. Szent-Györgyi's achievements mark the definite advent of this new biophysical view of the organism. By drawing the attention to the submolecular level of organisation and pointing out the biological importance of charge transfer and excited states of biomolecules, Szent-Györgyi created the field of bioelectronics. He suggested it was useful to explain the properties of living organisms in terms of the electrons which are responsible, ultimately, for the chemical reactions driving the organism. In his ground-breaking Koranyi lecture in 1941 Szent-Györgyi drew attention to the study of energy levels and the role of excitation in biochemistry, and suggested that the conduction-band model of solid-state physics could be applied to biological macromolecules which may possess semiconductive properties. Since then, semiconductive behaviour of organic molecules has been established. Important has also been his suggestion that metastable excited states involving triplet states may play an important role in biochemistry. The study of the role of electronic charge transfer in intermolecular interactions became an important field of biochemical research due to his suggestions. The quantum-theoretical framework for it was created in the early 1950's by R.S. Mulliken. Not less significant was that he pointed out the possibility of collective excitation processes in biological systems and their propagation in the form of excitons (excitation waves).

Szent-Györgyi's pioneering work in bioelectronics has not only led to the discovery of the unique electronic properties of biological materials which recently have become an inspiration for the new field of molecular computing, but have also brought new insights into energy transmission and information processing in biological systems. It is now clear that organisms use and process information by using molecules, electrons, and protons (hydrogen ions); they are able to direct electronic, ionic and proton currents along certain defined pathways (Bone & Zaba, 1992). We are also starting to understand the important role water molecules and their supramolecular organization play in this process, another fact that Szent-Györgyi was one of the first to point out. He remarked that like fish who are unable to see the most obvious fact of their life, "biology has forgotten water, or never discovered it" (Szent-Györgyi, 1971). In his view, water who has "the lowest energy and the highest ionization potential among all biological substances", was "at the bottom of nature", was "the matrix of life" – much more than a material medium filling the space between the structural elements in the cell and the intercellular tissues (Szent-Györgyi, 1957)."It is a mistake to talk about proteins, nucleic acids or nucleoproteins *and* water, as if they were two different systems. They form one single system which cannot be separated into its constituents without destroying their essence". Water greatly mediates and modifies energy transmission, reactivity and many other processes between molecules (Szent-Györgyi, 1960). Szent-Györgyi's work has also played an important role in "bioplasma theory", one of the modern biological field theories synthesizing the solid-molecular aspect and the field aspect of organisms (see below).

In the 1970's, Freeman Cope has also done very valuable work in the application of solid-state physics to biological systems (Cope, 1969, 1970, 1973, 1975, 1976, 1977).

3.4 Cosmic and Atmospheric Influences upon Biological Systems (Biometeorology, Cosmobiology)

As already Claude Bernard has emphasized, higher organisms have their existence in two environments at the same time: the inner environment (milieu interne) and the "milieu externe" of terrestrial and cosmic environment; their life is bound to and influenced by both environments (Holmes, 1963). "The conditions necessary to life are found neither in the organism nor in the outer environment, but in both at once" (Bernard, 1865).

The thesis that weather and climate played a significant role in the health and diseases of man is already expressed in the Hippocratic writings, but it was only seriously tested with the beginning of the 20th century, when such prerequisites as, for instance, regular meteorologic observations under standard conditions, reliable epidemiological records, adequate physiological understanding, and the invention of the climatic chamber for exposing human subjects to simulated weather conditions became available.

As one of the first modern scientists, Svante Arrhenius, one of the founders of physical chemistry who became Nobel laureate in Physics in 1903, tried in 1898 to show the impact of cosmic influences, including that of the moon, on the weather and on living organisms (Arrhenius,1898).

3.4.1 Willy Hellpach

An early pioneer of biometeorology (other names of the field are bioclimatology, medical climatology, and meteorobiology) was the German physician, professor of social psychology and politician, Willy Hellpach (1877-1955), also one of the first German physicians to champion Freud's sexual etiology of neuroses. He was also one of the founders of environmental psychology and treated the relation of man and environment in a very interdisciplinary way, with a broad knowledge about its scientific, psychological and social aspects (Hellpach, 1911, 1924). He mainly became famous through his work *"Die geopsychischen Erscheinungen"* (1911), called *"Geopsyche"* from the 5th edition, which experienced many printings and was translated into several languages. It was this unique book - even later there is nothing comparable to it – which first established biometeorology as a science. He first defined biometerology as the investigation of the influence of all meteorological factors and all atmospheric changes on living organisms, e.g., radiation, air pressure, temperature, humidity, weather fronts, atmospheric electricity, etc., especially the relation of weather and climate to certain health disturbances and illnesses, and the question if certain weather situations and factors can cause or trigger illness, and finally the question if different types of people (constitutional types) react differently to influences from weather and climate. Then Hellpach defined his own "geopsychical" approach. The natural environment was one of the three environmental spheres that man lived in, the others being the social-psychological one and the artificial environment created by man himself. According to Hellpach, the natural environment of weather, climate, soil and landscape had two modes of influence on man, namely the sensory "impression", the influences that are directly perceived by means of the senses, such as heat, cold, wind, rain, storminess, gloom or brightness, on the one hand, and the "tonic effects" of influences like air pressure, atmospheric electricity, the moderate oscillations of air pressure etc., that we cannot directly perceive, that are mediated by the vegetative system and have an even greater influence on us. Together with the influence of landscape and soil, the tonic effects of weather and climate determine our life tonus, vitality, freshness or flaccidity, readiness to act, and mood. "Geopsyche" is the total influence of these factors on our psychic life. In *"Psychologie der Umwelt"* (1924) Hellpach pointed out that general psychology has been conceived "environmentless"; it either is concerned with the internal life of the individual or with the psychological interaction between individuals (social psychology), but not with the third sphere of the natural environment. He pleaded for the inclusion of the two modes of environmental influence mentioned above into social psychology; until now it had been "almost exclusively interpersonally oriented" and "no attention had been paid in it to physical objects, in spite of their unmistakable personal and social significance".

Another instance of early modern research on cosmic influences on human physiology is the report on the correlation of the passage of sunspots over the meridian with sudden illnesses and deaths, based on the continuous observation of patients during 267 days, that the French physicians Faure and Sardou, together with the astronomer Vallot, presented in 1922 to the French Medical Academy (Tocquet, 1951). We have already described the work of one of the major early pioneers of cosmobiological research, the Russian biophysicist Alexander L.Chizhevsky, who in the time of 1920-1950 investigated the correlation of the sunspot cycle with human life with statistical methods (see above).

As for other fields, the 1930's were a very fruitful period for biometeorology in which many of the foundations of our modern knowledge were laid. In 1934, the first biometeorological journal was founded, the "Bioklimatische Beiblätter" (bioclimatological supplements) of the "Meteorologische Zeitschrift".

3.4.2 Bernard de Rudder

In the early 1930's, Bernard de Rudder (1894-1962), a pupil of Meinhard von Pfaundler - a pioneer of constitutional thinking in medicine - and director of the department of pediatrics of the University of Frankfurt, Germany, published the first textbook on biometeorology, *"Grundriss einer Meteorobiologie des Menschen"*(1931), an extensive review and critical analysis of the already vast amount of literature in the field. It helped considerably to overcome many misconceptions of his medical colleagues (Tromp, 1963). Rudder established the seasonality of disease, now a fundamental tenet of biometeorological concepts, and concluded that constitution was clearly a factor in weather sensitivity; weather-sensitive persons usually exhibited an unstable vegetative nervous system. However, he found that the usual constitutional studies of correlations of body types and other classifications with weather sensitivity were not convincing. The question of how important weather-induced vegetative stress was for the induction of illness needed further research. Besides biometeorological studies, de Rudder also did research on growth and development, child psychology, and constitutional anomalies, and examined the question of human cosmic rhythms (de Rudder, 1937, 1948). His broad outlook, interdisciplinarity and talent for synthesis, the ability to discern complex interrelations among phenomena of medical and natural science in a time of specialization, made him the Nestor of German biometeorology (Sargent, 1982).

3.4.3 William F. Petersen

Another important pioneer of clinical biometerorology was William F.Petersen (1887-1950), professor of internal medicine and head of the Department of Pathology and Bacteriology at the University of Illinois, Chicago (Petersen, 1935-37, 1947; Berg et al., 1940; Büttner, 1951; Sargent, 1982). He had been already one of the leading researchers in constitutional medicine, the non-specific resistance to disease and the pathophysiology of the autonomous nervous system, and was just writing the first draft of *"The Skin Reactions, Blood Chemistry and Physical Status of 'Normal' Men and of Clinical Patients"* (Petersen & Levinson, 1930), when in the late 1920's he discovered the hippocratic writings. They led him into intensive investigations of the relationship between weather and illness. In the late 1930's, he started to publish the results in the voluminous studies *"The Patient and the Weather"* (Petersen, 1935-37) and *"Man, Weather, Sun"* (Petersen, 1947). They are the most extensive and comprehensive investigations of man's ability to cope with weather and climate ever undertaken (Sargent, 1982).

In summary, Petersen found that the organismic response to environmental forces comprised a rhythmic pendulation of the autonomic nervous processes of the vegetative system, connected with endocrine and biochemical changes. He called this vegetative rhythm, discovered by him in 1930, the "ARS-COD rhythm", because of its connection with the two types of reactions to weather stimuli. One was the sympaticotonic reaction to cold fronts which he called "ARS phase" because it comprised anoxemia, reduction, spasm (vasoconstriction), and relative alkalosis, and the other was the vagotonic reaction – which could, but not always did, coincide with the succeeding warm front — of the "COD phase", a period of stimulation comprising catabolism, oxidation, dilation (vasodilatation), and relative acidosis. There were of course individual differences in these reactions at the cellular, organ and organismic levels, but "on the whole the entire population swings in a chemical and physiochemical rhythm that is identical" (Petersen, 1935).

Petersen's key contribution was his insight into the essential unity of man with his geophysical environment – he called man a "cosmic resonator" – , which he proved with the best tools available in his time and explained with a rational theory (Sargent, 1982). His model drew on constitutional medicine and neuroendocrine physiology and clearly explained the successes and failures in coping with the impacts of atmospheric influences. His understanding of the individuality and variability of man, which he successfully applied to the explanation of biometeorological influences, was amply confirmed by later studies of biochemical and physiological individuality and chronobiology (Williams, 1956; Halberg, 1969b; Luce, 1971).

3.4.4 Wladislaw Kopaczewski

In the late 1930's, the French physician, biochemist and physical chemist, Wladislaw Kopaczewski (1886-1953), also made an early attempt to produce a synthesis of the research in the field; he also tried to develop a theoretical foundation for it based on his work on biological colloids (Kopaczewski, 1938, 1939). His fundamentally biophysical conception shows when he writes that "we are now already in possession of a rich documentation which permits the codification of this new branch of pathology. (...) This codification is necessary, as there is already a deplorable tendency to attribute a dominating role to statistics and to clinical observation, when only physics and experimental physiology are able to give the development of this child of our century a happy impetus" (Kopaczewski, 1938).

3.4.5 Ellsworth Huntington

Another contemporary of Petersen, one of America's most distinguished geographers, Ellsworth Huntington (1876-1947), was mainly concerned with the occurrence of climatic cycles in historical time and their influence on human history, health and welfare, but also investigated the influence of seasonal cycles and weather (Huntington, 1915, 1923, 1924, 1930, 1945;

Sargent, 1982). He was one of the founders of the Ecological Society of America in 1916 and in the same year invented the term "human ecology" to distinguish the work of geographers from that of biologists (see below). In works like *"Civilization and Climate"* (1915), *"Earth and Sun"* (1923), *"Weather and Health"* (1930), and *"The Mainspring of Civilization"* (1945), he concluded from the statistically highly significant correlations found, that were also consistent with the findings of other biometeorologists, that physical environment, constitutional factors (including heredity), and cultural influences work together in determining the rate at which culture and civilization advance; of the physical environment, he concluded that the atmospheric factors had the greatest impact, and found that physical vigour, intellectual performance and health were the most sensitive indicators of that impact. He believed he could demonstrate that certain weather and climatic conditions were optimal for physical vigour and health, including temperature, moisture, and storminess. However, he found that weather factors could only account for one third of the variance in the physiological and pathological processes he studied.

In 1941, the Japanese professor of medicine, Maki Takata, found the influence of an unknown radiation from the sun on blood serum (Takata and Murasugi,1941; Takata, 1951). His findings inspired much of the later biochemical research, as did the work of Piccardi.

3.4.6 Giorgio Piccardi

Highly influential work on the influence of extraterrestrial forces on chemical and biological processes was done from the 1930s to the 1970s by Giorgio Piccardi (1895-1972), professor of physical chemistry at the University of Florence, Italy, in his study of "fluctuating phenomena" (Piccardi, 1962; Capel-Boute, 1974, 1983, 1985, 1990; Manzelli et al., 1994). It began in the late 1930's with his research on the removal of calcareous incrustations in water boilers, which he was able to prevent by adding water pretreated by the glow produced by stirring it with a glass sphere containing a drop of mercury in a low-pressure neon atmosphere. However, he made the strange observation that the action of such "activated water" was not uniformly effective, and he came to believe this was due not to chance, but to some extraterrestrial influence. Already at that time, he suspected that this may be a more general phenomenon, and that often the cause for the well-known non-reproducibility of chemical reactions resided not in the imperfection of experimental conditions, as traditionally assumed, but had their origin in the surrounding space. To test this hypothesis, he devised what he called "chemical tests", standardized reactions like the sedimentation rate of inorganic colloids, such as bismuthyl chloride (BiOCl) and other substances. In order to eliminate known environmental factors such as temperature, atmospheric pressure and humidity, the sedimentation rate of unshielded BiOCl in either activated or normal water was compared with that in shielded condition, in a series of experiments that were made three times daily for more than 22 years; similar series of experiments were made by Carmen Capel-Boute in Brussels and other researchers in other locations.

The effect of the activated water and/or of a metal screen on the sedimentation rate was found to vary with solar eruptions, the 11-year sunspot cycle, and bursts of cosmic rays. In his book *"The Chemical Bases of Medical Climatology"* (1962) he therefore proposed a solar origin for the changes observed, and concluded that "certain phenomena which take place in geophysical space and all of the phenomena which take place in solar space and astrophysical space act at a distance. No matter what the nature of the far-off spatial phenomena, their action is exercised by means of radiation of an electromagnetic or corpuscular nature, or by means of variations in the general field, electrical, magnetic, electromagnetic or gravitational. All of this may today be listed as being distant actions". However, the effects seemed also to depend on the position of the earth relative to the plane of its equator and on the speed of the earth on its motion through the galaxy. Piccardi also investigated the influence of the phases of the moon on his chemical test, and the influence of terrestrial, solar, and cosmic phenomena on biological reactions such as the blood sedimentation rate. He came to believe from these experiments that the basis of what is now called the "Piccardi Effect" is the influence these cosmic factors exert on the structure of water and aqueous solutions. He thought that such an effect would mainly occur in systems that are not in thermodynamic equilibrium, and believed that colloid systems are especially suited to demonstrate these effects. He assumed that even small changes in the intensities of electromagnetic fields were able to exert dramatic and long-lasting effects on the properties of water and aqueous solutions.

Piccardi's work was continued by Capel-Boute, by Tromp's International Society of Biometeorology, by the International Committee for the Research and Study of Environmental Factors (known by its French acronym, CIFA) founded by Capel-Boute, and by Simon E. Shnol' and his group at the Institute of Theoretical and Experimental Biophysics of the Russian Academy of Sciences in Pushchino, among others. CIFA's last meeting was the seminar "Cosmic Ecology and Noosphere" on October 4-9, 1999 at Partenit, Crimea, and was dedicated to the 30th anniversary of its foundation (see http://www.ccssu.crimea.ua/eng/conf/cen99/).

In the time of the 1950's and 1960's biometeorology developed rapidly, chiefly as a result of many environmental studies carried out mainly in Great

Britain and the USA. The deeper physiological mechanisms involved in the observations reported by earlier authors were extensively treated in Solco W.Tromp's *"Medical Biometeorology"* (1963). This was made possible through the creation of the International Society of Biometeorology, founded in 1956 at the occasion of a symposium held at the UNESCO headquarters in Paris, by Tromp, Frederick Sargent II, and many others. The now established discipline was defined at the symposium as follows: "Biometeorology comprises the study of the direct and indirect interrelations between the geophysical and geochemical environment of the atmosphere, and living organisms, plants, animals and man". With a membership of more than 600 scientists from more than 50 countries (1983), representing many disciplines, it was very active and had a high output of publications, such as the *International Journal of Biometeorology* (from 1957), and the book series *"Biometeorology"* (proceedings of the triannual conferences, from 1966) and *"Progress in Biometeorology"* (from 1970). (see Lieth, 1981, and Schnitzler & Lieth, 1983). The society is still active, but underwent a schism, with the partisans of the Piccardi type of work going their own way and founding CIFA. Tromp with his Biometeorological Research Center in Leiden, Netherlands, has performed much interesting research himself, mainly on the influence of weather and climate on blood pressure, blood composition and the physico-chemical state of the blood (see his review on effects of extraterrestrial stimuli on colloidal systems and living organisms, 1972).

For more recent work, see Tomassen (1990) and Shnol' (1996).

3.4.7 Human Ecology

The further development of biometeorology led to the establishment of the wider concept of a "Human Ecology" in which the comprehensive significance of the ecological viewpoint also implicit in quantum physics becomes clear (Sargent & Shimkin, 1965; Sargent, 1974, 1982, 1983). The notion of human ecology goes back to Huntington, who coined it in 1916 and in 1921 initiated a project on the "general study of human ecology relations" of the U.S. National Research Council, in which health is seen as a consequence of complex interactions between man and environment, and disease treated as a consequence of many, not a single factor acting on man, including microbes, but also meterological influences, socioeconomic conditions, and the nature of community health care. Jan Christiaan Smuts' holistic philosophy in *"Holism and Evolution"* (1926) and the ideas of Huntington and Petersen became the basis for the emerging field of "human ecology" as first expressed by botanist J.W.Bews in a book with the same title (Bews, 1935). Since then, many different disciplines have adopted the notion, but always with the common concept of considering man-environment relationships in a holistic and multicausal way, based on the study of the effects of the physical, biological, and socioeconomical environments on man and those of man on his environment; the modeling of the consequences of man-environment relations for health and disease is always treated in this general framework and therefore departs from conventional biomedical thinking.

3.5 Chronobiology

While the crystal model of the solid-state approach clearly brought a further advance in the physical understanding of living systems, it was not sufficient to ascribe only spatial order to the organism, which so obviously displays temporal order as well. Temporal organization is as important as spatial organization for the functioning of living organisms. Franz Halberg, one of the pioneers of this field, has stressed that at the physiological level, not only must the "right" amount of the "right" substance be at the "right" place, but also this must occur at the "right" time (Halberg, 1979). Also, Claude Bernard's concept of homeostatic constancy proved to be too static and had to be expanded by the idea of a rhythmic order. It was the discipline of chronobiology that developed from the discovery of the temporal dimension of biological organization.

Rhythmic movements of plant leaves of carob trees have already been reported by Androthenus, scientist in the retinue of Alexander the Great's military expedition to India in the 3rd century B.C. In 1729, the French astronomer Jean Jacques de Mairan discovered that the diurnal movements of mimosa leafs continued even when kept in the dark for several days. He suggested that the plant knew the time of the day because it possessed its own clock. Also in the 18th century, the Swedish botanist Carl von Linné studied the consistent opening and closing of of flower petals and in his *"Philosophia Botanica"* (1751) inaugurated the study of periodic phenomena in the development of organisms. In 1797, the German physician Christoph Wilhelm Hufeland, one of the protagonists of Romantic medicine, noticed the connection between the rhythmic organization of living organisms and that of our terrestrial and cosmic environment, when he remarked that "the 24 hour period which is transmitted by the regular rotation of our earth to all its inhabitants, is the unity, so to speak, of our natural chronology". In 1832, the Swiss botanist Auguste de Candolle continued the observations on mimosa and reported their adaptation to artificially changed lighting conditions such as shortening or phase shifting of the lighting period. At the same time, the first systematic investigations on the rhythmicity of human physiological functions were done by Grützmann in 1831, by Schweig in 1843, by Gierse in 1842 and by Bergmann in 1845. By experimenting on themselves, Davy in 1845 and Ogle in 1866 observed changes in their body temperatures, and assumed the existence of regular biological cycles in functions such as sleep and body temperature that are undisturbed by external cues. In his last years in the early 1880's, Charles Darwin studied the rhythmical phenomena in plants and animals and tried to relate them to the evolutionary process; he proposed that the daily movements of leaves had a survival value. In experiments on grass seedlings he showed that the tip of the seedlings was the part of the plant that is responsible for the plant's response to the changing light.

One of the most popular versions of biorhythmology was created in 1887 by the Berlin physician and biologist, Wilhelm Fliess, a friend of Sigmund Freud. In his book *"Der Ablauf des Lebens – Grundlegung zur exakten Biologie"* (Fliess, 1906) he postulated that every human had male and female cycles of 23 and 28 days which controlled the physiology of every cell and the tides of bodily and psychological vitality. He even believed the length of individual life and the day of death could be predicted on this base, and proposed a simple mathematical formula to predict favourable and adverse days. However, Fliess' biorhythmology proved to be too simplistic, as many more cycles were discovered in the course of time. Today we know that they cover a spectrum of 10 orders of magnitude, and occur on all levels of organization, from biochemical processes, cellular components and cells, to cell populations, organs, organisms, and populations. In 1898, Svante Arrhenius, in a paper that was also an important milestone for biometeorology (see there) stated that the clock of an organism constituted an open system and that in constant conditions the timing of the periods is derived from the external environment.

The beginning of modern research in chronobiology could be connected with the 1915 hypothesis of an autonomous internal clock by German botanist Wilhelm Pfeffer. After the first observations on time memory of bees by Auguste Forel in 1910, the investigations of Ingeborg Beling done on the instigation of Karl von Frisch in the late 1920's, gave definite proof of a memory of time and place, with elimination of the influence of external Zeitgebers like light, temperature, humidity and atmospheric electricity. At the same time, the Swedish physicians E.A.Forsgren and J.Möllerström investigated the diurnal rhythmicity of bile production in the rabbit and of the erythrocyte sedimentation rate in humans. In 1937 these studies led to the foundation of the medically oriented first International Society for the Study of Biological Rhythms in Stockholm.

The 1930's were very fruitful for the study of chronobiology. The zoologist Erich von Holst studied the interelationships of various rhythms and came to an understanding of the ways they are coupled and lead to a coordination of movements. The physiologist Albrecht Bethe investigated the mechanisms of biological periodicity and stated that rhythmicity was a fundamental property of life and that seesaw oscillations could serve as a model for understanding them. Hoagland postulated the existence of an internal clock and suggested that physiological time may depend on the rate of certain chemical reactions.

However, the most influential hypothesis, based on the actual state of chronobiological research, came from plant physiologist Erwin Bünning (1936, 1958), who took up Arrhenius' suggestion of external (cosmic or atmospheric) influence on biological processes. He did not succeed in showing an influence of electrical charges in the atmosphere on rhythmical movements of leaves, but found that the plants kept in constant darkness deviated from the 24 hour rhythm. He took the periodicity shift as a proof for the endogenous character of the biological clock. Further experiments formed the basis of his hypothesis on the mechanism of timing in plants, namely that the organism undergoes a change of a photophilic and of a photophobic phase of 12 hours each. Light stimuli given in the first phase accelerate the flowering, but stimuli given in the second phase inhibit it. This hypothesis became the starting point of a considerable rise of chronobiological research, which however came to a climax only in the late 1940's, when the hypothesis of the internal clock became more widely known.

It was in the late 1940's and in the 1950's that most of the great pioneers of modern chronobiology, among them Franz Halberg, Frank A.Brown, jr. (1954), Jürgen Aschoff (1965), and A.Sollberger (1965), started their systematic researches that have established the discipline. Franz Halberg and his associates (Halberg, 1953, 1960, 1962, 1969,1979, 1998) started their pioneering chronobiological research with the investigation of eosinophil levels in blood, and with detecting and documenting circadian rhythms in many endocrine functions. They also tracked such rhythms in cell division. Until the late 1940's, most physiologists had believed the meaning of the body's homeostasis was that its normal state was one of constancy. This assumption influenced the norms of medical diagnosis, treatment, and laboratory research, and still continues to influence many branches of physiology, biology and medicine, despite ample evidence to the contrary. Halberg also coined the notions of "circadian" (roughly 24 hours), "ultradian" (periods shorter than circadian, 90-100 minutes) and "infradian" (longer than circadian, i.e. weeks or months) cycles. Frank A.Brown started his investigations with research on the endocrine systems of fiddler crabs (Brown, 1954) which led him into chronobiology. In 1959, he postulated that biological rhythms are imposed by external rhythms. He suggested that diurnal fluctuations in air pressure and gravity associated with rotation of the earth in relation to sun and moon may be responsible. This geographical effect, however, was refuted by K.C.Hammer in the 1960's on the base of experiments at the South Pole.

For a long time, the field was characterized by a controversy between the advocates of exogenous origin of biological rhythmicity and those of an endogenous timing. The former, with Brown as their leading proponent, held that the organism was dependent on environmental rhythms. All known properties of clock-timed rhythms can be explained by entrainment through external "Zeitgebers", rhythmic changes in the organism's ambient physical environment, including light cycles, temperature changes, pulsating magnetic, electric, and extremely low frequency electromagnetic fields (Brown, 1976). The advocates of internal timing, most prominently Erwin Bünning and C.S.Pittendrigh, believed that the organism was independent of environmental rhythms, and possesses an inherited autonomous, self-sustained, temperature-compensated internal clock-timing system (Edmunds, 1976). However, as Edmunds has pointed out, all characteristics of circadian clocks – ubiquity, approximate 24-hour period, entrainability, persistence, phase shiftability, and temperature-compensation – are accounted for by either hypotheses.

If we consider that in the course of evolution all organisms have developed in an environment rhythmically structured by the planetary and cosmic radiation fields, we may assume that originally biological rhythms have been established by the integration of the temporal structure of environmental stimuli into the evolving organisms. Later, they probably have detached themselves from these Zeitgebers and have become autonomous pulsations, but still are in resonance with the external cycles. Anyway, the search for linear monocausal relationships which underlies these controversies must very probably be given up in favour of ecological concepts that take the interwovenness of all phenomena into account.

Such a concept has been developed by Rainer Sinz, physiologist at the University of Leipzig, in the late 1970's (Sinz, 1978, 1980). His concept of the "dynamical multioscillatory functional order of the organism" represents the perfect connection between chronobiological research and integrative biophysics. Based on the developments from Claude Bernard and Walter B. Cannon to Erich von Holst's central-nervous coordination and Werner Rudolf Hess' "functional organization of the vegetative nervous system" (Hess, 1948), Sinz' proposes a dynamical systems theory of the organization of regulation, coordination, and oscillation in the organism, based on the functional unity of regulatory processes, functional coordination and rhythmicity. It integrates the insights of the non-equilibrium thermodynamics of irreversible ordering processes with the knowledge on the hierarchical coordination of nonlinear regulatory and oscillatory systems in the organism. The spatial organization of the organism is correlated to the endogenous time structures arising from the nonlinear processes of regulation and oscillation of dissipative structures and their coordination at various levels of organization, which, according to Sinz, have been optimized by phylogenetic adaptation to the dominating environmental rhythms. If we consider this "oscillator model of the organism" in the light of our field model, the living organism with its innumerable rhythmical processes basically is a highly complex resonating system of oscillating fields coupled nonlinearly by their phase-relations.

Today, chronobiology (the name was coined in 1960) appears as a well founded and highly significant, but still somewhat marginal science – common fate of all the "border sciences" (Tromp) that do not fit into the classical disciplinary scheme and challenge the conventional scientific world view (for reviews and recent work see Luce, 1971, Halberg, 1979, 1998; Edmunds, 1984; Gutenbrunner et al., 1993).

3.6 Non-Equilibrium Thermodynamics and Self-Organization

One of the most fundamental conceptual shifts making the new biophysics possible certainly was that from classical equilibrium thermodynamics to non-equilibrium thermodynamics. Although generally associated with the name of Ilya Prigogine, who formalized non-equilibrium thermodynamics in the late 1940's and in 1977 received the Nobel prize in chemistry for this achievement, the concept of living organisms as non-equilibrium (open) systems was first developed by Alexander G.Gurwitsch, Ervin S.Bauer, Ludwig von Bertalanffy, and Erwin Schrödinger.

3.6.1 Alexander G. Gurwitsch

Gurwitsch, also known as discoverer of mitogenetic radiation (now called biophoton emission), suggested in the early 1920's that the molecular aggregates in living cells and tissues existed in a state of non-equilibricity which in his view was due to the action of biological fields (Gurwitsch, 1923; Gurwitsch & Gurwitsch, 1959). These excited "unequilibrated molecular constellations" were the foundation of the aliveness of the cell, and in their decay, the potential energy stored in their conformation would be released in the form of mitogenetic radiation (biophotons) or transformed into kinetic energy which then would cause the directed movement of substance. Aliveness was based on the constant change between the excitation of molecules by the field and the relaxation of the molecular binding energy which was retransformed into field energy. According to Gurwitsch, the individual atomic and molecular composition of living substance was of no import; it was their spatial organization that was decisive. As this was solely determined by the biological field, it was the latter that became the central element of biological understanding.

With these concepts Gurwitsch became one of the pioneers of modern biophysical thinking. The fact that he designated the "molecular constellations" as "states of mutual alignment and orientation" of molecules, makes him the first one to postulate what is now known as "collective states" or "cooperative phenomena" as the basis of life processes. He also was the first to realize that the distribution of energy in the living system was of paramount importance. The cooperative behavior of many molecules "greatly enhances the chances of energy transmission from one molecule to the next" and also "the propagation of chain reactions is only possible due to the regular arrangement of the reacting particles". He pointed out that the ability of the protoplasm to shift easily between its "immobile" (solidified) and fluid states was understandable as a "result of the spatial rearrangement of the regular molecular order and therefore as a result of a redistribution of energy" (Gurwitsch & Gurwitsch, 1959).

3.6.2 Ervin S. Bauer

Very similarly, Bauer, Russian theoretical biologist of Hungarian origin, already in the 1920's and then 1935 set forth three fundamental postulates, from which all biological phenomena may be derived (Bauer, 1920, 1935). (1) All living systems are always in stable non-equilibrium condition and therefore have energy to perform work. They use this energy to sustain the non-equilibrium state, therefore they are purposeful. From this postulate follows homeostasis. (2) Living systems are performing internal work, but also external work (interaction with environment, obtaining food etc.). The postulate states that the proportion of the energy used in external work to that used in internal work increases in development. (3) Non-equilibricity in living systems consists of molecular compounds in whose conformation the metabolic energy used to form them is stored.

As already described above, some of the most decisive work about the non-equilibrium nature of living systems was done in the 1940's and 1950's by the Austrian-Canadian biologist and founder of "General System Theory", Ludwig von Bertalanffy (Bertalanffy 1940, 1949, 1950, 1953). Together with A.C.Burton (Burton, 1939), he introduced the concept of the living organism as an "open system" in the sense of thermodynamics.

3.6.3 Erwin Schrödinger

In his small, but influential book *"What is Life ?"* (1944), Erwin Schrödinger (1887-1961), one of the founders of quantum physics and Nobel laureate of 1933, also engaged in the discussion of the thermodynamics of living systems. After discussing the nature of heredity and predicting the structure of the then still hypothetical gene, calling it an "aperiodic crystal" which he suggested was storing information as a code in its structure, he discussed the problem of how organisms could retain their highly improbable (from the standpoint of equilibrium thermodynamics) ordered structure in the face of the Second Law of Thermodynamics (the entropy principle). Schrödinger pointed out that organisms retained order by feeding on "negative entropy" from the environment. They liberate themselves from the entropy which they cannot avoid producing by exporting it into their environment. The term "negative entropy" Schrödinger coined (like its later abbreviation into "negentropy" by Brillouin), has not been well received by other scientists. He used it instead of talking of the free energy, the non-equilibrium of the environment, upon which organisms feed, a fact that had already been pointed out by F.G.Donnan in 1928 (Donnan, 1929). It was this new principle, "order from order", Schrödinger said, that was they key for understanding life. He explained there were two principles for creating order in nature – the statistical one which creates "order from disorder", and the new "order-from-order"-mechanism, which was much simpler and much more able to explain the behaviour of living matter. He stated the new one was not foreign to physics – it was nothing else than the principle of quantum theory.

Schrödinger, like Niels Bohr, was intrigued by the idea that understanding living systems may involve "other physical laws" beyond the "known law of physics", but without evading these. He stated that "the living organism seems to be a macroscopic system which behaves in some of his aspects like matter close to absolute zero, where molecular disorder is removed". Thus, the new laws Schrödinger was looking for were of the nature of the intermolecular forces of solid state physics.

3.6.4 Ilya Prigogine

One of the most fundamental contributions to this field is certainly the work of Ilya Prigogine, Nobel laureate in chemistry of 1977, on the thermodynamics of irreversible processes and on the theory of self-organization (Prigogine 1947, 1961; Glansdoff & Prigogine, 1971; Nicolis & Prigogine, 1977). Although there are earlier contributions to the theory of self-organization, such as the classical works by Henri Poincaré, Alexander M.Lyapunov and later Alan M.Turing, it was Prigogine's work that definitely established non-equilibrium thermodynamics and allowed its use in the explanation of life processes. Using the earlier experiments of the Germans F.Ferdinand Runge and Raphael E. Liesegang, the French physicist H.Bénard, and the Russians Boris P.Beloussov and Anatol M.Zhabotinsky on self-organization in chemical reactions (Kuhnert & Niedersen, 1987), Prigogine established that these nonlinear reactions are just inorganic examples of the more general phenomenon of "dissipative structures". The work of Beloussov probably would not have been noticed, if not Simon E.Shnol' had recognized the general significance of the phenomenon (Shnoll et al., 1981, 1982). He then commissioned Zhabotinsky to do a more detailed investigation, who extended the reaction considerably (Kuhnert & Niedersen, 1987). These pattern-forming, oscillating chemical reactions, used by Prigogine as a model for the physical understanding of biological order, exhibit the properties of cooperative behaviour, sensitivity to small perturbations, and memory. When a continuous and sufficiently strong flow of energy is fed into a collection of disordered molecules, spontaneous order arises in them and is maintained as long as the inflow is kept up. This new class of order "far from the thermal equilibrium" arises from the sudden collective behaviour of a great number of particles, triggered by small fluctuations which in thermal equilibrium do not have such an effect. Thus, non-equilibrium thermodynamics can explain how biological systems use nutrition as an external pump to establish a stable state far from equilibrium. The stability of living systems is based on what Popp has called the main optimization principles of physics, the First and Second Law of Thermodynamics, applied to the common system of both biological systems and their surrounding, which together can be treated as closed systems.

A first textbook on non-equilibrium in biophysics was published in 1967 by Aharon Katchalsky and P.F.Curran (Katchalsky & Curran, 1967).

3.6.5 Herbert Fröhlich

While the physico-chemist Prigogine uses the concept of coherence to characterize the mutual spatial alignment and the "collective", holistic behaviour of a large number of particles, Herbert Fröhlich (1905-1991), British theoretical physicist of German origin, directed his attention to the electromagnetic interactions of these particles. In contrast to Prigogine, he not only referred to thermodynamics which is concerned with the statistical behaviour of large numbers of particles, but based his work on the quantum theory of solids and liquids. Before introducing the concept of coherence into biology, Fröhlich had created the foundations for understanding supraconductivity. However, the 1972 Nobel prize for this achievement was not awarded to him, but to John Bardeen, Leon Cooper and John Schrieffer. Superconducting behaviour is due to the establishment of long-range coupling forces binding particles together in pairs separated by hundreds of atomic diameters. Fröhlich's application of quantum field theory to the physics of solids revolutionized this field not less than his introduction of coherence changed biophysics.

In the fundamental paper of 1968 in which he did this, he writes that biological systems should show longitudinal vibrations (so-called "optical phonons") of frequencies between 10^{11} and 10^{12} Hz (microwaves) arising from the dipole properties of cell membranes, macromolecular bonds such as hydrogen bonds and probably also from delocalized electrons. The non-equilibrium state created by the continuous supply of a certain minimum of energy should then bring about a stationary state in which the energy is not completely thermalized, but stored in an very orderly way. The strong excitation of a particular frequency of these phonons leads to long-range phase-correlations. Fröhlich pointed out that this phenomenon had a strong similarity to the condensation of a Bose gas at very low temperatures, the so-called Bose condensation (Fröhlich, 1968).

Fröhlich's concept of coherence corresponds in all essential points to Prigogine's dissipative structures. In both concepts nonlinear, autocatalytic (self-amplifying) processes lead to the system's stabilization far from thermodynamic equilibrium and in the form of a dynamic order, provided that energy beyond a certain minimum is continuously fed into the system. Like Prigogine, Fröhlich concludes that long-range correlations and coherent structures must play a fundamental role in living organisms, by regulating biological processes like the function of enzymes, the storage of energy, growth and the establishment of order.

However, in contrast to Prigogine, Fröhlich does not consider the molecular structure of the organism as most essential in this process, but the electromagnetic fields coupled to the molecules, which in his view are responsible for the higher order of the non-equilibrium state. Fröhlich's model is more fundamental and far-reaching, because electromagnetic interactions are the most fundamental couplings in living systems. All binding forces between atoms and molecules are of electromagnetic nature, and as a consequence, all chemical reactions are basically electromagnetic processes. Fröhlich has also suggested to use the heat bath as an idealization of the external world, and the chemical potential to describe the interaction between the surroundings and the biological system.

Popp has pointed out that the energy distribution over the spatio-temporal structure has to be considered, which in the form of the equal spectral distribution of electronic excitations over all possible modes (wavelengths) constitutes the non-equilibrium threshold of a system, i.e., thermodynamic non-equilibrium corresponds to the predominance of excited species over non-excited ones, and thus to the lasing state.

However important the thermodynamic aspects of life may be, the mechanisms of establishing coherent states, considered by Popp as the fundamental biophysical principle of life, cannot be explained by thermodynamics (or electrodynamics), but concern quantum theory (Popp, 1994).

3.7 Biological Fields

The "building-stone" view of equilibrium thermodynamics which underlies classical physics and chemistry, has, up to now, been the world-view upon which biophysics has, like the rest of the biological sciences, based its picture of the organism. One of the first and most important steps in the development of a holistic biophysics must therefore be not only to complement the classical view with the field aspect, but even to build its model of the organism completely on the field picture. The electromagnetic field theories of Faraday and Maxwell proposed in the second half of the 19th century have inspired biologists already from 1900 onwards to develop biological field theories. As I have shown elsewhere, the concept of the field has occupied a central place in the school of "organismic" or "holistic" biologists in the first half of the 20th century (Bischof, 1998). Leading biologists of the time, such as Hans Spemann, Ross Harrison, Paul A.Weiss, Joseph Needham, Conrad H.Waddington, and Alexander G.Gurwitsch (Lipkind, 1998), used the hypothesis of a biological, or morphogenetic, field, introduced by Gurwitsch and Weiss in the early 1920's, as a tool for understanding the phenomena of development, regeneration and morphogenesis and to make predictions for experimental testing. They generally considered the biological field as a purely heuristic concept and left the exact nature of the fields open. The time (and electromagnetic science) was not yet ripe for the notion of real electrical, electromagnetic or otherwise physical, fields of long-range force. At this time, there was neither enough experimental evidence for the existence of bioelectromagnetic fields nor for the biological effects of EM fields. Another reason for the late acceptance of the electromagnetic aspect of organisms was the success of the biochemical approach to life.

There are several reasons why the concept of the biological field has almost completely vanished from biology in the 1950's (Gilbert et al., 1996). The main reason was the rise of genetics as an alternative program to explain development. In the 1930's, the biological field had been a clear alternative to the gene as the basic unit of ontogeny and phylogeny, and this was seen as a threat to the rise of genetics as the leading field of biology. This was the reason why Thomas Hunt Morgan, himself originally a developmental biologist, actively fought the concept of the morphogenetic field and tried to ridicule it in all possible ways. He denounced field theorists, and embryologists in general, as old-fashioned, metaphysical, holistic, mystical and bad scientists, not to be taken seriously in the new gene-based reductionist biology (Mitman & Fausto-Sterling, 1992). However, Morgan never presented any evidence against fields or gradients (Gilbert et al., 1996), and at that time genes were not less abstract and metaphysical than morphogenetic fields. On the other hand, most embryologists also did not feel they needed to take genes seriously (Gilbert et al., 1996), which helped to push embryology out of the "New Synthesis" of genetic-evolutionary theory. By the late 1930's, evidence started to come for the role of genes in embryogenesis, but then, embryologists still had reason to doubt the geneticists' claim to be able to explain development, as it was not possible to explain how identical genes in each cell created different cell types and no genes active in early development had been identified. However, with the breakthrough of genetically oriented molecular biology through Watson and Crick's 1953 model of DNA, field theories were definitely out (Bischof, 1996) and the predominance of the molecular approach in biology began, as we know it today.

More recently, however, a group of leading U.S. biologists, Scott F.Gilbert, John M.Opitz, and Rudolf A.Raff, have challenged the adequacy of genetics for alone explaining evolution and the devaluation of morphology, and have demanded the rehabilitation of the biological field (Gilbert et al, 1996). They suggest that evolutionary and developmental biology should be reunited by a new synthesis, in which morphogenetic fields are proposed to mediate between genotype and phenotype. As an alternative to the solely genetic model of evolution and development, the new synthesis should be based on the morphogenetic field as a major unit of ontogenetic and phylogenetic change. Gene products should be seen as first interacting to create morphogenetic fields in order to have their effects; changes in these fields would then change the ways in which organisms develop.

3.8 Bioelectromagnetics

Of course, investigations of bioelectromagnetism have a long history, and electrophysiological studies have played a major role in early biophysics, starting from the Berlin school and the French school of Claude Bernard (Bischof, 1994). Among the less known pioneers of bioelectromagnetic research I would like to mention the early French school of biophysics, with Bernard's pupil Arsène d'Arsonval (Delhoume, 1939; Chauvois, 1941; Kopaczewski, 1947), Fred Vlès and his collaborators, Georges Lakhovsky, and Charles Laville. Arsonval, who in 1894 succeeded his teacher Bernard on the chair of physiology at the Collège de France, founded in 1882 the first official laboratory of biophysics in France at the École des Hautes Études.

Various bioelectromagnetic field concepts were proposed in the first decades of the 20th century, e.g., by Rudolf Keller, George W.Crile, Charles Laville, Harold S.Burr, Elmer J.Lund, and Georges Lakhovsky (Bischof, 1994). Laville, French electrical engineer and biologist, proposed in the 1920's that muscles behave like capacitors and that the cell was an electrical engine; he pointed out the significance of electronegativity for biology and proposed that cancer is an "electromagnetic disturbance of the cell". He was convinced that the organization of living matter was due to the structure of underlying force fields, and ascribed changes in it to changes in the distribution of energy (Laville, 1925, 1928, 1932, 1950). Lund concluded from decades of work on the bioelectric potentials of plants and animals that all polar systems are surrounded by and possess interpenetrating electric fields generated by each one of the constituent polar cells in order to maintain electrical correlation within the system (Lund, 1947). He hypothesized that the fields constitute a primitive type of integrating mechanism that plays an important role in the spatial organization of metabolic processes and coordinates growth and possibly other processes. The hypothesis of Burr and Northrop (Burr and Northrop, 1935), based on Burr's work on bioelectric potentials (Burr, 1972), is summarized in the following quote:

> *The pattern of organization of any biological system is established by a complex electro-dynamic field, which is in part determined by its atomic physico-chemical components and which in part determines the behaviour and orientation of those components. This field is electrical in the physical sense and by its properties it relates the entities of the biological system in a characteristic pattern and is itself in part a result of the existence of those entities. It determines and is dertermined by the components. More than establishing pattern, it must maintain pattern in the midst of a physico-chemical flux. Therefore, it must regulate and control living things, it must be the mechanism the outcome of whose activity is "wholeness", organization and continuity. The electro-dynamic field then is comparable to the entelechy of Driesch, the embryonic field of Spemann, the biological field of Weiss"* (Burr and Northrop, 1935).

Lakhovsky (1963), French engineer of Russian origin, postulated that each living being is simultaneously a broad band emitter and receptor of EM radiation, and that every cell of the organism is an electrical resonator. According to Lakhovsky, the chromosomes of the cell nucleus constitute oscillatory circuits and are surrounded by a weak electromagnetic field which interacts with the ambient EM fields. Perturbations of the natural frequency of the cell can induce illness; they may be caused by changes in the composition of the medium in the cell, by the presence of bacteria or viruses, which are supposed to emit their own EM fields, or by unusual fluctuations of solar or cosmic radiation.

However, these proposals were premature, and the breakthrough and beginning of modern EM field theories in biology came only in 1970, with Alexander M.Presman's report of the pioneering work of Soviet bioelectromagnetics researchers, which also contained a first outline of a holistic EM field theory of the organism and his relationships to the environment (Presman, 1970). Since then, there is ample evidence for bioeffects of EM fields and endogenous EM fields (Bischof, 1994, 1998). It is now established that organisms react sensitively to the impact of electromagnetic fields, including very weak ones; effects of various types of endogenous physical fields on cellular organization and morphogenesis are very likely. We also know that several kinds of electromagnetic fields, including microwaves and optical frequencies (biophotons), are emitted from living beings. There is also evidence that weak endogenous electrical currents are involved in regeneration and growth of new tissue; the role of ionic currents in morphogenesis and development has also been demonstrated. Communication by electromagnetic fields is established for fishes and insects, which suggests this may be a more general phenomenon.

Presman's idea of the role of EM fields in the organism is still valid. Based on his review of the Russian and Western work on biological effects of EM fields and on biological EM radiations, he argued that environmental EM fields have played some, if not a central, role in the evolution of life and also are involved in the regulation of the vital activity of organisms. Living beings behave as specialized and highly sensitive antenna systems for diverse parameters of weak fields of the order of strength of the ambient natural ones. According to Presman, EM fields serve as mediators for the interconnection of the organism with the environment as well as between organisms, and EM fields produced by the organisms themselves play an important role in the coordination and communication of physiological systems within living organisms. In all these biological functions, in addition to energetic interactions, informational interactions play a significant (if not the main) role. These bioinformational capacities of electromagnetic fields occur, thus Presman's conclusion from experimental evidence, in their fullest state of development only in the organism as a whole, and are either not present, or not present in comparable form, at the molecular level. As these capacities seem to be a function of the complexity of organisms, it may be useless to attempt to investigate EM interactions with biological systems on the molecular level. In healthy organisms, the internal EM information systems are reliably shielded from interference by natural EM fields, but in pathological states spontaneous variations of EM fields can upset the regulation of physiological processes.

4. CONTEMPORARY ASPECTS OF INTEGRATIVE BIOPHYSICS

4.1 The Quantum Nature of Life

Although the concept of a "quantum biology" has already been proposed in the 1930's by Pascual Jordan, one of the founders of quantum theory (Jordan, 1934, 1942; Beyler, 1996), and was taken up by Friedrich Dessauer, one of the founders of modern biophysics, in the 1950's (Dessauer & Sommermeyer, 1964), and carried in the name of the famous "Cold Spring Harbor Symposia on Quantum Biology" that were involved in the birth of molecular biology, full use of the consequences of quantum physics has not yet been made in biology (Dürr, 1997). The philosophical implications of quantum mechanics have played an important role at the very origin of molecular biology, but today the Schroedinger equation is merely used for calculating molecules and their interactions, while until recently, its philosophical implications rarely have been taken into account.

The philosophical contribution to the rise of molecular biology (Culotta, 1974) originated in Niels Bohr's speech *"On Light and Life"* at the International Congress on Light Therapy at Copenhagen, August 15, 1932 (Bohr, 1932), where he postulated that a new physics was required for interpreting life; life was not reducible to atomic physics. His suggestions were taken up by Pascual Jordan, who in his "amplifier theory of life" proposed that living systems may be able to amplify weak signals, even single photons, in such a way that they could trigger macroscopic events (Jordan, 1938, 1948). However, in this farsighted suggestion he did not take the nonlinearity of the amplification process into account.

Bohr's views also strongly influenced some of the key figures in early molecular biology, mainly Max Delbrück and Erwin Schrödinger (Kay, 1985). Schrödinger took them up in his famous small book *"What is Life ?"* (Schrödinger, 1944), which inspired many young physicists traumatized by the wartime use of physical expertise, to go into biology and had a strong influence on the development of molecular biology (Murphy & O'Neill, 1995).

However, the book certainly was more influential by interpreting the genetic viewpoint of Herman Joseph Muller, Thomas Hunt Morgan and Max

Delbrück in physicist's terms and by backing it with the prestige of physics, than by convincing biophysicists of the Bohr-Schrödinger hope of discovering new physical laws through biophysical investigation of biological phenomena (Keller, 1990). These latter ideas – which have to be seen in the context of the rather strong holistic tendency of the biology of the time – did not exert any lasting influence on molecular biology. On the other hand, they have been, and still are, a seminal influence for the later emergence of a holistic biophysics.

Today, molecular biology proposes itself as the manageable project of refashioning life and redirecting the course of evolution that some of its early pioneers like H.J.Muller and Warren Weaver had envisioned (Carlson, 1971; Kay, 1993). Because of the enormous technological and social power promised by molecular biology, even the increasing awareness of the bad science on which it is based in many respects, such as the numerous weak points in genetic and evolution theory (Ho, 1999), does not prevent it to carry us into such immature and dangerous projects as genetic engineering biotechnology in agriculture and medicine, and the *"Human Genome Project"*, whose deeper nature is revealed by the military epithet of a *"Manhattan Project of the life sciences"*.

The noted quantum chemist, Hans Primas, agrees with this fundamental criticism (Primas, 1990; Primas et al., 1999). He writes that molecular biology, as it exists today, is in fact engineering, not science. It is pragmatic, instrumental knowledge which aims at the power over nature, but not at understanding. It does not constitute a scientific theory of life able to give us orientation to live rationally with nature, but only provides technological control over life. Contemporary molecular biology has become a scientific technology which has lost contact with the epistemological sciences. Aspects of life that cannot be treated or understood from the molecular viewpoint, such as morphology, are glossed over. The assessment Robert Rosen made in 1967 is actually still valid*: "...it must be pointed out that the older problems* [he refers to the questions that preoccupied an older generation of biologists] *have merely been displaced and not solved by the recent developments at the molecular level. These problems involve the very core of biological organization and development: homeostasis, ontogenesis, phylogenesis"* (Rosen, 1967).

4.1.1 The Intrinsic Holism of Quantum Theory

For Primas (1990), the primary shortcoming of molecular biology is that the holistic character of the physical world now recognized in quantum theory is either not acknowledged by the bioengineers or rejected as irrelevant. He emphasizes that molecular biology, though well grounded in empirical knowledge, has no foundation whatsoever in the principles of quantum theory, contrary to a widely held belief to the opposite. It uses the methods and technologies of quantum mechanics, but its way of thinking is still committed to the classical physics of the 19th century and has not taken notice of the fundamental insights of quantum mechanics on the structure of the material world.

According to Primas, on whose statements the following is mainly based (1981, 1982, 1985, 1990, 1992, 1995, 1999), the atomistic-molecular view of matter and the reductionist-mechanist philosophy have no longer any scientific foundation, according to the actual understanding of quantum theory. The description of reality by isolated, context-independent, elementary systems such as quarks, electrons, atoms, or molecules is only permissible under certain specific experimental conditions, and these entities cannot in any way be considered as "fundamental building stones" of reality. Besides the molecular one, there are other, fundamentally different descriptions, complementary to the molecular one, which are quantum-theoretically equivalent and equally well founded. Quantum theory is much richer in possibilities than is admitted in the worldview of molecular biology.

In Primas' view, the feature of quantum theory that is most significant for biology is its *intrinsic fundamental holism*. Already Max Planck said that "the conception of wholeness must (...) be introduced into physics, as in biology, to make the orderliness of nature intelligible and capable of formulation" (Planck, 1931). Since then it has been made clear that for quantum mechanics, the scientific theory most widely recognized as fundamental and best confirmed by experiment, material reality forms an unbroken whole that has no parts. These holistic properties of reality are mathematically precisely defined by the Einstein-Podolsky-Rosen (EPR) correlations which are experimentally well defined. Primas postulates that, by virtue of this, quantum mechanics constitutes the first and up to now only logically consistent, universally valid, mathematically formulated holistic theory. In quantum mechanics, it is never possible to describe the whole by the description of parts and their interrelations.

With this view of quantum mechanics Primas follows Bohr and the school of Heisenberg (Heisenberg, 1976; Dürr, 1968), while quark physics as founded by Murray Gell-Mann continues to cultivate democritean atomism with their clinging to the concept of elementary particles (Kanitscheider, 1979). Similar holistic views of quantum theory are the *"bootstrap theory"* of Geoffrey Chew (1970), David Bohm's *"Causal Quantum Theory" or "Holographic Theory of Reality"* (Bohm, 1980; Bohm & Hiley, 1993), and others advocated by Henry Stapp (1997), Amit Goswami (1989, 1993, 1994), Menas Kafatos & Robert Nadeau (1990), Norman Friedman (1997), David Peat, Fritjof Capra, Harald Atmanspacher

(1996), and many others. This holistic view of quantum theory, although the phenomena on which it is based are not yet completely understood theoretically, cannot be rejected anymore because the strange EPR quantum correlations of non-interacting and spatially separated systems have been amply demonstrated in many experiments (Aspect & Grangier, 1986; Duncan & Kleinpoppen, 1988; Hagley et al., 1997). Therefore the worldview of classical physics, atomism and mechanistic reductionism, definitely cannot anymore be the basis of our worldview, and of biophysics. Quantum mechanics has established the primacy of the unseparable whole.

Another important epistemological consequence of quantum mechanics, complementarity, is also connected to its holism. As Primas writes, there is no single description, such as the molecular-reductionistic one, which can alone represent the whole reality of the subject of a scientific investigation, or is better or "truer" than any other. Nature is extremely diverse and stratified; each description comprehends only a minute partial aspect of its unfathomable multiplicity. Any scientific description of a natural phenomenon is only possible if we renounce the description of its complementary aspects. Quantum theory can only be applied if we abstract from certain aspects and thereby break the holistic symmetry. However, the kind of abstraction we use is not prescribed by the first principles of the theory, such that quantum mechanics allows, and even requires, many different, but equivalent, complementary descriptions of nature. As an important postulate for future science, Primas (1990) therefore emphasizes that we will have to learn to work simultaneously with several complementary descriptions of nature.

In this perspective, the molecular view is legitimate and important and should not be abandoned; molecular biologists can be rightly proud of their successes. It should be cultivated, but not at the expense of other viewpoints. It is its extreme one-eyedness that must be criticised. However, as Primas points out, *"biology is more than molecular biology"*. He postulates that science must now redirect its attention to the wholeness of nature, and therefore will have to ask radically new questions. It has to develop a concept of reality which does not exclude any part of it. Those properties which belong to organisms only as wholes must remain within the scope of science. Therefore, it will be necessary to consider the phenomena as well from "bottom-up", as in mechanistic understanding, as from "top-down", as in vitalistic and holistic understanding. According to Primas, the notion that the latter is not legitimate or secondary is a prejudice that must be overcome. From the viewpoint of the quantum-theoretical worldview, both are completely equivalent, but lead to fundamentally different research agendas and insights. Even the criteria according to which the scientist decides what is scientifically defendable and interesting, are completely different from these two viewpoints. Also, according to quantum theory functional and teleological explanations are completely legitimate and equivalent to causal ones; even the primacy of causality has no foundation in the first principles of physics. Primas points out that it is not possible to distinguish between causal and final processes by purely mechanical means and that such a distinction only makes sense for irreversible processes. As to the existence of the hypothetical vitalistic forces, Primas states that modern physics is well able to integrate new forces into its system.

4.1.2 The Hidden Domain of Potentiality and Connectedness

Thus, matter has become "dematerialized" by modern quantum theory, and this property of "thinglessness" in the quantum worldview is closely connected to the property of "interconnectedness". The emphasis is no longer on isolated objects, but on relations, exchanges, interdependences, on processes, fields, and wholes. Quantum theory is a non-local theory (Stapp, 1997). It is important to see that it is not sufficient to retain the classical world of objects and only add the interconnectedness as a supplementary property of these objects.

They are two of the complimentary descriptions or aspects of reality which Primas has alluded to and cannot be used simultaneously; thus they rather should be considered as different dimensions of reality. The holistic interpretation of quantum theory in fact may also be taken as implying a multidimensional structure of reality (Shacklett, 1991; Friedman, 1997). In this view, there are, besides the world of objects, one or several more fundamental levels of reality where interconnectedness rather than separatedness dominates. Fields certainly belong into this category; however, apart from electromagnetic and other physical fields which are still among the phenomena considered as belonging to the four fundamental forces of the observable world, we must assume the existence of additional field-like levels of reality not directly observable at present, which may be beyond space-time and represent the realm of potentiality (Heisenberg, 1958), or of the "noumena", the realm behind the phenomena assumed by Newton, in contrast to the actuality of the observable. The Schroedinger wave function of quantum theory actually describes this hidden domain of potentiality, of the non-observable, unmanifested, pre-physical world of non-local correlations and superluminal, instantaneous connections, rather than the world of observable phenomena (Friedman, 1997). Only with the act of measurement this infinity of potentialities, described in the Schroedinger equation as a superposition of all possible quantum states, is "collapsed" into one single actuality. Connected to the concept of potentiality is the concept of "entanglement" which describes the characteristic of interconnectedness (Shimony, 1988). In the absence of any interaction (such as a measurement), two systems are in an entangled state in which neither system by itself can be said to be in a "pure state", i.e., can be fully specified without reference to the other.

This hidden domain can be considered as a fundamental dimension of reality, a domain of dynamical connectivity, from which the patterns of the physical world arise. According to some authors, this realm of pre-physicality is not only the basis of the physical world and of matter, but also seems to be connected to, consciousness, which some see as the fundamental field underlying it (Bohm, 1980; Hagelin, 1987; Goswami, 1989, 1993; 1994; Shacklett, 1991; Gough & Shacklett, 1993; Laszlo, 1995, 1996; Friedman, 1997; Grandpierre, 1997). In physics, it is treated by the various models of the physical vacuum. Its possible relevance to biophysics as a basis for a true quantum biology (Josephson & Pallikari-Viras, 1991; Zeiger, 1998; Zeiger & Bischof, 1998; Thaheld, 1998, 2001a,b) seems obvious to us, as we will explain later. Therefore we postulate the development of a *"vacuum biophysics"*.

The "hidden domain" of connectivity has characteristics completely different from those of the classical, macroscopic world of separated objects. For a long time, the quantum description that reveals the properties of phenomena belonging to this domain, or arising from it, was taken to apply only to the microscopic world of atoms and molecules, while the world of macroscopic phenomena of our experience was considered to be purely classical and not to manifest quantum properties. However, today we know that this is not true, and that there are many macroscopic quantum manifestations, although our knowledge about them is still limited (Leggett, 1986, 1992, 1996; De Martino et al., 1997; Sassaroli et al., 1998). Biological systems obviously possess the characteristics of macroscopic quantum systems.

4.1.3 Macroscopic Quantum Phenomena

The usual assumption is that quantum properties disappear at macroscopic scales. But already in the 1960's, Herbert Fröhlich pointed out that low-temperature phenomena in fluids, like liquid helium, had shown "macroscopic quantization of a more subtle nature", and that these observations made it necessary to assume the existence of "a new kind of order based on the concept of phase correlations" (Fröhlich, 1969). Fröhlich, one of the foremost protagonists of the superconductivity and superfluidity research of the time, was convinced that this concept could be applied beyond low-temperature physics, "in particular in systems that are relatively stable but not near thermal equilibrium, and which show an organized collective behaviour which cannot be described in terms of an obvious (static) spatial order", i.e., biological systems. He pointed out that macroscopic concepts expressed in terms of microscopic (particle) behaviour were well developed for systems near thermal equilibrium, but not so much for explaining the existence of a macroscopic spatial order which is not of such an obvious nature as that of crystal structures, such as in the phenomena of superconductivity and superfluidity (Fröhlich, 1969).

As described above, Fröhlich introduced the concept of coherence into biology in 1968 by suggesting that the kind of macroscopic quantum behaviour first observed in low-temperature physics, namely the so-called Bose condensation, may also be observed in biological systems (Fröhlich, 1968, 1969, 1975). The concept of Bose (-Einstein) condensation originated in 1924, when Albert Einstein and his Indian colleague Satyendra Nath Bose predicted that atoms at extremely low temperatures would not dissipate as in an ordinary gas, but fall into their lowest energy state and collectively oscillate in a common coherent state. For technical reasons, this assumption could not be confirmed until 1995, when Eric A. Cornell and Carl E. Weiman of the University of Colorado (Andersen et al., 1995) and Wolfgang Ketterle of Boston's MIT (Davis et al., 1995) first succeeded with their research teams in producing Bose condensates of rubidium and sodium atoms; the three physicists received the 2001 Nobel prize for this achievement. Fröhlich's suggestion that this new, macroscopic, non-equilibrium quantum state of matter, very different from all known physical states, played a fundamental role in living systems, was later taken up and developed by Bhaumik et al. (1976), Mishra et al. (1979) and others, and is a central element of biophoton theory and integrative biophysics.

4.2 Quantum Electrodynamics and Optics

4.2.1 Quantum Coherence

The bases of our knowledge of quantum coherence have been established by in the 1960's by Roy J.Glauber of Harvard University in Cambridge, USA (Glauber, 1963a,b, 1964), based on an early suggestion by Erwin Schrödinger (Schrödinger, 1926). Starting from the theory of single-photon detection, Glauber developed the concept of "coherent states" which transcends the classical concept of coherence. Coherent states – in terms of the Schrödinger equation, "eigenstates of the annihilation operator" – are not just characterized by the ability for interference ("coherence of the second order"), but fulfil the ideal relations of a coherence of an arbitrarily high order. They are wave packets of which neither position nor momentum are exactly known and which have minimal uncertainty. They integrate the otherwise unbridgeable opposites of the wave and the particle nature of light into a new and higher unity, and simultaneously partake of the localized nature of particles and of the long-range delocalization of waves, and in coherent states these two modes are feedback-coupled to each other in such a way that one never can annihilate the other completely. As already Heisenberg has shown (1930), the state of minimal uncertainty is the most stable state and can carry the highest amount of information of all states.

As Popp has pointed out, this situation of coherent states between particle and wave, coherence and incoherence, corresponds to their existence at the laser threshold and enables them to act as homeostatic regulators allowing flexible switching of the field between coherence and incoherence. Because of their extremely low quantum fluctuation and therefore low noise they have a very high signal-to-noise ratio. According to Popp, coherent states are fundamental for biological systems; they enable them to optimize themselves concerning organization, information quality, pattern recognition, etc.

4.2.2 The Dicke Theory of Coherent Emission of Coupled Multiatomic Emitters

In 1954, Richard H.Dicke, later an eminent astrophysicist, pointed out that antenna systems, e.g., groups of molecules, with distances smaller than the wavelength of the radiation they emit, always will emit coherently (Dicke, 1954). The reason for this is that in this case they cannot emit independently from each other, but are coupled to each other by a common radiation field. The field consists of interference patterns reflecting the structure of the antenna system to which it is feedback-coupled. The emission can happen in two ways, either explosively in the form of intense "superradiance" flashes, or in the slowed down and attenuated form of weak "subradiance". The smaller the distances between the emitters compared to the wavelength of the radiation, the higher is the coherence of the emission. DNA perfectly fulfils the "Dicke condition" in the optical region, as the distances of 3 Ångstrom between base pairs are very much smaller than the wavelength of visible light (around 5000 Å). Dicke's predictions have been extensively tested in the past decade and have led to the establishment of the new field of "Cavity Quantum Electrodynamics" (Haroche & Kleppner, 1989; Haroche, 1991; Berman, 1994; Benedict et al., 1996). Independently, in the last few years many experimental physicists have followed up and confirmed the related suggestion of V.S Letokhov (1968) that amplification and superradiant emission could take place in strongly scattering media such as gases, dye solutions, powders or collections of styrofoam balls (Lawandy et al., 1994; Genack & Drake, 1994).

These findings refute a frequent objection against the assumption of a function for light in the organism, namely that tissues and cells are optically dense (opaque) and therefore light could not transmit information. Popp has pointed out that Emil Wolf in 1963 has predicted an increase of the coherence of light in highly dispersive media with increasing thickness of the layer (Wolf, 1963), and has shown in absorption experiments (Popp et al., 1984) that biophoton emission – in opposition to artificial light – is able to penetrate disperse media like sand and soybean cultures of a thickness of thousands of cells. The optical transparency of biological tissue has also been demonstrated by Mandoli & Briggs (1982, 1984) and others (e.g., van Brunt et al, 1964; Vogelman, 1989). Cavity Quantum Electrodynamics makes it clear that just *because* tissues are optically dense, they interact with the vacuum by cavity formation and thus create special coherence effects.

4.2.3 Nonclassical Light

Recent findings in the new field of "nonclassical light" confirm that these phenomena of quantum coherence are the more pronounced, the lower the intensities of the light, or in many cases, only are possible at low intensities (Pike & Walther, 1988; Meystre & Walls, 1991). Nonclassical light is light at very low intensities (single photons), which shows properties, such as antibunching and squeezed states, not explainable in the framework of classical optics. For this reason, it can only be treated nonclassically. Recently it has become clear that squeezed states actually are a more general concept than coherent states, and the latter should be considered as special cases of the former (Li, this volume). While in coherent states the uncertainty of momentum and that of position are equal, in squeezed states either the momentum uncertainty is bigger than position uncertainty, or vice versa. Therefore, in squeezed states, we still have coherence, but not a coherent state. Squeezed light can both be focused narrowly to a spot of atomic dimension, or extended over the whole volume of an organism. When the phenomena predicted and experimentally verified by this very young field of quantum optics became known, they perfectly seemed to fit the findings of biophoton research; recently the earlier proposition of Li (1992), Gu (1998) and Bajpai et al. (1998) that squeezed light existed in living systems was confirmed by Popp et al. (2001).

4.2.4 Coherence Space-Time

Finally, in a series of fundamental papers the theoretical physicist Ke-Hsueh Li of the Chinese Academy of Sciences has demonstrated that the concept of coherence is really rooted in Heisenberg's uncertainty principle, and that coherence space-time is actually equivalent to uncertainty space-time (Li, 1992, 1994, 1995). As Li has shown, the insight that the uncertainty relation is only an alternative approach to describe coherence properties of fields and particles can be traced back to Heisenberg himself (1958b). Li points out that the interference between different probability amplitudes, hence the coherence properties of probability packets, constitute the essential point of quantum theory. Coherence time is the time within which interference is kept up; coherence length or coherence volume is the distance or space within which this happens. The latter corresponds to the breadth of the wave function which is the region within which matter (in the case of matter fields) or radiation is statistically distributed. Any field, including chaotic ones, has a coherence space-time.

Coherent states exist only within the coherence space-time; it is the region where the phase is defined. Outside, phase information is lost. Within the coherence volume – and *only* within it -, interference patterns are formed and a particle loses its classical picture; the particles and fields within it must be considered as an indivisible whole. As a consequence, it is not possible to assign the emission of a system situated within the coherence volume to any particular emitter within it. The field, even if consisting only of single photons, is delocalized within the whole system, and can be trapped (stored) within it for a long time. According to Li the relationship to the Dicke theory is evident: Inside the coherence volume, the Dicke principle applies; outside the coherence volume, when particles are separated by a distance larger than the coherence length, they are independent of each other. Therefore, Planck's radiation law is only valid in cases not within the coherence volume, while the Dicke principle applies to the cases within it. Li's insight concerns matter waves as well as electromagnetic waves. In the case of atoms or molecules, the electrons are always within the coherence volume, which is the reason why it is not possible to assign any determined orbits to them. Equally, the Pauli principle and the covalent bonding between two atoms or molecules appear in a new light, and can be explained by the destructive interference of the electron wave packets within the uncertainty/coherence volume. Finally, Li points out that Schrödinger's principle of the production of "order from order" (Schrödinger, 1945, see above) finds a solid physical basis in this mechanism, which appears to be more fundamental than the statistical mechanism producing order from disorder conventionally used by physics as a basis for deriving its laws.

Li emphasizes that the coherence volume is a region with unusual properties where other laws rule than in ordinary space-time. Space and time within it are different than those outside. All usual measurements of physics are done outside it, and up to now, physics has been concerned only with the properties of space-time outside the coherence volume. The recent experiments on the Einstein-Podolsky-Rosen (EPR) effect and on quantum tunneling that indirectly have shown the existence of superluminal velocities, can easily be understood in terms of coherence space-time, because velocities "higher than the speed of light" only occur within the coherence volume, or more precisely, space and time are not defined within coherence space-time. In this volume, all phenomena happen at the same time, and are interconnected; they form an "unbroken whole without any parts". The coherence volume can become macroscopic; actually coherent states have unlimited coherence space and time, or rather, these cannot be defined for coherent states. Thus, coherence space-time is of general significance for physics, beyond optics, and of course very important for biophysics. In the organism, many different coherence volumes can exist simultaneously, can overlap, and can change, while technical systems like lasers have always fixed coherence volumes.

4.3 Modern Bioelectromagnetic Field Theories

Modern electromagnetic field theories could only be created as a result of the recent developments in thermodynamics, quantum electrodynamics and quantum optics described above, and especially not without the theory of quantum coherence and Fröhlich's work on coherence. These achievements became important elements in the *"biophoton theory"* developed by Popp and his group (Popp et al, 1989, 1992; Bischof, 1995; Chang et al., 1998; Beloussov et al., 2000). Like most modern field-theoretical proposals, it tries to reconcile particle and field approaches. Based on the modern evidence for the coherent emission of ultraweak luminescence by organisms first discovered in the 1920's by Alexander Gurwitsch (Bischof, 1995; Lipkind, 1998), it conceptualizes organisms as biological lasers of optically coupled emitters and absorbers operating at the laser threshold. The solid part of the organism is coupled with a highly coherent, holographic biophoton field which is proposed to be the basis of communication on all levels of organization. The components of the organism are seen to be connected in such a way by phase relations of the field, that they are instantly informed about each other at all times. The biophoton field is also postulated to be the basis of memory and of the regulation of biochemical and morphogenetic processes.

A related approach, *"bioplasma theory"* (Inyushin et al., 1968; Inyushin, 1970, 1977, 1983; Sedlak, 1967, 1976, 79, 1980, 1982, 1984; Wolkowski et al., 1983; Zon, 1980;Wnuk & Bernard, 2001), was developed from early suggestions by Szent-Györgyi, who pointed out that biomolecules in the organism are predominantly present in the excited state, and that the energetics of living systems are based on excitation-deexcitation dynamics which are also the basis of chemical bonding. Biological plasma is described as a "cold" plasma of highly structured collective excitations produced by the dielectric polarization of biological semiconductors, which functions as a single unit. The collective excitations of the molecules propagate in the form of excitons. The complex aggregates and configurations formed by the plasma particles serve as an energy network in the organism. External and intrinsic radiation is stored in the bioplasma in the form of trapped cavity oscillations which form the biological field; it has a complex broadband holographic wave structure of great stability. The biological effects of external radiation are ascribed to resonance properties of the whole system, and not to any of its parts. Like biophoton theory, the bioplasma concept implies non-equilibrium and electronic population-inversion, and therefore laser-like processes, as postulated by Inyushin in the early 1970's.

Pribram's *"holographic theory of perception and memory"*, first proposed in 1971 (Pribram 1971; Pribram et al, 1974), has also been an important contribution to modern biological field theory. It proposes that information from the sensory input is enfolded by Fourier-like transformations and stored in the brain in the form of holographic interference patterns, i.e., coherent EM fields. For reading it out in remembering, it is unfolded again by inverse Fourier transformation. In 1975 Pribram synthesized his model with the more general *"holographic theory of reality"* proposed by Bohm in 1971-73 (Bohm, 1980). It suggests that the organization of reality itself may be holographic, the world of objects we perceive (the "explicate order") being a second-order manifestation of the more fundamental *"implicate order"* or *"holomovement"* forming the basis of the world's fundamental unbroken wholeness.

The *"holographic concept of reality"* proposed by Miller et al. (1979, 1992) is a useful attempt to sketch the outlines of a possible synthesis of field models emphasizing the particle aspect, like bioplasma theory, and those who focus on the connecting and/or underlying fields, like biophoton theory and the holographic theories. At the same time, it tries to elucidate the significance of bioelectric phenomena and physico-chemical parameters like the acid-base and electrolytic balances and redox potentials within the bioelectromagnetic fields.

Today, the field view of the organism and its interactions is finding increasing acceptance in biology, biophysics and medicine (see e.g., Tiller, 1977; Tiller et al., 2001; Braud, 1992; Bistolfi, 1991; Dossey, 1992; Goodwin, 1987; Gough & Shacklett, 1993; Ho, 1993, 1997; Jerman & Stern, 1996; Morton & Dlouhy, 1989; Rubik, 1996; Sit'ko & Gizhko, 1991; Welch, 1990, 1992; Wolkowski, 1985, 1988, 1995; Savva, 1999; Oschman, 2000; Thaheld, 2001a,b).

4.4 Biophysics and Consciousness

Already the German founders of biophysics in the early 19th century had recognized consciousness as the *"ultimate problem of biology"*, but science at that time was not prepared to include consciousness into biophysics, although the necessity was not denied (Culotta, 1974). Today, the situation is different. Since around the turn of the last century, Freud and his followers have made the problem of consciousness a central topic of a broad research effort whose results widely influenced Western society, but were hardly taken seriously by the natural sciences.

The last few years have changed this: consciousness has ceased to be a "non-subject" and is now definitely on the scientific agenda (Gray, 1992; Chalmers, 1995). An increasing number of authors are emphasizing the necessity of introducing consciousness into the scientific worldview, and some believe it should even become its very foundation (Cowan, 1975; Bohm, 1980; Bohm & Hiley, 1993; Goswami, 1989, 1993, 1994; Atmanspacher, 1996; Hagelin, 1987; Jahn & Dunne, 1987; Elitzur, 1989; Harman, 1992; Rubik, 1996; Radin, 1997; Amoroso et al., 2000). As to the consequences of a "biology without consciousness", Efron (1967) has pointed out the disastrous epistemological confusion the exclusion of consciousness from biology has caused.

Although the highly animated discussion in consciousness research is characterized by widely divergent standpoints, the decades-long dispute about the inclusion of the observer and the possible role of consciousness in quantum mechanics (Shimony, 1963; Walker, 1970, 1974, 2000) has certainly been a significant influence. As can be seen from the new discipline of *"Quantum Neurodynamics"* (Pribram, 1993; Jibu & Yasue, 1993, 1995), Pribram's and Bohm's holographic theories, together with Eccles' suggestion that fields analogous to the probability fields of quantum theory could be responsible for the coupling between consciousness processes and neural events (Eccles, 1986), have also been of considerable influence. Thus, an important segment of the most recent efforts in consciousness research is based on the hypothesis that consciousness may have a field-like nature, and/ or that fields may play a mediating role between consciousness and the biological organism. This hypothesis has already a considerable history and is more widely held than commonly is known (Bischof, 1998b).

4.4.1 Consciousness as a Field

It was probably William James who first introduced the concept of a field of consciousness into modern psychology in 1890. Carl Gustav Jung's *"Collective unconscious"*, first proposed in 1917, is conceived as a deeper, fundamental field-like unitary psychophysical reality *("unus mundus")* occasionally producing *"synchronicity"* effects. Gestalt Theory, initated by Christian v.Ehrenfels and Wolfgang Köhler, postulates an isomorphism between psychological and psychophysiological processes, mediated by fields analogous to Maxwell's electromagnetic fields and not bound to the nervous substrate, whose geometrical structure mirrors that of the perceived stimuli (Ash, 1995). In the 1930's Kurt Lewin proposed in his field theory of social psychology that social interactions are best understood by a field model (Lewin, 1936, 1951). In the 1940's Gardner Murphy developed a field concept of the organism, of personality, and of communication and suggested the existence of an interpersonal field which he thought to be part of a wider universal field (Murphy, 1945, 1947, 1961). Murphy also explained Psi effects in groups as a loosening of the usual interpersonal barriers and opening up to the interpersonal field. Paul Schilder demonstrated the existence of a "body-image", a 3-dimensional picture of the "perceived body", different from the body of anatomy and physiology, a kind of a constantly reorganized field that is constructed from the visual, tactile, kinesthetic, postural etc. experiences of an individual's lifetime (Schilder, 1950). In 1964, Aron Gurwitsch proposed a field theory of experiential organization in the tradition of Gestalt theory, Kurt Lewin's work and Edmund Husserl's phenomenology (Gurwitsch, 1964). The Russian mathematician and philosopher, Vassily V.Nalimov, has recently proposed a theory of the *"semantic vacuum"*, according to which there is a deeper, unobservable process from which ordinary, reflexive consciousness emerges which he calls the semantic vacuum, in deliberate analogy to the concept of the physical vacuum (Nalimov, 1981, 1982, 1989). Today, the field view of consciousness is proposed by many authors who suggest it may possess the property of quantum "nonlocality" (Dossey, 1992, 1999; Thaheld, 1998, 2001a,b).

4.4.2 Quantum Neurodynamics

A school of thought that has found wide interest and has led to a new branch of neuropsychology, has its origin in the work of Pribram and Eccles. In the early 1970's Karl H.Pribram had proposed that coherent holographic fields mediate between consciousness and neurological processes; John Eccles postulated in 1977 that consciousness has an existence independent of the brain, and that the self interacts with the body and the material world using the brain as an instrument (Popper & Eccles, 1977). The physicist Henry

Margenau suggested in 1984 that mind may be a unique type of non-material field, analogous to quantum probability fields; this suggestion was taken up in 1986 by Eccles who proposed that this field may modify the probability of emission of neurotransmitters at the dendritic synapse (Eccles, 1986). This finally led to the formation of the new field of Quantum Neurodynamics, based on the hypothesis that brain processes are to be understood on the basis of quantum field theory and are based on quantum fields, or potentials (Pribram, 1993; Jibu & Yasue, 1995). Long-time memory is conceived as a structured complex of vacuum states; remembering as the emission of coherent biophoton signals from the vacuum state. The coupling of neurophysiology with the "quantum sea" of the vacuum is assumed to be the basis of brain processes (Jibu & Yasue, 1993).

Quantum neurodynamics illustrates the many recent efforts to find approaches bridging consciousness as an entity which is not directly measurable, and the solid material aspect of the organism, with the hypothesis of a mediating field domain. Similar efforts have been made in the last few decades in many areas, not least in connection with the scientific investigation of Eastern medical systems, such as acupuncture, of contactless healing and various other phenomena.

4.5 Interpersonal Fields and "Subjective Anatomy"

The existence of electromagnetic fields emitted by living organisms, including humans, is now well established, even if there is not much secure knowledge about their biological functions. On the other hand, man has a long history of subjective experience of field-like interpersonal connections which usually are relegated to the realm of imagination by the scientifically minded. More recently, however, a number of scientific experiments have to some extent given evidence for the physical reality of these field observations (Bischof, 1998b).

As to the observations, an important example are the studies of nonverbal behaviour that have shown a synchrony of the body motion of speakers and listeners with the speech pattern (McBroom, 1966; Condon & Sander, 1974; Hatfield et al., 1994), which probably serves to establish empathic resonance. Such a synchronization between mother and child may be the basis and origin of human bonding and communication (Condon & Sander, 1974; Bullowa, 1975). A related phenomenon is the well established phenomenon of *"emotional contagion"* (Hatfield et al., 1994). Psychiatrists and psychotherapists have been familiar since decades with the *"praecox feeling"*, the field-like aura displayed by their patients announcing impending psychosis or schizophrenic episodes, and have been well aware of the contagious nature of these states (Deane, 1961; Ihle, 1962). The phenomenon of *"transference"* between therapist and patient is equally well known and has led a number of authors to the hypothesis of an *"interpersonal field"* (Schwartz-Salant, 1988; Mansfield & Spiegelman, 1995). In mutual hypnosis two persons create a common psychic field which in the more advanced stages can turn into a shared hallucinatory or dreamlike reality (Tart, 1969). Families may, according to some psychotherapists, possess a common unconscious and shared emotional field (Taub-Bynum, 1984).

A number of recent experimental studies give evidence that such interpersonal field effects may have some physical basis. Already in 1965, Duane & Berendt have shown brain-wave correlations between the brains of identical twins (Duane & Berendt, 1965). In order to test the many anecdotal reports that illness or trauma in one of a pair of identical twins affects the other, even when the twins are far apart, they altered the EEG pattern of one twin to see if this would produce a similar response in the other twin. In 2 of 15 pairs of twins tested, eye closure in one twin not only produced an immediate alpha rhythm in his own brain, but also in the brain of the other twin, even though he kept his eyes open and sat in a lighted room. This effect was reproduced in several experiments. In 1981, Jean Millay demonstrated in an EEG study of 22 individuals designed to test the old hypothesis that telepathy involves a mental resonance, that the ability to identify and increase interhemispheric EEG phase synchronisation and the ability of 11 pairs formed from the same individuals to identify and increase EEG phase coherence between their right hemispheres (interpersonal brain-wave synchronisation) corresponded to their ability for telepathic communication (Millay, 1981).

Conducted under much more stringent conditions, studies of empathically bonded pairs by Jacob Grinberg-Zylberbaum have shown interhemispheric and interpersonal EEG coherence and the appearance of transferred (evoked) potentials in the unstimulated partner after separation by a Faraday cage (Grinberg-Zylberbaum, 1987, 1992, 1994). A series of experiments to replicate the Greenberg-Zylberbaum studies and to investigate the possibility that states of consciousness can exert biological effects at a distance, are currently under way at Bastyr University and the University of Washington in the United States (Richards & Standish, 2000; Thaheld, 2001a,b). As a second neurophysical measure of the transfer of information between the two brains they are also using functional Magnetic Resonance Imaging (fNMR). According to preliminary results, the experiments have been successful so far. These studies are being conducted under a 2 year grant from the National Institutes of Health (Thaheld, 2001a,b). Thaheld (2001a,b) proposes that the experiments of Duane & Berendt and of Grinberg-Zylberbaum demonstrate the possibility of controllable biological quantum nonlocality and imply superluminal communication between individuals. Since neither neural nor electromagnetic energy could have penetrated the two Faraday cages used in the Grynberg-Zylberbaum experiments, Thaheld holds that they also demonsterated that mental events can influence or control neuronal events, a question that in present consciousness research is referred to as the "reverse direction" problem. Thaheld (1998, 2001b) has proposed several variations of these experiments to test the hypothesis of biological nonlocality and the interpersonal transfer of states of consciousness, which may also be used to address the question of animal consciousness.

The *"field-REG experiments"* done by the Princeton Engineering Anomalies (PEAR) Laboratory demonstrate anomalous influence of group events with a "high degree of subjective resonance between participants" on the random output of portable random event generators (REG) that suggest the presence of a field within such groups (Nelson et al., 1996, 1998; Radin et al., 1996; Bierman 1996; Radin 1997). Experiments on distant mental influence on living systems (DMILS) show that persons are able to exert direct mental influences upon various distant biological systems shielded from all conventional informational and energetic influences (Braud & Schlitz, 1983, 1991; Braud, 1992).

While these experiments suggest the possible non-electromagnetic nature of the studied fields, the measurements of the "Copper-Wall Project" performed by Elmer Green demonstrate that in healing sessions, exceptional subjects, such as healers and sensitives, are able to generate anomalous voltage surges in electrical body potential which are transmitted to and measured by electrometers attached to the four highly polished copper walls surrounding them in some distance (Green, 1990, 1991).

The former category of fields fits somewhere between the more physiology-related electromagnetic fields and the consciousness field of the organism; this field dimension of the organism belongs to both the "objective anatomy" and to what some have called "subjective anatomy". Their investigation may help in the future to understand and integrate the "subjective anatomy" of Schilder's "body image" and the "felt body" or "experienced body" investigated by the phenomenologists, experienced by most of us and most important for the understanding of the symptoms reported by the patient to the doctor (Schilder, 1950; Schutz, 1975; Fisher, 1986; Uexküll, 1997). One may also ask if the "phantom sensations" of amputees could not be more than just representations of lost limbs in the brain, perceptuomotor memories of a once functional body part. Since Brugger et al. have recently shown that phantom sensations are also experienced by persons with congenitally absent limbs (Brugger et al., 2000), we may all have an internal blueprint of a fully developed organism. This blueprint may have the form of a field instead of being just represented in material structures of the brain – a field that we even occasionally may experience as a "felt body". Some authors also feel that the "chi" of Chinese medicine and other life-energy concepts of non-Western medicine, including the "subtle bodies" of Oriental anthropology, possibly could be substantiated by research in the area of "subtle energies" that becomes increasingly important by the name of "energy medicine" (Morton & Dlouhy, 1989; Oschman, 2000; Bischof, 2000c).

4.6 The Physical Vacuum

In some of the above described experiments showing the existence of interpersonal field phenomena (Grinberg-Zylberbaum and DMILS) electromagnetic fields have been excluded; therefore we must consider the possibility that some kind of probably unknown, non-electromagnetic field(s) may be involved. Recently, a number of authors have suggested that electromagnetic potentials (vector and scalar potentials) may play a role in living systems (Tiller, 1993, 1995, 1999; Tiller et al., 2001; Taylor, 1996, 1998; Zeiger & Bischof, 1998; Bischof, 1998), and a series of preliminary experiments (which still have to be reproduced independently) seems to show biological effects of vector potentials different from those of ordinary electromagnetic fields (Rein, 1989, 1991, 1993, 1997, 1998; Ho et al., 1994; Smith, 1994; for a summary see Bischof, 1998). It has been proposed that there may be a whole class of non-electromagnetic fields underlying electromagnetic phenomena, which have been called "subtle energies" by some authors, following a suggestion by Einstein (Tiller, 1993, 1999; Taylor, 1998).

In fact, while the potentials have long been considered mere mathematical conveniences without physical reality, the reevaluation of their significance made possible by the groundbreaking paper by Aharonov & Bohm (1959), is now opening up a new field of electromagnetic research which we suspect may turn out to be highly significant for bioelectromagnetics and biophysics in general. A number of recent attempts to formulate extensions of electromagnetic theory point to the existence of an additional, hitherto unsuspected dimension of electromagnetism, which seems to be able to interact with the very structural fabric of space and time (Lakhtakia, 1993; Barrett & Grimes, 1995). Aharonov & Bohm (1959) have shown that in certain cases the potentials act as real physical fields and must even be considered more fundamental than the electric and magnetic forces; in the experiment they proposed the potentials exert an physical effect on charged particles in a field-free volume but not in the way force-fields do – they only influence the phase, and thus are fields of information.

However, the reason for the now well proven Aharonov-Bohm (AB) effect has only become evident in the wake of its analysis and generalization by Wu & Yang (Wu & Yang, 1975; Wu, 1975). Terence W. Barrett (1990; 1993) has given evidence that in the AB experiment the electromagnetic field, normally of U(1) symmetry, is "conditioned" into SU(2) form (in other cases even higher symmetries can be obtained) by the geometrical constraints of the experiment, which adds a degree of freedom to the field allowing an interaction with the space-time metric (neutrino network) and its topological structure, and endows it with a gravitation-like essence and form. Barrett also has shown that the AB effect is only one of a whole class of effects where this is the case.

According to William Tiller (1993) potentials have the important function of mediating between electromagnetic fields and the macroscopic quantum states of solid matter on the one hand, and the physical vacuum on the other hand, because of their property of controlling the phase of electromagnetic fields. He suggests that the subtle energy" fields of the vacuum domain, belonging to a higher dimension beyond space-time, organize the structure of space-time, which in turn, by the intermediate of the potentials, generates the corresponding electromagnetic fields. These finally give rise to the observed processes in space and time.

This hypothesis is supported by the work of Barrett (1990, 1993) on the conditioning of the electromagnetic field. In this process, the phase-controlling property of potentials is central. This is highly significant for biophysics, not only because of the coherence of biolectromagnetic fields; its importance can also be illustrated by the fact that the living organism with its many rhythmical processes basically is a complex system of oscillating fields coupled nonlinearly by their phase-relations.

Apart from potentials, a number of further non-electromagnetic fields have been forwarded in the various proposals for extensions of the Maxwell theory, as possible elements of an intermediate subtle realm" between particles and force fields and the vacuum, or as elements or aspects of the vacuum itself, for instance, *torsion fields"* (Akimov & Moskowski, 1992, Akimov & Tarasenko, 1992, Akimov, 1995; Akimov & Shipov, 1997, 1998; Shipov, 1998) and the B(3) ghost field" of longitudinal magnetic polarization (Evans, 1992, 1994; Evans & Vigier, 1994, 1995; Evans et al., 1996, 1998; Evans, 1998).

4.7 The Concept of Vacuum Biophysics

We are convinced that it is one of the central tasks of biology and biophysics, as it is of physics itself, to investigate the process of becoming and of manifestation, the arising of actuality from potentiality. It is clear that this is not yet completely realized in quantum physics, although the recent discussions about the interpretation of quantum theory and the alternatives to the Copenhagen interpretation have shown that it is groping in this direction. The same tendency can be found in some recent attempts at developing unfied theories of all physical interactions.

In the various unification programs of physics the concept of the physical vacuum occupies a central place. *The vacuum is fast emerging as the central structure of modern physics"* (Saunders & Brown, 1991). We postulate that it also merits such a central place in biophysics (Zeiger, 1998; Zeiger & Bischof, 1998; Bischof, 1998). It has in fact already been used in a number of recent models, e.g. by Conrad (1989a,b), Grandpierre (1997), Laughlin (1996), Laszlo (1995, 1996), Jibu & Yasue (1993, 1995), Shacklett (1991), and Tiller (1993, 1995, 1999; Tiller et al., 2001). We would not be surprised if it would turn out to be the very foundation a holistic quantum biophysics needs. The holistic quantum logic of biological processes and structures may not be sufficiently understood without the explicit inclusion of the vacuum concept into biophysics. Biophysics should be able to explain how, in the generation and development of organisms, pre-physical potentialities are transformed into physical realities. For practical reasons, it should also be interested in improving our knowledge on the more subtle, early levels of biological manifestation, where we may have access, for instance, to the formation of preconditions for illness, and which may also be the interface for the influence of the mind on the body. The assumption of a pre-physical dimension of potentiality is a prerequisite for the full understanding of life. To quote the Heisenberg pupil Hans-Peter Dürr: *"Living systems prove that actuality (factuality) is not all there is, but potentiality is also important. Like all macroscopic quantum systems, they are emergences of potentialities into factuality"* (Dürr, 1998).

We postulate that the concept of the vacuum is the appropriate framework to model the fundamental quantum-mechanical domain of potentiality. The vacuum is the "ground of being" from which the information for the structured development and regeneration of inorganic as well as living forms arises. All the features of the unbroken wholeness of reality implicit in quantum theory – non-separability, non-locality, fundamental connectedness – which are so fundamental for biological understanding, are an expression of the properties of the vacuum. The vacuum is the origin of microscopic and macroscopic coherence, an essential feature of living organisms. And, finally, the understanding of the vacuum may provide the crucial insights on the role of consciousness in physical reality, and in the various stages in which the creativity of the ground of existence unfolds on its way from pure potentiality and information to physical manifestation.

The concept of of the quantum vacuum may provide an important tool in developing both, a holistic terminology, and a holistic methodology for the integration of biology and physics necessary for the emergence of a holistic biophysics, or quantum biophysics. Especially significant may be its usefulness as a suitable framework for the treatment of organisms as macroscopic quantum systems (cf. the significance of vacuum degeneracy).

However, it must be clarified that we are not only talking about the electromagnetic vacuum of zero-point fluctuations, but of a more inclusive and fundamental unified vacuum of all four interactions.

4.8 Superfluid Vacuum Model of the Organism

In biophoton research first considerations on the possible role of the vacuum have been made in the 1980's in connection with the stability and optical properties of DNA and the optimal signal/noise ratio in the information transfer by biophotons. The central role that the vacuum plays in the Dicke theory and in Cavity Quantum Electrodynamics is well known. In 1985, Popp has suggested that biophoton emission as measured may arise from a non-measurable, virtual, delocalized, highly coherent field within the tissue, denoted by him as the realm of *"potential information"* in the organism; he conjectured it may be a kind of vacuum state (Popp, 1986; Bischof, 1995).

More recently, Zeiger (Zeiger, 1998; Zeiger & Bischof, 1998) has developed a *"superfluid vacuum model"* for understanding the biophoton emission of seeds and its connection to their viability. According to this model, seed vitality and biophoton emission are two parallel expressions of the same underlying reality: the superfluid Bose-condensate of photons. He proposes that the radiation field coupled to biological systems has to be understood on the basis of a twofold ground state. It consists, on the one hand, of a non-perturbative, collective-coherent state responsible for stability, internal communication and photon storage, which endows the organism with a quiet background field connecting all its components by long-range phase relations with each other and with the environment. The second ground state, a perturbative, fluctuating-coherent state consisting of the excitations of the collective-coherent state, is responsible for flexibility, adaptation and external communication, and from it the observed biophotons are emitted. The two states are separated by an energy gap which controls the behaviour of the system and is a basic measure for the overall state of the organism. It is a parameter that promises to become an important new tool in biophysics providing additional information on the living system. Zeiger's model may be a significant step in the biophysical modelling of the process of the emergence of "becoming" from the potentiality of the "ground of being".

Thus, if we speak of the electromagnetic field, or biophoton field, of cells, tissues or of the whole organism, as opposed to the biophoton emission measured, we may actually be dealing not with an electromagnetic field in the usual sense, but with a virtual field, or vacuum state. This has actually been proposed by Thomas E. Bearden (1991, 1993, 1995) and is partially supported by the work of Barrett and others, already mentioned, on one or more deeper level(s) of electromagnetism.

If we define vacuum physics as that branch of physics concerned with the fundamental pre-physical level of potentiality from which matter and fields arise and which contains the information for their dynamic structuring processes, I suggest that the corresponding field of biophysics concerned with the investigation of the biological role of the physical vacuum and of the mediating role of potentials and other nonelectromagnetic fields between the vacuum on the one hand and force fields and solid matter on the other hand, should be called "vacuum biophysics". It may become an important theoretical element and research subject of the new biophysics.

5. CONCLUSION: A MULTIDIMENSIONAL MODEL OF THE ORGANISM

Under the above assumptions a hierarchy of levels of biological function, or regulation systems, based on fields, may be envisaged as a working hypothesis, where we have (for the human case), between the solid physical body on the one hand, and consciousness on the other hand, the intermediate levels of the primitive holistic regulation systems and physiological-biochemical regulation, bioenergetic (EM) fields, bioinformation fields, and finally the "subjective anatomy" of the "experienced body".

The model is not only based on the inclusion of consciousness and the subjective aspects of our existence, but also on a tentative reversal of the usual primacy of matter over consciousness. In the diversity of standpoints in contemporary consciousness research it is now again possible to hold a view like that of Popper and Eccles, who have proposed that consciousness may not be a (secondary) product of (primary) material brain structures, but rather an independent entity that uses the brain as an instrument to interact with the body and the material world. It may be useful to remember what James Clerk Maxwell said to his friend, Professor Hort, on his deathbed: "What is achieved by the so-called Me, is in reality achieved by something greater than myself in me" (Norretranders, 1998). This view also roughly corresponds to that of some of the anthropologies of non-Western cultures and our own past, which assume that the innermost core of man consists of a "self" or a "soul", i.e., a consciousness, but one without activity and without an object, and therefore to differentiate from the mental (thoughts, imaginations etc.) arising from this deeper level. This self is regarded as the highest instance of the person, the origin of all intentions, and as the "place" of the wholeness of our psychic and corporeal existence, and therefore as the "highest regulatory instance" for psyche and body. Following the phenomenologists, and in accordance with non-Western models, this concept has to be supplemented by the insight that our body has not only an objective, but also a subjective aspect. Therefore in the spectrum of the dimensions of human existence the "experienced body" forms an intermediate state between the domain of the "objective body" of the biosciences and the "subjective anatomy" of the psychosocial and spiritual worlds. Consequently, we obtain the following model of a "multidimensional anatomy of man", whose dimensions are at the same time levels of self-regulation, levels of regulatory disturbance, and levels of external (therapeutic) intervention:

Table 1. Multidimensional Model of the Organism. Hierarchy of regulation levels, regulatory disturbances and regulatory interventions

PHYSICAL BODY	Solid-Physical Body	morphological changes, reversible and irreversible "lesions", « medical evidence », repair and replacement
	Primitive, Non-specific Systems and Regulations	
	holistic regulation systems Friedrich Kraus' "vegetativum": functional unity of humoral system of body liquids (hormones, electrolytes, acid-base balance, redox potentials etc.), ground regulation system of connective tissue, and autonomous nervous system	« Terrain » predisposition to disease functional disturbances, latent and larvated early forms of disease, premorbid states prevention healthy life style

ELECTROMAGNETIC FIELD-BODY	Bioenergetic Fields electromagnetic (force-) fields energetic interactions distribution of energy biophotons in the narrow (optical frequencies) and in the wider sense (all wavelengths)	energetic imbalances field interventions, electromagnetic medicine phototherapy etc.
NON-ELECTROMAGNETIC FIELD-BODY	Bioinformation Fields Quantum potentials, scalar waves etc. « higher » dimensions of EM field made possible by extensions of EM theory, may interact with space-time and consciousness possibly interface to consciousness "vacuum biophysics"	dysregulations on the level of the infor-mation fields subtle field interventions homeopathy etc.

Subjective Anatomy

SENSATE BODY **SOMATIC UNCONSCIOUS** **MENTAL BODY** **EMOTIONAL BODY** **DREAM BODY**	"Experienced Body" Subjective Condition Emotions, Thoughts, Moods "subtle bodies" body sensation body image (Schilder) streaming sensations "life energy" ? only partly conscious closely connected with "atmosphere" of environ-ment and with transper-sonal fields and moods of others no sharp separation of I and world	 unhealthy thoughts emotional instability negative feelings and moods stress social integration good relationships psychotherapy placebo effects

SELF	Deepest Core of Person Soul consciousness level of wholeness of the organism and of interconnectedness with the world highest regulatory instance	unbalanced mental and spiritual atti-tudes and intentions, often unconscious meditation metanoia mysticism

A consequence of this model of a hierarchy of regulatory levels is the idea of a gradual pathogenesis and corresponding therapeutic interventions which I would like to formulate as a conclusion. I start from the premise that upon transgression of a certain range of tolerance, of fluctuations which the organism just can tolerate, dysregulations on one level will lead to disturbances on the next lower level. If all intentions originate primarily from the deep self, then inadequate or unbalanced mental and spiritual attitudes or intentions would represent the first level of dysregulation. If the imbalance is not corrected on this level, then it will cause dysregulation on the next level, e.g. unhealthy thoughts, emotional instability, negative feelings and moods, and stress. Without a correction of these imbalances, they will again cause dysregulations on the next level of bioinformation fields and the electromagnetic body, on whose properties we know not enough to describe their exact nature. Without adjustment, again these energetic imbalances will lead to functional disturbances in the primitive non-specific systems. And only now, if no correction on the level of the physiological "terrain" is done, we get the "medical evidence", i.e., the morphological lesions and changes in the tissues, that make the whole chain of dysregulations manifest at its very end.

REFERENCES

Abir-Am, P.G. (1987) The Biotheoretical Gathering: Trans-disciplinary authority and the incipient legitimation of molecular biology in the 1930's. *History of Science*, **25**, 1-70.

Aharonov, Y., Bohm, D. (1959) Significance of electromagnetic potentials in the quantum theory. *Physical Review*, 2nd Ser., **115 (3)**, 485-491.

Akimov, A.E., Moskovskii, A.V. (1992) *Quantum Non-Locality and Torsion Fields*. Moscow: Center of Intersectorial Science, Engineering and Non-Conventional Venture Technologies (CISE-VENT).

Akimov, A.E., Tarasenko, V.Ya. (1992) Models of polarized states of the physical vacuum and torsion fields. *Soviet Physics Journal* **35 (3)**, 214-222.

Akimov, A.E., ed. (1995) *Consciousness and the Physical World*. Collected Papers. Moscow: Publishing Agency "Yakhtsmen".

Akimov, A.E., Shipov, G.I. (1997) Torsion fields and their experimental manifestations. *Journal of New Energy* **2 (2)**, 67.

Akimov, A.E., Shipov, G.I. (1998) Experimental manifestation of torsion fields and the torsion model of consciousness. *Consciousness and Physical Reality*, **1 (1)**, 42-63.

Amoroso, R., Antunes, R., Coelho, C., Farias, M., Leite, A., and Soares, P. (eds.) (2000) *Science and the Primacy of Consciousness*. Orinda, CA., Noetic Press.

Anderson, M.H., Ensher, J.R., Matthews, M.R., Weiman, C.E. and Cornell, E.A. (1995) Observation of Bose-Einstein condensation in a dilute atomic vapor. *Science*, **269**, 198-201.

Arrhenius, S. (1898) Die Einwirkung kosmischer Einflusse auf physiologische Verhältnisse. *Skandinavisches Archiv für Physiologie*, **8**, 376-416.

Arsonval, A. d', Chauveau, J.B.A., et al. (1901-1903) *Traitè de Physique Biologique*. Paris, Masson. 2 vols.

Aschoff, J. (1965) *Circadian Clocks*. Amsterdam, North-Holland.

Ash, M.G.(1995) *Gestalt Psychology in German Culture 1890-1967. Holism and the Quest for Objectivity*. Cambridge, Cambridge University Press.

Aspect, A., Dalibard, J., and Roger, G. (1982) Experimental test of Bell's inequalities using time-varying nalyzers. *Physical Review Letters*, **49**, 1804-1807.

Aspect, A., Grangier, P. (1986) Experiments on Einstein-Podolsky-Rosen-type correlations with pairs of visible photons. In *Quantum Concepts in Space and Time*, R.Penrose and C.J.Isham (eds.) Oxford, Clarendon Press, 1-15.

Atmanspacher, H. (1996) Erkenntnistheoretische Aspekte physikalischer Vorstellungen von Ganzheit. *Zeitschrift für Parapsychologie und Grenzgebiete der Psychologie* **38 (1/2)**, 20-45.

Bailes, K.E. (1990) *Science and Russian Culture in an Age of Revolutions. V.I.Vernadsky and His Scientific School, 1863-1945*. Bloomington, Indiana University Press.

Bajpai, R.P., Kumar, S., Sivadasan, V.A. (1998) Biophoton emission in the evolution of a squeezed state of frequency stable damped oscillator. *Applied Mathematics and Computation*, **93**, 277-288.

Barcroft, J. (1934) *Features in the Architecture of Physiological Function*. Cambridge: Cambridge University Press. Reprinted: New York, Garland Publishing, 1988.

Barrett, T.W. (1976) Mechanoelectrical transduction in hyaluronic acid salt solution is an entropy-driven process. *Physiological Chemistry and Physics*, **8**, 125-134.

Barrett, T.W. (1990) Maxwell's theory extended. Part 1: *Annales de la Fondation Louis de Broglie* 15 (2), 143-183. Part 2: Theoretical and pragmatic reasons for questioning the completeness of Maxwell's theory. *Annales de la Fondation Louis de Broglie*, **15 (3)**, 253-283.

Barrett, T.W. (1993) Electromagnetic phenomena not explained by Maxwell's equations. In: Lakhtakia, Aklesh (ed.):*Essays on the Formal Aspects of Electromagnetic Theory*. Singapore, World Scientific Publishing, 6-86.

Barrett, T.W., Grimes, D.M. (eds.) (1995) *Advanced Electromagnetism: Foundations, Theory & Applications*. Singapore World Scientific.

Bauer, E. (1920) *Die Grundprinzipien der rein naturwissenschaftlichen Biologie und ihre Anwendungen in der Physiologie und Pathologie*. In W.Roux (ed.), Vorträge und Aufsätze über Entwicklungsmechanik der Organismen, No. XXVI. Berlin, Julius Springer Verlag.

Bauer, E. S. (1935) *Theoretical Biology*. Moscow. Reprinted Budapest: Akademiai Kiadó, 1982 (in Russian, with abridged English translation and commentaries by B.P.Tokin and S.E.Shnol).

Beall, P.T., Cailleau, R.M., and Hazlewood, C.F. (1967) The relaxation times of water protons and division rate in human breast cancer cells: a possible relationship to survival. *Physiological Chemistry and Physics*, **8**, 281-284.

Bearden, T.E. (1991) *Gravitobiology*. Ventura, California: Tesla Book Company.

Bearden, T.E. (1993) Information content of the field: Dynamic EM structuring of the scalar potential and its use by biological systems. Paper presented at the *2nd Estrian Workshop on "The Organization of Information Transfer in Living Systems"*, August 26-28, 1993, University of Sherbrooke, Canada.

Bearden, T.E. (1995) Vacuum engines and Prioré's methodology: The true science of energy-medicine. Parts I and II. *Explore!* **6 (1)**, 66-76 and **6 (2)**, 50-62.

Beier, W. (1965) *Einführung in die theoretische Biophysik*. Stuttgart: Gustav Fischer.

Bell, J.S. (1987) *Speakable and Unspeakable in Quantum Mechanics*. Cambridge, Cambridge University Press.

Beloussov, L.V. (1988) The problem of biological morphogenesis: General approaches and mechano-geometrical models. In Kull, K. and Tiivel, T. (eds.), *Lectures in Theoretical Biology*. Tallin, Estonia, Valgus, 55-64.

Beloussov, L.V. (1993) Transformation of morphomechanical constraints into generative rules of organic evolution. *World Futures*, **38**, 33-42.

Beloussov, L.V. (1998) *The Dynamic Architecture of a Developing Organism. An Interdisciplinary Approach to the Development of Organisms*. Dordrecht, Kluwer Academic Publishers.

Beloussov, L.V., Popp, F.A., Voeikov, V., and Wijk, R. van (eds.) (2000) *Biophotonics and Coherent Systems*. Moscow, Moscow University Press.

Benedict, M.G., Ermolaev, A.M., Malyshev, V.A., Sokolov, I.V., Trifonov, E.D. (1996) *Super-Radiance - Multiatomic Coherent Emission*. (Optics and Optoelectronics Series). Bristol and Philadelphia, Institute of Physics Publishing.

Bénézech, C. (1962) *L'Eau, Base Structurale et Fonctionelle des Êtres Vivants*. Paris, Masson.

Berg, M., Mayne, A., and Petersen, W.F. (1940) Variability of blood pH and its association with meteorological factors. *American Journal of Physiology*, **130**, 9-21.

Bergmann, G.von (1932) *Funktionelle Pathologie*. Berlin, Julius Springer,.

Bergsmann, O. (1994a) *Bioelektrische Phänomene und Regulation in der Komplementärmedizin*. Vienna, Facultas.

Bergsmann, O. (1994b) *Struktur und Funktion des Wassers im Organismus*. Vienna, Facultas.

Bergsmann, O. (1998) *Chronische Belastungen*. Vienna, Facultas.

Berman, P.R. (1994) *Cavity Quantum Electrodynamics*. (Advances in Atomic, Molecular, and Optical Physics, Supplement 2). San Diego and London, Academic Press.

Bernard, C. (1859) *Leçons sur les propriétés physiologiques et les altérations pathologiques des liquides de l'organisme*. Paris , Baillière.

Bernard, C. (1865) *Introduction à l'étude de la médecine expérimentale*. Paris, Baillière. English ed.: *An Introduction to the Study of Experimental Medicine*. New York, Dover, 1957.

Bertalanffy, L.v. (1927a) Studien über theoretische Biologie. *Biologisches Zentralblatt*, **47**.

Bertalanffy, L.v. (1927b) Über die Bedeutung der Umwälzungen in der Physik für die Biologie. *Biologisches Zentralblatt*, **47**.

Bertalanffy, L.v. (1940) Der Organismus als physikalisches System betrachtet. *Naturwissenschaften*, **28**, 521-531.

Bertalanffy, L.v. (1945) Zu einer allgemeinen Systemlehre. *Deutsche Zeitschrift für Philosophie*, **18** (3/4).

Bertalanffy, L.v. (1948) *Theoretical Biology*. Ann Arbor: J.W.Edwards.

Bertalanffy, L.v. (1949) Open Systems in Physics and Biology. *Nature*, **163**, 384 ff.

Bertalanffy, L.v. (1950) The Theory of Open Systems in Physics and Biology. *Science*, **111**, 23-29.

Bertalanffy, L.v. (1968) *General System Theory*. New York, George Braziller.

Bertalanffy, L.v. (1975) Perspective on General System Theory. New York, George Braziller.

Bertalanffy, L.v. ,Beier, W. and Laue, R. (1977) *Biophysik des Fließgleichgewichts*. 2nd ed. Braunschweig, Vieweg.

Bews, J.W. (1935) *Human Ecology*. London: Oxford University Press.

Beyler, Richard H. (1996) Targeting the Organism – The scientific and cultural context of Pascual Jordan's Quantum Biology. *History of Science*, 87, 248-273.

Bhaumik, D., Bhaumik, K., Dutta-Roy, B. (1976) On the possibility of Bose condensation in the excitation of coherent modes in biological systems. *Physics Letters*, 56A, 145-148.

Bierman, D.J. (1996) Exploring correlations between local emotional and global emotional events and the behavior of a random number generator. *Journal of Scientific Exploration*, 10 (3), 363-373.

Bigu, J. (1976) On the biophysical basis of the human "aura". *The Journal of Research in Psi Phenomena*, 1 (2), 8-43.

Bischof, M. (1992) "Wasser (water)", in *Dokumentation der besonderen Therapierichtungen und natürlichen Heilweisen in Europa,*Vol.II, Wissenschaftliche Grundlagen der besonderen Therapierichtungen und natürlichen Heilweisen, Essen, Verlag für Ganzheitsmedizin VGM (co-authored by F.Rohner), 91-148.

Bischof, M. (1994) The history of bioelectromagnetism. In *Bioelectrodynamics and Biocommunication*, M.W.Ho, F.A.Popp and U.Warnke (eds.), Singapore, World Scientific.

Bischof, M. (1995a) *Biophotonen – das Licht in unseren Zellen*. Frankfurt, Zweitausendeins; 11th ed. 2001.

Bischof, M. (1995b) Vitalistic and Mechanistic Concepts in the History of Bioelectromagnetics. In *Biophotonics - Non-Equilibrium and Coherent Systems in Biology, Biophysics and Biotechnology*. L.V.Belousov and F.A.Popp (eds.), 3-14. Moscow, Bioinform Services.

Bischof, M. (1996) Some Remarks on the History of Biophysics and its Future. In: *Current Development of Biophysics*, C.L. Zhang, F.A. Popp and M. Bischof (eds.), Hangzhou, China, Hangzhou University Press.

Bischof, M. (1998a) Holism and Field Theories in Biology - Non-Molecular Approaches and their Relevance to Biophysics. In: *Biophotons*, J.J.Chang, J.Fisch and F.A.Popp (eds), Dordrecht, Kluwer Academic Publishers, 375-394.

Bischof, M. (1998b) *The fate and future of field concepts – from metaphysical origins to holistic understanding in the bio-sciences*. Invited lecture given at the Fourth Biennial European Meeting of the Society for Scientific Exploration, Valencia, Spain, October 9-11.

Bischof, M. (1998c) Skalarwellen und Quantenfelder als mögliche Grundlage biologischer Information. *Erfahrungsheilkunde* 47 (5), 295-300.

Bischof, M. (2000a) Field concepts and the emergence of a holistic biophysics, in Beloussov, L.V., Popp, F.A., Voeikov, V.L., and Van Wijk, R. (eds.) *Biophotonics and Coherent Structures*. Moscow, Moscow University Press, 1-25.

Bischof, M. (2000b) Wege zu einer integralen Heilkunde. In *Salutive*, ed. Gesundheitsakademie e.V. Frankfurt, Mabuse.

Bischof, M. (2000c) Energiemedizin - Heilkunst der Zukunft. *Esotera*, **8**, 16-21, and **9**, 20-25.

Bohm, D. (1980) *Wholeness and the Implicate Order*. London, Routledge.

Bohm, D.; Hiley, B.J. (1993) *The Undivided Universe. An Ontological Interpretation of Quantum Theory*. London, Routledge.

Bohr, N. (1932) Light and Life. *Nature* **131**, 421-423, 457-459.

Bone, S., Zaba, B. (1992) *Bioelectronics*. Chichester, Wiley.

Bose, J.C. (1902) *Response in the Living and Non-Living*. London, Longmans, Green & Co..

Bose, J.C. (1907) *Comparative Electro-Physiology*. London, Longmans, Green & Co.

Bose, J.C. (1915) *Plant Autographs and Their Revelations*. Washington.

Bose, J.C. (1920) *Collected Physical Papers*. London, Longmans, Green & Co.

Bose, J.C. (1923) *The Physiology of the Ascent of Sap*. London, Longmans, Green & Co.

Braud, W.G. (1992) Human interconnectedness: Research indications. *ReVision*, **14 (3)**, 140-148.

Braud, W., Schlitz, M.J. (1983) Psychokinetic influence on electrodermal activity. *Journal of Parapsychology*, **47 (2)**, 95-119.

Braud, W., Schlitz, M.J. (1991) Consciousness interactions with remote biological systems: Anomalous intentionality effects. *Subtle Energies*, **2 (1)**, 1-46.

Brown, F.A., jun. (1954) Biological clocks and the fiddler crab. *Scientific American*, **190**, 34-37.

Brown, F.A., jun. (1976) Evidence for external timing of biological clocks. In Palmer, J.D., *An Introduction to Biological Rhythms*. New York, Academic Press, 209-279.

Brugger, P. et al. (2000) Beyond re-membering : Phantom sensations of congenitally absent limbs. *Proceedings of the National Academy of Science USA*, **97 (11)** 6167-6172.

Brunt, E.E. van , Shepherd, M.D., Wall, J.R., Ganong, W.F., Clegg, M.T. (1964) Penetration of light into the brain of mammals. *Annals of the New York Academy of Sciences*, **117**, 217-227.

Bullowa, M. (1975) When infant and adult communicate, how do they synchronize their behaviors ? In: Kendon, A., Harris, R.M., Key, M.R. (eds.), *Organization of Behaviour in Face-To-Face Interaction*. The Hague, Mouton, 95-129.

Burns, J.T. (1997) *Cosmic Influences on Humans, Animals, and Plants. An Annotated Bibliography.* (Magill Bibliographies). Lanham, MD, Scarecrow Press and Pasadena, CA, Salem Press.

Bünning, E. (1958) *Die physiologische Uhr.* English ed. *The Physiological Clock.* 3rd rev. ed. Berlin, Springer 1973.

Burr, H.S. (1947) Field theory in biology. *The Scientific Monthly,* **64 (March)**, 217-225.

Burr, H.S., Northrop, F.S.C. (1935) The electrodynamic theory of life. *Quarterly Review of Biology,* **10**, 322-333.

Burr, H.S. (1972) *Blueprint for Immortality - The Electric Patterns of Life.* London, Neville Spearman.

Buttersack, F. (1912) *Latente Erkrankungen des Grundgewebes, insbesondere der serösen Häute.* Stuttgart, Ferdinand Enke Verlag.

Büttner, K. (1951) W.F.Petersen †. *Archiv für Meteorologie, Geophysik und Bioklimatologie,* Series B, **2**, 291.

Cannon, W.B. (1929) *Bodily Changes in Pain, Hunger, Fear, and Rage.* 2nd. ed. New York, Appleton.

Cannon, W.B. (1932*) The Wisdom of the Body.* Philadelphia, W.W.Norton.

Capel-Boute, C. (1974) L'oeuvre scientifique de G.Piccardi. *Medicina Termale e Climatologia,* **22**, 69-74.

Capel-Boute, C. (1983) Fluctuating phenomena in physical chemistry and biology. *Cycles,* **34 (3)**, 49-53.

Capel-Boute, C. (1985) Water as receptor of environmental information? *Journal of Biometeorology,* **29**, Suppl.2, 71-88.

Capel-Boute, C. (1990) Water a receptor of environmental information: A challenge to reprodcibility in experimental research – The Piccardi scientific endeavour. In Tomassen, G.J.M. (eds.), *Geo-Cosmic Relations – The Earth and its Macro-Environment.* Wageningen, Pudoc, 75-91.

Carlson, E.A. (1971) An unaknowledged founding father of molecular biology, H.J.Muller's contribution to gene theory, 1910-1936. *Journal of the History of Biology,* **4 (1)**, 149-170.

Carrel, A. (1935) *Man – the Unknown.* New York.

Cat, J. (1998) The physicists' debates on unification in physics at the end of the 20th century. *Historical Studies in the Physical and Biological Sciences,* **28**, part 2, 253-299.

Chalmers, D.J. (1995) Facing up to the problem of consciousness. *Journal of Consciousness Studies* **2 (3)**, 200-219.

Chauvois, L. (1941) *D'Arsonval – une vie, une époque 1851-1940*. Paris, Librairie Plon.

Chew, G. (1970) Hadron bootstrap: triumph or frustration ? *Physics Today* **23**, 23-28.

Chizhevsky, A.L. (1930) *Epidemic Catastrophes and the Periodic Activity of the Sun* (in Russian). Moscow.

Chizhevsky, A.L. (1968) *The Earth in the Universe*. V.V.Fedynsky, ed. NASA TT F-345 TT 66-51025.

Chizhevsky, A.L. (1973) *The Terrestrial Echo of Solar Storms*. Moscow: Mysl (written in 1936).

Clark, W. M. (1925) *The Determination of Hydrogen Ions*. Baltimore, Williams & Wilkins.

Clark, W. M. (1928) Studies on oxidation-reduction. *Hygienic Laboratory Bulletin*, No.151. Washington, United States Public Health Service.

Clegg, J.S. (1981) Intracellular water, metabolism, and cell architecture. Part I: *Collective Phenomena*, **3**, 289-312.

Clegg, J.S. (1983) Intracellular water, metabolism, and cell architecture. Part II. In: Fröhlich, H., Kremer, F. (eds.), *Coherent excitations in biological systems*. Berlin, Springer, 162-177.

Clegg, J.S. (1984) Properties and Metabolism of the Aqueous Cytoplasm and its Boundaries. *American Journal of Physiology*, **246**, R133-R151.

Clegg, J.S. (1987) On the physical properties and potential roles of intracellular water. In G.R.Welch and J.S.Clegg (eds.), *The Organisation of Cell Metabolism*. (Nato ASI Series.Series A: Life Sciences, 127). New York, Plenum, 41-55.

Clegg, J.S. (1992) Cellular infrastructure and metabolic organization. *Current Topics in Cellular Regulation*, 33, 3-14. New York, Academic Press.

Condon, W.S., Sander, L.A. (1974) Neonate movement is synchronized with adult speech. *Science*, **183**, 99-101.

Conrad, M. (1989) Physics and Biology: Towards a Unified Model. *Applied Mathematics and Computation* **32**, 75-102.

Conrad, M. (1989) The Fluctuon Model of Force, Life, and Computation: A Constructive Analysis. *Applied Mathematics and Computation* **56**, 203-259.

Cope, F.W. (1969) Nuclear magnetic resonance evidence using D_2O for structured water in muscle and brain. *Biophysical Journal*, **9 (3)**, 303-319

Cope, F.W. (1970) The solid-state physics of electron and ion transport in biology. *Advances in Biological and Medical Physics*, **13**, 1-42.

Cope, F.W. (1973) Supramolecular biology: A solid state physical approach to ion and electron transport. *Annals of the New York Academy of Sciences*, **204**, 416-433.

Cope, F.W. (1975) A review of the applications of solid state physics concepts to biological systems. *Journal of Biological Physics* **3**, 1-41.

Cope, F.W. (1976) A primer of water structuring and cation association in cells. I. Introduction: The big picture. *Physiological Chemistry and Physics*, **8**, 479-483; II. Historical notes, present status, and future directions. *Physiological Chemistry and Physics*, **8**, 569-574.

Cope, F.W. (1977) Pathology of structured water and associated cations in cells (the tissue damage syndrome) and its medical treatment. *Physiological Chemistry and Physics*, **9**, 547-553.

Cowan, D.A. (1975) *Mind Underlies Spacetime – A Idealistic Model of Reality*. San Mateo, CA, Joseph Publishing.

Cranefield, P.F. (1957) The organic physics of 1847 and the biophysics of today. *Journal of the History of Medicine*, **12**, 407-423.

Culotta, C.A., (1974) German biophysics, objective knowledge, and romanticism. In *Historical Studies in the Physical Sciences*, ed. R. McCormmach. **4**, 3-39.

Damadian, R. (1971) Tumor detection by nuclear magnetic resonance. *Science*, **171** (3976), 1151-1153.

Dasgupta, S. (1999) *Jagadis Chandra Bose and the Indian Response to Western Science*. New Delhi, Oxford University Press.

Davis, K.B. et al. (1995) Bose-Einstein condensation in a gas of sodium atoms. *Physical Review Letters*, **75**, 3969-3973.

Deane, W.N. (1961) The reactions of a nonpatient to a stay on a mental hospital ward. *Psychiatry*, **24**, 61-68.

Delhoume, L. (1939) *De Claude Bernard à d'Arsonval*. Paris, Baillière.

Delore, P. (1926) *Facteur acide-base et tuberculose pulmonaire. Étude physiologique du terrain dans la tuberculose*. Paris Doin.

De Martino, S., De Siena, S., De Nicola, S., Fedele, R., Miele, G., (eds.) (1997) *New Perspectives in the Physics of Mesoscopic Systems - Quantum-like Descriptions and Macroscopic Coherence Phenomena*. Singapore, World Scientific.

Dessauer, F., Sommermeyer, K. (1964) *Quantenbiologie*. 2nd ed. Berlin, Springer.

Dicke, R.H. (1954) Coherence in Spontaneous Radiation Processes. *Physical Review*, **93**, 99-110.

Donnan, F.G. (1929) The Mystery of Life. *Report of the British Association for the Advancement of Science*, Glasgow, 96th meeting, 660.

Dossey L (1992) Era III Medicine: The Next Frontier. *ReVision* **14** (3), 128-139.

Dossey, L. (1999) *Reinventing Medicine*. San Francisco, HarperCollins.

Drost-Hansen, W. , Clegg, J.S., (eds.) (1979) *Cell-Associated Water*. New York, Academic Press.

Duane, T.D., Behrendt, T. (1965) Extrasensory electroencephalographic induction between identical twins. *Science*, **150**, 367.

Dubrov, A.P. (1978) *The Geomagnetic Field and Life - Geomagnetobiology*. New York, Plenum Press.

Duncan, A.J., Kleinpoppen, H. (1988) The experimental investigation of the Einstein-Podolsky-Rosen question and Bell's inequality. In Selleri, F. (ed.), *Quantum Mechanics versus Local Realism – The Einstein-Podolsky-Rosen Paradox*, New York, Plenum Press, 175-218.

Dürr, H.P. (1968) Dynamik der Elementarteilchen. In Süssmann, G. and Fiebiger, N. (eds.), *Atome, Kerne, Elementarteilchen*, Frankfurt a.M., 229 ff.

Dürr, H.P.: (1997) Ist Biologie nur Physik ? *Universitas*, No.607, 1.

Dürr, H.P. (1998) Paper read at Gelterswoog symposium on biophysics, March 14th, Kaiserslautern, Germany.

Eccles, J.C. (1986) Do mental events cause neural events analogously to the probability fields of quantum mechanics ? *Proceedings of the Royal Society ,London*, **B 227**, 411-428.

Eccles, R. (1993) *Electrolytes, Body Fluids and Acid Base Balance*. London, Edward Arnold.

Edmunds, L.N., jr. (1976) Models and mechanisms for endogenous time keeping. In Palmer, J.D. *An Introduction to Biological Rhythms*, New York, Academic Press, 209-279.

Edmunds, L.N., (ed.) (1984) *Cell Cycle Clocks*. New York, Marcel Dekker.

Efron, R. (1967) Biology without consciousness – and its consequences. *Perspectives in Biology and Medicine*, **11(1)**, 9-36.

Einstein, A., Podolsky, B, and Rosen, N. (1935) Can quantum-mechanical description of physical reality be considered complete ? *Physical Review*, 47, 777-780.

Elitzur, A.C. (1989) Consciousness and the incompleteness of the physical explanation of behaviour. *The Journal of Mind and Behaviour* **10**, 1-19.

Endler, P.C. and Schulte, J., (eds.) (1994) *Ultra-High Dilution – Physiology and Physics*. Dordrecht, Kluwer Academic.

Eppinger, H. (1949) *Die Permeabilitäts-Pathologie als die Lehre vom Krankheitsbeginn*. Vienna, Springer-Verlag.

Ernst, E., Scheffer, L. (1928) *Pflügers Archiv f. d.ges. Physiol.*, **220**, 655.

Ernst, E. (1963) *Biophysics of the Striated Muscle*. Budapest, Akadémiai Kiadó.

Evans, M.W. (1992) The elementary static magnetic field of the photon. *Physica B* **182**, 227-236.

Evans, M.W. (1994) Classical relativistic theory of the longitudinal ghost fields of electrodynamics. *Foundations of Physics* **24 (11)**, 1519-1542.

Evans, M.W.; Vigier, J.P. (1994) *The Enigmatic Photon.* Vol.1: *The Field B^3*. Fundamental Theories of Physics, Vol.64, Dordrecht, Kluwer Academic Publishers.

Evans, M.W.; Vigier, J.P. (1995*)* *The Enigmatic Photon.* Vol.2: *Non-Abelian Electrodynamics.* Fundamental Theories of Physics, Vol.68, Dordrecht, Kluwer Academic Publishers.

Evans, M.W.; Vigier, J.P.; Roy, S.; Jeffers, S. (1996) *The Enigmatic Photon.* Vol.3: *Theory and Practice of the $B^{(3)}$ Field.* Fundamental Theories of Physics, Vol.77, Dordrecht, Kluwer Academic Publishers.

Evans, M.W., Vigier, J.P., Roy, S., Hunter, G. (1998) *The Enigmatic Photon.* Vol.4: *New Directions.* Fundamental Theories of Physics, Vol.90, Dordrecht, Kluwer Academic Publishers.

Evans, M.W. (1998) Electrodynamics as a Non-Abelian Gauge Field Theory. *Frontier Perspectives* **7 (2)**, 7-12.

Fick, A. (1856) *Die medizinische Physik.* Braunschweig, Vieweg.

Fick, A. (1904) *Gesammelte Schriften*, vol.3, pp.492 and 767, Würzburg.

Findl, E. (19xx) Bioelectrochemistry - Electrophysiology - Electrobiology. In Bockris, J.O'M., Conway, B.E., White, R.E. (eds.), *Modern Aspects of Electrochemistry.* New York, Plenum Press, 509-555.

Fischer, M.H., Moore, G. (1907) *American Journal of Physiology*, **20**, 330.

Fischer, M.H. (1910) *Oedema: A Study of the Physiology and the Pathology of Water Absorption by the Living Organism.* New York, John Wiley.

Fischer, M.H., Hooker, M.O. (1933) *The Lyophilic Colloids (Their Theory and Practice).* Springfield, Illinois, Charles C. Thomas.

Fischer, M.H., Suer, W.J. (1938) *Archives of Pathology*, **26**, 51.

Fischer, M.H., Suer, W.J. (1951) *Der kolloide Aufbau der lebenden Substanz.* Darmstadt Steinkopff.

Fisher, S. (1986) *Development and Structure of the Body Image.* Hillsdale, NJ, Lawrence Erlbaum, 1986. 2 vols.

Flexner, S, Flexner, J.T. (1941) *William Henry Welch and the Heroic Age of American Medicine.* New York, Viking Press.

Fliess, W. (1906) *Der Ablauf des Lebens – Grundlegung zur exakten Biologie.* Leipzig-Vienna, Franz Deuticke.

Fougerousse, A. (1992) L'approche bio-électronique de Vincent. *Sciences du Vivant*, **No.4** (2ème trimestre), 63-79.

Fox-Keller, E. (1983) *A Feeling for the Organism*. San Francisco W.H.Freeman.

Friedman, N. (1997) *The Hidden Domain*. Eugene OR, Woodbridge Group.

Fröhlich, F., Hyland, G. (1995) Fröhlich coherence at the mind-brain interface. In King, J. and Pribram, K.H. (eds.), *Scale in Conscious Experience; Is The Brain Too Important To Be Left To Specialists To Study*. Mahwah, N.J., Lawrence Erlbaum Associates, 405-438.

Fröhlich, H. (1968) Long range coherence and energy storage in biological systems. *International Journal of Quantum Chemistry*, **2**, 641-649.

Fröhlich, H. (1969) Quantum mechanical concepts in biology. In Marois, M. (ed.), *Theoretical Physics and Biology*, Amsterdam, North-Holland, 13-22.

Fröhlich, H. (1975) Evidence for Bose condensation-like excitation of coherent modes in biological systems. *Physics Letters*, **51A**, 21-22.

Gabler, R. (1978) *Electrical Interactions in Molecular Biophysics*. New York, Academic Press.

Gallagher, R. (1984) Riemann and the Göttingen school of physiology. *Fusion*, **6 (3)**, 24-30.

Garnier, G. (1959) *Wladislas Kopaczewski - le savant et son oeuvre*. Lunéville.

Gasan, A.I. / Maleev, V.Ya. / Semenov, M.A. (1990) Role of water in stabilizing the helical biomacromolecules DNA and collagen. *Studia Biophysica*, **136 (2-3)**, 171-178.

Geddes, P. (1930) Life and Work of Sir J.C.Bose. New York: Benjamin Blom 1971.

Genack, A.Z., Drake, J.M. (1994) Scattering for super-radiation. *Nature*, **368**, 400-401.

Gilbert, S.F., Opitz, J.M., and Raff, R.A. (1996) Resynthesizing evolutionary and developmental biology. *Developmental Biology*, **173**, 357-372.

Glansdorff, P, Prigogine, I., *Thermodynamics of Structure, Stability and Fluctuations*. London, Wiley, 1971.

Glasser, O. (ed.) (1944, 1950, 1960) *Medical Physics*. 3 vols. Chicago, Yearbook Publishers.

Glauber, R. J. (1963a) The quantum theory of optical coherence. *Physical Review*, **130 (6)**, 2521-2539.

Glauber, R. J. (1963b) Coherent and incoherent states of the radiation field. *Physical Review*, **131 (6)**, 2766-2788.

Glauber, R. J. (1964) Quantum theory of coherence. In Grivet, P. and Bloembergen, N. (eds.), *Quantum Electronics*. Paris, Dunod / New York, Columbia University Press, **1**, 111-120.

Goodwin, B.C. (1987) Developing organisms as self-organizing fields. In *Self-Organizing Systems*, 167-180, ed. Yates, F.E (ed.), New York, Plenum.

Goodwin, B.C. (1994) Toward a science of qualities. In *New Metaphysical Foundations for Modern Science*, Harman, W. (ed.), Sausalito, CA, Institute of Noetic Sciences, 215-250.

Gortner, R.A. (1938) *Outline of Biochemistry*. 2nd. ed. New York, Wiley & Sons.

Goswami, A. (1989) The Idealistic Interpretation of Quantum Mechanics. *Physics Essays* 2, 385-400.

Goswami, A. (1993) *The Self-Aware Universe*. J.P.Tarcher/Putnam.

Goswami, A. (1994) *Science Within Consciousness – Developing a Science Based on the Primacy of Consciousness*. Causality Issues in Contemporary Science, Research Report # CP-7, Sausalito CA, Institute of Noetic Sciences.

Gough, W.C., Shacklett, R.L. (1993) The science of connectiveness. *Subtle Energies* **4 (1)**, 57-76, **4 (2)**, 99-123, **4 (3)**, 187-214.

Grandpierre, A. (1997) The physics of collective consciousness. *World Futures* **48**, 23-56.

Grandpierre, A. (2000) The nature of man-universe connections. In Amoroso, R., Antunes, R., Coelho, C., Farias, M., Leite, A., and Soares, P. (eds.), *Science and the Primacy of Consciousness*, Orinda, CA., Noetic Press, 203-223.

Gray, J. (1992) Consciousness on the scientific agenda. *Nature* 358, 277.

Green, E. E. (1990) *Consciousness, Psychophysiology, and Psychophysics: An Overview*. Copper Wall Research Technical Note, No.1, November 21, 1990, Voluntary Controls Program, Topeka, Kansas, The Menninger Clinic.

Green, E. E. (1991) Copper Wall Research – Psychology and Psychophysics. *"Subtle Energy and Energy Medcine: Emerging Theory and Practice"*. Proc. 1st Ann. Conf. Int.Soc. for the Study of Subtle Energies and Energy Medcine (ISSSEEM), June 21-25, Boulder, Colorado. Golden, Colorado.

Grinberg-Zylberbaum, J. (1987) Patterns of interhemispheric correlation during human communication. *International Journal of Neuroscience*, **36**, 41-53.

Grinberg-Zylberbaum, J., et al. (1992) Human communication and the electrophysiological activity of the brain. *Subtle Energies*, **3 (3)**, 25-43.

Grinberg-Zylberbaum, J., et al. (1994) The Einstein-Podolsky-Rosen Paradox in the brain: The transferred potential. *Physics Essays*, **7 (4)**, 422-428.

Gu, Q. (1998) Biophotons and nonclassical light. In Chang, J.J., Fisch, J., Popp, F.A. (eds.), *Biophotons*, Dordrecht, Kluwer Academic Publishers, 299-321.

Gulyaev, Yu.V.; Godik, E.E. (1987) The physical fields of biological objects. In *Cybernetics of Living Matter*, Makarov, I.M. (ed.), 244-250, Moscow, Mir Publishers.

Gurwitsch, A. (1964) *The Field of Consciousness*. Pittsburgh, Duquesne University Press.

Gurwitsch, A.G. (1923) Versuch einer synthetischen Biologie. *Schaxels Abhandlungen zur theoretischen Biologie*, **17**, 1-87.

Gurwitsch, A.G., Gurwitsch, L.D. (1959) *Die mitogenetische Strahlung*. Jena, Gustav Fischer.

Gutenbrunner, C., Hildebrandt, G., and Moog, R., (eds.) (1993) *Chronobiology & Chronomedicine: Basic Research and Applications*. Frankfurt, Peter Lang.

Hagelin, J.S. (1987) Is consciousness the unified field ? A field theorist's perspective. *Modern Science and Vedic Science* **1 (1)**, 29-87.

Haken, H. (1978) *Synergetics*. New York, Springer.

Haken, H. (1980) *Dynamics of Synergetic Systems*. New York, Springer.

Hadjamu, J. (1974) *Professor Dr.med.Heinrich Schade, Begründer der Molekularpathologie, 1876-1935, Leben und Werk*. Düsseldorfer Arbeiten zur Geschichte der Medizin, Heft 39, Düsseldorf, Michael Triltsch Verlag.

Hagley, E. et al. (1997) Generation of Einstein-Podolsky-Rosen pairs of atoms. *Physical Review Letters* **79 (1)**, 1-5.

Halberg, F. (1953) Some physiological and clinical aspects of 24-hour periodicity. *Lancet (USA)*, **73**, 20-32.

Halberg, F. (1960) Temporal coordination of physiological function. *Cold Spring Harbor Symoposia on Quantum Biology*, **25**, 289-310.

Halberg, F. (1969a) Chronobiologie – rhythms et physiologie statistique. In Marois, M. (ed.), *Theoretical Physics and Biology*. Amsterdam, North-Holland, 347-393.

Halberg, F. (1969b) Chronobiology. *Ann. Rev. Physiol.*, **31**, 675-725.

Halberg, F. (1979) Physiologic 24-hour periodicity. *Zeitschrift für Vitamin-, Hormon-, und Fermentforschung*, **10**, 225-296.

Halberg, F. (1998) Chronomedicine. In: Armitage, P., Colton, T., (eds.), *Encycloedia of Biostatistics*, Vol.1. Chichester, Wiley, 642-649.

Harman, W. (1992) *Reconciling the experience of consciousness with the scientific worldview*. Presented at the HSRC Conference on Science and Vision, Pretoria, South Africa, January 31.

Harman, W., Sahtouris, E. (1998) *Biology Revisioned*. Berkeley, North Atlantic Books.

Haroche, S., Kleppner, D. (1989) Cavity Quantum Electrodynamics. *Physics Today*, **42, No.1**, 24-30.

Haroche, S. (1991) Cavity Quantum Optics. *Physics World*, **4 (3)**, 33-38.

Hatfield, E., Cacioppo, J.T., and Rapson, R.L. (1994) *Emotional Contagion*. Cambridge, England, Cambridge University Press.

Hauss, W.H., Losse, H. (1960) *Struktur und Stoffwechsel des Bindegewebes*. Stuttgart, Thieme.

Hauss, W.H., Junge-Hülsing, G. (1961) Über die universelle unspezifische Mesenchymreaktion. *Deutsche Medizinische Wochenschrift*, **86**, 763.

Hauss, W.H., Junge-Hülsing, G., and Gerlach, U. (1968) *Die unspezifische Mesenchymreaktion*. Stuttgart: Gustav-Thieme-Verlag.

Hay, E.D. (1983) *Cell Biology of Extracellular Matrix*. 2nd ed. Plenum Press, New York and London 1983.

Hazlewood, C.F., Nichols, B.L., Chang, D.C., and Brown, B. (1971) On the state of water in the developing muscle. *Johns Hopkins Medical Journal*, **178**, 117.

Hazlewood, C.F. (1979) A view of the significance of and understanding of the physical properties of cell-associated water. In Drost-Hansen, W., Clegg, J.S., (eds.), *Cell-Associated Water*, New York, Academic Press.

Hazlewood, C.F. (1991) Insight into the organization of water in living cells. In Tigyi, J., Kellermayer, M., and Hazlewood, C.F., (eds.) *The Physical Aspect of the Living Cell*. Budapest, Akademiai Kiadó, 107-118.

Heine, H., Anastasiadis, P., (eds.) (1992), *Normal Matrix and Pathological Conditions*. Stuttgart, Gustav Fischer.

Heine, H. (1997) *Lehrbuch der biologischen Medizin*. 2nd rev. ed. Stuttgart, Hippokrates.

Heisenberg, W. (1958a) *Physics and Philosophy*. New York, Harper & Row.

Heisenberg, W. (1958b) *Physikalische Prinzipien der Quantentheorie*. Mannheim, Bibliographisches Institut.

Heisenberg, W. (1976) Was ist ein Elementarteilchen ? *Naturwissenschaften* **63 (1)**, 1-7.

Hellpach, W. (1911) Die *geopsychischen Erscheinungen*. Publ. from 4th ed. under the title *Geopsyche*, Leipzig, Wilhelm Engelmann.

Hellpach, W. (1924) Psychologie der Umwelt. In, E.Abderhalden (ed.), *Handbuch der biologischen Arbeitsmethoden*, Abt.VI, part C1, 109-218, Vienna, Urban & Schwarzenberg.

Henderson, L.J. (1909) Das Gleichgewicht zwischen Basen und Säuren im tierischen Organismus. *Ergebnisse der Physiologie*, L.Asher and K.Spiro (eds.), **8**, 254-325.

Henderson, L.J. (1913) *The Fitness of the Environment*. New York, Macmillan.

Henderson, L.J. (1913) The regulation of neutrality in the animal body. *Science*, **37 (950)**, 389-395.

Henderson, L.J. (1928) *Blood: A Study in General Physiology*. New Haven.

Héricourt, J. (1927) *Le Terrain dans les Maladies*. Paris, Flammarion.

Hess, W.R. (1948) *Die funktionelle Organisation des vegetativen Nervensystems*. Basel, Schwabe.

Hill, A.V. (1930) *Proceedings of the Royal Society (London)*. Ser.B., 106, 477.

Hill, A.V., Kupalov, P.S. (1930) *Proceedings of the Royal Society (London)*. Ser.B., 106, 445.

Ho, M.W., Popp, F.A., and Warnke, U. (eds.) (1994) *Bioelectromagnetics and Biocommunication*. Singapore, World Scientific.

Ho, M.W. (1993) *The Rainbow and the Worm - The Physics of Organisms*. Singapore: World Scientific.

Ho, M.W. (19) Toward an indigenous Western science: Causality in the universe of coherent space-time structures. In *New Metaphysical Foundations for Modern Science*, Harman, W. (ed.) Sausalito, CA, Institute of Noetic Sciences, 179-213.

Ho, M.W., French, A., Haffegee, J., Saunders, P.T. (1994) Can weak magnetic fields (or potentials) affect pattern formation ? In *Bioelectrodynamics and Biocommunication*, Ho, M.W., Popp, F.A. and Warnke U. (eds.), Singapore World Scientific, 195-212.

Ho, M.W. (1996) Organisms as polyphasic liquid crystals. *Bioelectrochemistry and Bioenergetics*, **41**, 81-91.

Ho, M.W. (1997) Towards a theory of the organism. *Integrative Physiological and Behavioral Science*, 32 (4), 343-363.

Ho, M.W. (1998) Organism and psyche in a participatory universe. In Loye, D. (ed.) *The Evolutionary Outrider*, Westport, CT, Praeger, 49-65.

Ho, M.W. (1999) *Genetic Engineering – Dream or Nightmare ?* 2nd rev. ed. Dublin Gateway.

Holmes, F.L. (1963) Claude Bernard and the milieu interieur. *Archives Internationales d'Histoire des Sciences*, **16**, 369-376.

Hoppe, W., Lohmann, W., Markl, H., and Ziegler, H., (eds.) (1982) *Biophysics*. 2nd compl. rev. ed., New York-Berlin, Springer.

Huntington, E. (1915) *Civilisation and Climate*. New Haven, Yale University Press.

Huntington, E. (1923) *Earth and Sun*. New Haven, Yale University Press.

Huntington, E. (1930) *Weather and Health*. Bulletin of the National Research Council, Washington, National Academy of Science.

Huntington, E. (1945) *Mainsprings of Civilisation*. New York, John Wiley & Sons.

Ihle, G. (1962) Das "Praecoxgefühl" in der Diagnostik der Schizophrenie. *Archiv für Psychiatrie*, **203**, 385-406.

Inyushin, V.M., Grishchenko, V.S., Vorobyov, N.A., Shuyski, N.N., Fyodorowa, N.N. and Gibadulin, F.F. (1968) *On the Biological Character of the Kirlian Effect – The Concept of Biological Plasma* (in Russian). Alma-Ata Kazakh State University.

Inyushin, V.M. (1970) Biological plasma of human and animal organisms. *International Journal of Paraphysics*, **5 (1/2)**, 50-53.

Inyushin, V.M. (1977) Bioplasma: The fifth state of matter ? In White, J., Krippner, S. (eds.) *Future Science*, Garden City, N.J., Anchor-Doubleday, 115-120.

Inyushin, V.M. (1983) Resonance, biostimulation and the problem of bioplasma. In Wolkowski, Z.W. (ed.) *Proceedings Int. Symposium of Wave Therapeutics*, Créteil, 123-129.

Jantsch, E. (1980) *The Self-Organizing Universe*. Oxford Pergamon Press.

Jibu, M.; Yasue, K. (1993) The basics of quantum brain dynamics. In Pribram, K.H., (ed.) (1993) *Rethinking Neural Networks, Quantum Fields and Biological Data*. Hilldale N.J., Erlbaum.

Jibu, M., Yasue, K. (1995) *Quantum Brain Dynamics and Consciousness: An Introduction*. Philadelphia, John Benjamins Publishing.

Jordan, P. (1934) Quantenphysikalische Bemerkungen zur Biologie und Psychologie. *Erkenntnis*, **4 (3)**, 215-252.

Jordan, P. (1938) Die Verstärkertheorie der Organismen in ihrem gegenwärtigen Stand. *Naturwissenschaften*, **26**, 537.

Jordan, P. (1942) Begriff und Umgrenzung der Quantenbiologie. *Physis* (Stuttgart), **No.1**, 13-26.

Jordan, P. (1948) *Die Physik und das Geheimnis des organischen Lebens*. Braunschweig, Friedrich Vieweg & Sohn.

Josephson, B.D.; Pallikara-Viras, F. (1991) Biological utilization of quantum nonlocality. *Foundations of Physics*, **21 (2)**, 197-207.

Kafatos, M., Nadeau, R. (1990) *The Conscious Universe – Part and Whole in Modern Physical Theory*. New York, Springer.

Kanitscheider, B. (1979) *Philosophie und moderne Physik*. Darmstadt, Wissenschaftliche Buchgesellschaft.

Katchalsky, A., Curran, P.F. (1967) *Nonequilibrium Thermodynamics in Biophysics*. Cambridge, Mass., Harvard University Press.

Katz, J.R. (1917) The role of swelling. *Kolloidchemische Beihefte*, **9**, 1-182.

Kay, L.E. (1985) The secret of life: Niels Bohr's influence on the biology program of Max Delbrück. *Rivista di Storia della Scienza,* **2 (3)**, 487-510.

Kay, L.E. (1993) *The Molecular Vision of Life – Caltech, the Rockefeller Foundation, and the Rise of the New Biology.* New York, Oxford University Press.

Keller, E.F. (1990) Physics and the emergence of molecular biology: A history of cognitive and political synergy. *Journal of the History of Biology* **23**, 389-409.

Kollath, W. (1968) *Regulatoren des Lebens - vom Wesen der Redox-Systeme.* Heidelberg, Karl F.Haug Verlag.

Kopaczewski, M.W. (1921-22) La tension superficielle en biologie. Parts 1-3. *Archives de Physique Biologique,* **1 (4) (1921)**, 145-167; **2 (2) (1922)**, 1-20; **2 (3) (1922)**, 1-12.

Kopaczewski, M.W. (1923) *Théorie et pratique des colloides en biologie et en médecine.* Paris, Vigot Frères.

Kopaczewski, M.W. (1926a) *Introduction à l'étude des colloides. L'état colloidal et ses applications.* Paris, Gauthier-Villars.

Kopaczewski, M.W. (1926b) *Les ions d'hydrogène. Signification, mesure, applications, données numériques.* Paris, Gauthier-Villars.

Kopaczewski, M.W. (1930-38) *Traité de Biocolloidologie.* 2eme éd. Paris, Gauthier-Villars. vol. 1: Pratique des colloides (1930), tome 2, Biocolloides, vol. 3: Phénomènes colloidaux (1932-33), vol. 4: État colloidal en biologie (1933-36), vol. 5: État colloidal en médecine (1937-38).

Kopaczewski, M.W. (1933) Role de la tension superficielle en biologie. *Protoplasma,* **19**, 255-292.

Kopaczewski, M.W. (1936) Le terrain, le microbe et l'état infectieux: Claude Bernard ou Pasteur? *Revue Scientifique,* **74 (14)**, 417-425.

Kopaczewski, M.W. (1938) Meteoropathologie. *La Nature* (Paris), **66 (3024)**, 272-276.

Kopaczewski, M.W. (1939) *Essai de météoropathologie - physique, clinique, thérapeutique.* Paris, Baillière.

Kopaczewski, M.W. (1947) *D'Arsonval et la Biophysique.* Rabat.

Kracmar, F. (1971) Zur Biophysik des vegetativen Grundsystems. *Physikalische Medizin und Rehabilitation,* **12 (6)**, 120-122.

Kraus, F. (1919) *Allgemeine und spezielle Pathologie der Person. Klinische Syzygiologie.* Thieme, Leipzig; 2nd ed.1926, Vol.1: Allgemeiner Teil. Vol.2: Besonderer Teil: Tiefenperson.

Kühnau, J. (1941) Redoxpotentiale und Redoxsysteme in der Medizin. *Therapie der Gegenwart,* **82**, 35 and 375.

Kuhnert, L., and Niedersen, U., (eds.) (1987) *Selbstorganisation chemischer Strukturen – Arbeiten von F.F.Runge, R.E.Liesegang, B.P.Belousov und A.M.Zhabotinsky.* Ostwalds

Klassiker der exakten Wissenschaften, 272, Leipzig, Akademische Verlagsgesellschaft Geest & Portig.

Lakhovsky; G. (1963) *The Secret of Life. Cosmic Rays and Radiations of Living Beings.* 3rd rev. ed. Rustington, Sussex, England, Health Science Press.

Lakhtakia, A., ed. (1993) *Essays on the Formal Aspects of Electromagnetic Theory.* Singapore, World Scientific Publishing.

Laszlo, E. (1995) *The Interconnected Universe – Conceptual Foundations of Transdisciplinary Unified Theory.* Singapore, World Scientific.

Laszlo, E. (1996) *The Whispering Pond – A Personal Guide to the Emerging Vision of Science.* Rockport, Mass., Element Books.

Laughlin, C.D. (1996) Archetypes, neurognosis and the quantum sea. *Journal of Scientific Exploration* **10**, 375-400.

Laville, C. (1925) *Electrodynamique du muscle.* Paris.

Laville, C. (1928) *Le cancer - dérangement électrique de la cellule.* Paris, Dunod, 1928.

Laville, C. (1932) *La négativation électrique.* Montrouge: Imprimerie de l'Industrie, Later ed. Paris, Masson, 1934.

Laville, C. (1934) De la notion du pH considérée comme révélatrice de l'intimité des mécanismes vitaux. *pH*, juillet 1934.

Laville, C. (1950) *Mécanismes Biologiques.* Paris, Dunod.

Lawandy, N.M., et al. (1994) Laser action in strongly scattering media. *Nature*, **368 (6470)**, 436-438.

Lawrence, C. and Weisz, G., (eds.) (1998) *Greater Than the Parts - Holism in Biomedicine 1920-1950.* New York, Oxford University Press.

Leggett, A.J. (1986) The superposition principle in macroscopic systems. In *Quantum Concepts in Space and Time*, Penrose, R. and Isham C.J. (eds.) Oxford, Clarendon Press, 228-240.

Leggett, A.J. (1992) *Quantum Tunnelling in Condensed Media*, Yu Kagan & Leggett, A.J. (eds). Elsevier Science.

Leggett, A.J. (1996) The current status of quantum mechanics at the macroscopic level. *Foundations of Quantum Mechanics in the Light of New Technology*, eds. Nakajima S., Murayama, Y., and Tonomura, A. (eds.), Singapore: World Scientific, 257-267.

Letokhov, V.S. (1968) Generation of light by a scattering medium with negative resonance absorption. *Soviet Physics JETP*, **26**, 835-839.

Lewin, K. (1936) *Principles of Topological Psychology.* New York, McGraw-Hill.

Lewin, K. (1951) *Field Theory in Social Science*. New York, Harper & Brothers.

Li, K.H. (1992) Coherence in physics and biology. In Popp, F.A., Li, K.H., Gu, Q. (eds.), *Recent Advances in Biophoton Research and its Applications*, Singapore, World Scientific Publishing, 113-155.

Li, K.H. (1994) Uncertainty Principle, coherence, and structures. In R.K.Mishra, D. Maass, and E.Zwierlein (eds.), *On Self-Organization*, Berlin, Springer, 245-255.

Li, K.H. (1995) Coherence - A Bridge between Micro- and Macro-Systems. *Biophotonics - Non-Equilibrium and Coherent Systems in Biology, Biophysics and Biotechnology.*, ed. L.V.Belousov, and Popp, F.A. (eds.), 99-114, Moscow, Bioinform Services.

Lichtwitz, L., Liesegang, R.E., and Spiro, K. (1935) *Medizinische Kolloidlehre*. Dresden, Steinkopff.

Lieth, H.H. (1981) 25 years of International Society of Biometerology. *International Journal of Biometerology*, **25 (3)**.

Ling, G.N. (1960) The interpretation of selective ionic permeability and cellular potential in terms of fixed charge/induction hypothesis. *Journal of General Physiology*, **43**, 149.

Ling, G.N. (1962) *A Physical Theory of the Living State: The Association-Induction Hypothesis*. Waltham, Mass., Blaisdell.

Ling, G.N. (1992) *A Revolution in the Physiology of the Living Cell*. Malabar, Florida, Krieger Publishing.

Ling, G.N. (2001) *Life at the Cell and Below-Cell Level. The Hidden History of a Fundamental Revolution in Biology*. New York, Pacific Press.

Lipkind, M. (1998) Alexander Gurwitsch and the concept of the biological field. Part 1. *21st Century Science & Technology*, **11 (2)**, 36-51.

Lipkind, M. (1998) Alexander Gurwitsch and the concept of the biological field. Part 2. *21st Century Science & Technology*, **11 (3)**.

Loofbourow, J.R. (1940) Borderland problems in biology and physics, *Rev.Mod.Phys.*, **12 (4)**, 267.

Luce, G.G. (1971) *Biological Rhythms in Human and Animal Physiology*. New York, Dover.

Lund, E.J. (1947) *Bioelectric Fields and Growth*. Austin/Texas University of Texas Press.

Mandoli, D.F., Briggs, W.R. (1982) Optical properties of etiolated plant tissues. *Proceedings of the National Academy of Sciences of the USA*, **79**, 2902.

Mandoli, D.F., Briggs, W.R. (1984) Fiber optics in plants. *Scientific American*, **251**, 90-98.

Mansfield, V. (1995) On the physics and psychology of transference as an interactive field. *http://lightlink.com/vic/field.htm*.

Manzelli, P., Masini, G., and Costa, M. (1994) *I Segreti dell'Acqua. L'opera scientifica di Giorgio Piccardi*. Roma, Di Renzo Editore.

Marois, M., ed. (1969) *Theoretical Physics and Biology*. Amsterdam, North-Holland Publishing.

Marois, M., (ed.) (1971) *From Theoretical Physics to Biology*. Paris, Editions du CNRS.

Marois, M., (ed.) (1973) *From Theoretical Physics to Biology*.Basel , S.Karger.

Marois, M., (ed.) (1975) *From Theoretical Physics to Biology*. Amsterdam, North-Holland.

Marois, M. (1997) *Documents of History: Life and Human Destiny. The Institut de la Vie*. Paris, Rive Droite.

McBroom, P. (1966) Body Dances to Speech. *Science News*, **89**, 483.

Meystre, P., Walls, D.F. (1991) *Nonclassical Effects in Quantum Optics*. Key Papers in Physics, Bristol, Adam Hilger, American Institute of Physics.

Michaelis, L. (1914) *Die Wasserstoffionenkonzentration*. Berlin, Julius Springer.

Michaelis, L. (1933) *Oxydations- und Reduktionspotentiale*. Berlin, Julius Springer.

Mickulecky, M., (ed.) (1993) The Moon and Living Matter. Bratislava, Slovak Medical Society.

Mickulecky, M., (ed.) (1994) *Sun, Moon and Living Matter*. Bratislava, Slovak Medical Society.

Mickulecky, M., (ed.) (1997) Chronobiology and its Roots in the Cosmos. Bratislava, Slovak Medical Society.

Milburn, M.P. (1994) Emerging relationships between the paradigms of oriental medicine and the frontiers of Western biological science. *American Journal of Acupuncture*, **22 (2)**, 145-157.

Milburn, M.P. (2001) *The Future of Healing – Exploring the Parallels of Eastern and Western Medicine*. Freedom, California: The Crossing Press.

Millay, J. (1981) Brainwave synchronization: A study of subtle forms of communication. *Humanistic Psychology Institute Review*, **3**, 9-40.

Mishra, R.K. (1975) Occurrence, fluctuations and significance of liquid crystallinity in living systems. *Molecular Crystals and Liquid Crystals*, **29**, 201-224.

Mishra, R.K., Bhaumik, K., Mathur, S.C., and Mitra, P.P. (1979) Excitons and Bose-Einstein condensation in living systems. *International Journal of Quantum Chemistry*, **16**, 691-706.

Mishra, R.K. (1990) The living state - the matrix of self-organization. In Mishra, R.K.(ed.), *Molecular and Biological Physics of Living Systems*, Dordrecht, Kluwer Academic Publishers, 215-237.

Mitman, G., Fausto-Sterling, A. (1992) Whatever happened to Planaria? C.M.Child and the physiology of inheritance. In Clarke A.E. and Fujimura J.H. (eds.), *The Right Tool for the Right Job: At Work in Tweentieth-Century Life Sciences*. Princeton, Princeton University Press, 172-197.

Moore, G., Roaf, H.E. (1908) *Biochemical Journal*, 3, 55.

Morton, Michael E., Dlouhy, Carrie (eds.) (1989) *Energy Fields in Medicine. A Study of Device Technology Based on Acupuncture Meridians and Chi Energy*. Kalamazoo, Michigan, John E.Fetzer Foundation.

Murphy, G. (1945) Field theory and survival. *Journal of the American Society for Psychical Research*, **39 (4)**, 181-209.

Murphy, G. (1947) *Personality – A Biosocial Approach to Origins and Structure*. New York, Harper & Brothers.

Murphy, G. (1961) Toward a field theory of communication. *Journal of Consciousness*, **11**, 196-201.

Murphy, M.P., O'Neill, L.A.J., (eds.) (1995) *What is Life ? The Next Fifty Years. Speculations on the Future of Biology*. Cambridge, Cambridge University Press.

N.N. (1999) Can physics deliver another biological revolution ? *Nature* **397**, 476-77.

Nalimov, V.V. (1981) *In the Labyrinths of Language*. Philadelphia, ISI Press.

Nalimov, V.V. (1982) *Realms of the Unconscious*. Philadelphia, ISI Press.

Nalimov, V.V. (1989) *Spontaneity of Consciousness*. Moscow, Prometheus Publishers (in Russian).

Nassonov, D.H. (1959) *The Local Reaction of the Protoplasm and the Spreading of Excitation* (in Russian). Moscow, Academy of Sciences.

Nelson, R.D., Bradish, G.J., Dobyns, Y.H., Dunne. B.J., and Jahn, R.G. (1996) FieldREG anomalies in group situations. *Journal of Scientific Exploration*, **10 (1)**, 111-141.

Nelson, R.D., Jahn, R.G., Dunne. B.J., Dobyns, Y.H., and Bradish, G.J. (1998) FieldREG II: Consciousness field effects: Replications and explorations. *Journal of Scientific Exploration*, **12 (3)**, 425-454.

Newton, R., and Gortner, R.A. (1922) A method for estimating hydrophilic colloid content of expressed plant tissue fluids. *Botanical Gazette*, **74**, 442-446.

Nicolis, G., Prigogine, I. (1977) *Self-Organization in Non-Equilibrium Systems*. New York, Wiley.

Ninham, B.W. (1982) Surface forces in biological systems. In Franks, F., Mathias, S.F., (eds.), *Biophysics of Water*, Chichester: Wiley, 105-119.

Norretranders, T. (1998) *The User Illusion. Cutting Consciousness Down to Size*. New York, Viking.

Nuccitelli, R., (ed.) (1986) *Ionic Currents in Development*. New York, Alan R. Liss, 1986.

Olmsted, E.H. (1967) Historical phases in the influence of Bernard's scientific generalizations in England and America. In Grande, F. and Visscher, M.B. (eds.), *Claude Bernard and Experimental Medicine*. Cambridge, Massachusetts, Schenkman, 24-34.

Országh, J. (1990) Réactions d'oxydo-réduction et acido-basiques - vers une approche théorique et expérimentale plus cohérente. *Sciences du Vivant*, **No.1 (3ème trimestre)**, 23-35.

Országh, J. (1992) Quelques aspects physio-chimiques des coordonnées bioélectroniques. *Sciences du Vivant*, **No.4 (2ème trimestre)**, 45-62.

Oschman, J.L. (1996) The nuclear, cytoskeletal and extracellular matrices: a continous communication network. Center for Advanced Studies in the Space Life Sciences Workshop on "*The Cytoskeleton: Mechanical, Physical, & Biological Interactions*", Nov.15-17, 1996, Marine Biological Laboratory, Woods Hole, Mass., Poster Session. http://www.mbl.edu/CASSLS/Oschman.poster.html.

Oschman, J.L. (2000) *Energy Medicine – The Scientific Basis*. Edinburgh, Churchill Livingston.

Palmer, J.D. (1976) *An Introduction to Biological Rhythms*. New York, Academic Press.

Petersen, W.F., Levinson, S.A. (1930) *The Skin Reactions, Blood Chemistry and Physical Status of 'Normal' Men and of Clinical Patients*, Archives of Pathology, 9, published as a separate monograph.

Petersen, W.F. (1936) *The Patient and the Weather*. Ann Arbor, Michigan: Edwards. 4 vols.

Petersen, W.F. (1947) *Man Weather Sun*. Springfield, Ill., C.C.Thomas.

Piccardi, G. (1962) *The Chemical Basis of Medical Bioclimatology*. Springfield, Ill., C.C.Thomas.

Pienta, K.J., and Coffey, D.S. (1991) Cellular harmonic information transfer through a tissue tensegrity-matrix system. *Medical Hypotheses*, **34**, 88-95.

Pierce, G.J. (1927) Sir J.C.Bose and his latest book. *Science*, **66 (1721)**, 621-622.

Pike, E.R., Walther, H., (eds.) (1988), *Photons and Quantum Fluctuations*. Bristol and Philadelphia, Adam Hilger.

Pischinger, A. (1991) *Matrix and Matrix Regulation – Basis for a Holistic Theory in Medicine*. Brussels, Haug International.

Planck, M. (1931) *Nature*, April **18**, 1931.

Popp, F.A., Nagl, W., Li, K.H., Scholz, W., Weingärtner, O., Wolf, R. (1984) Biophoton emission, New evidence for coherence and DNA as a source. *Cell Biophysics*, **6**, 33-51.

Popp, F.A. (1986) *Bericht an Bonn*. Essen, Germany, VGM.

Popp, F.A., Warnke, U., König, H.L. and Peschka, W., (eds.) (1989) *Electromagnetic Bio-Information*. 2nd ed. Munich-Baltimore, Urban & Schwarzenberg.

Popp, F.A., Li, K.H., Gu, Q., (eds.) (1992) *Recent Advances in Biophoton Research and its Applications*. Singapore, World Scientific.

Popp, F.A. (1994) Electromagnetism and Living Systems. In Ho, M.W., Popp, F.A., and Warnke, U. (eds.), *Bioelectromagnetics and Biocommunication*. Singapore, World Scientific, 33-80.

Popp, F.A., Chang, J.J., Herzog, A., Yan, Z, and Yan, Y. (2002) Evidence of non-classical (squeezed) light in biological systems. *Physics Letters A*, **293**, 98-102.

Popper, K.R., Eccles, J.C. (1977) *The Self and Its Brain*. New York, Springer International.

Porter, K.R., Tucker, J.B. (1981) The ground substance of the living cell. *Scientific American*, **244**, 57-67.

Porter, K.R. (1984) The cytomatrix: A short history of its study. *Journal of Cell Biology*, **99**, 3s-12s.

Presman, A.S. (1970) *Electromagnetic Fields and Life*. New York and London, Plenum Press. (Original Russian edition: Nauka, Moscow 1968).

Pribram, K.H., (ed.) (1993) *Rethinking Neural Networks: Quantum Fields and Biological Data*. Hillsdale N.J., Erlbaum.

Prigogine, I. (1947) *Étude Thermodynamique des Phénomènes Irréversibles*. Liège, Desoer.

Prigogine, I. (1961) *Introduction to Thermodynamics of Irreversible Processes*. New York´, Wiley.

Primas, H. (1981) *Chemistry, Quantum Mechanics, and Reductionism*. Lecture Notes in Chemistry, Vol.24, Berlin, Springer.

Primas, H. (1982) Chemistry and complementarity. *Chimia* **36**, 293-300.

Primas, H. (1985) Kann Chemie auf Physik reduziert werden ? *Chemie in unserer Zeit* **19 (4)**, 109-xx and **19 (5)**, 160-166.

Primas, H. (1990) Biologie ist mehr als Molekularbiologie. In Fischer, E.P., Mainzer, K., (eds.), *Die Frage nach dem Leben*, Piper, Munich, 63-92.

Primas, H. (1992) Umdenken in der Naturwissenschaft. *GAIA* 1 **(1)**, 5-15.

Primas, H. (1993) Ein Ganzes, das nicht aus Teilen besteht – Komplementarität in den exakten Naturwissenschaften. In *Mannheimer Forum 92/93*, ed. E.P.Fischer. Mannheim: Boehringer.

Primas, H., Atmanspacher, H. and Amann, A. (eds.) (1999) *On Quanta, Mind and Matter. Hans Primas in Context.* Lecture Notes in Chemistry, 24, Dordrecht, Kluwer.

Radin, D., Rebman, J.M. and Cross, M.P. (1996) Anomalous organization of random events by group consciousness: Two exploratory experiments. *Journal of Scientific Exploration*, 10 (1), 143-168.

Radin, D. (1997) The Conscious Universe - The Scientific Truth of Psychic Phenomena. San Francisco, Harper, San Francisco.

Rashevsky, N. (1960) *Mathematical Biophysics*. 3rd. ed. New York, Dover.

Rashevsky, N. (1962) *Physicomathematical Aspects of Biology*. New York, Academic Press.

Rein, G. (1989) Effect of non-hertzian scalar waves on the immune system. *Journal of the U.S. Psychotronics Association*, 1, 15.

Rein, G. (1991) Utilization of a cell culture bioassay for measuring quantum potentials generated from a modified caduceus coil. *Proceedings of the 26th Intersociety Energy Conversion Engineering Conference, August 4-9, 1991, Boston*, 4, 400-403.

Rein, G. (1993) Modulation of neurotransmitter function by quantum fields. In Pribram, K.H., ed. *Rethinking Neural Networks: Quantum Fields and Biological Data*, Hillsdale N.J., Erlbaum, 377-388.

Rein, G. (1997) A bioassay for negative Gaussian fields associated with geometric patterns. In *Proceedings of the Fourth International Symposium of New Energy, Denver, Colorado, May 23-26, 1997*. Fort Collins, Colorado, International Association for New Science.

Rein, G. (1998) Biological effects of quantum fields and and their role in the natural healing process. *Frontier Perspectives* 7 (1), 16-23.

Reiss, P. (1926) *Le pH intérieur cellulaire*. Paris.

Reiss, P. (1940) L'action du potentiel d'oxydation-réduction du milieu sur l'autolyse protéique du cristallin et du corps vitré. *Archives de Physique Biologique*, 15 (4), 39-40.

Reiss, P. (1942/43) Actions du potentiel d'oxydation-réduction du milieu sur l'activité de differentes protéinases: Hydrolyse et condensation. *Archives de Physique Biologique*, 16, suppl.

Reiss, P. (1943/44) Les mesures de potentiels d'oxydation-réduction dans le sang. Revue bibliographique. *Archives de Physique Biologique*, 17, suppl.

Reiss, P., Lemair, R. (1943/44) Évolution du potentiel d'oxydo-réduction du sang humain in vitro. *Archives de Physique Biologique*, 17, suppl.

Richards, T.L., Standish, L.J. (2000) EEG coherence and visual evoked potentials : Investigation of neural energy transfer between human subjects. Abstract No.393. *Tucson 2000 Consciousness Conference*, http://www.imprint.co.uk/Tucson2000.

Ricker, G. (1905) *Entwurf einer Relationspathologie*. Jena, Gustav Fischer.

Ricker, G. (1924) *Pathologie als Naturwissenschaft. Relationspathologie*. Berlin, Julius Springer.

Rosen, R. (1958-59) A relational theory of biological systems. *Bulletin of Mathematical Biophysics*, **20**, 245-260, and **21**, 109-128.

Rosen, R. (1967) *Optimality Principles in Biology*. New York, Plenum Press / London, Butterworth.

Rosen, R. (1970) *Dynamical Systems in Biology*. New York, Wiley-Interscience.

Rosen, R. (1972) *Relationale Biologie*. Fortschritte der experimentellen und theoretischen Biophysik (ed.), W.Beier,Vol.15. Leipzig, Georg Thieme.

Rosen, R. (1974) The role of quantum theory in biology. *International Journal of Quantum Chemistry: Quantum Biology Symposia*, **No.1**, 229-232.

Rowlands, S. (1985) Condensed matter physics and the biology of the future. *Jourrnal of Biological Physics*, **13**, 103-105.

Rubik, B. (1996) *Life at The Edge of Science*. Philadelphia, Institute for Frontier Science.

Rudder, B. de (1931) *Grundriss einer Meteorobiologie des Menschen*. Berlin, Springer.

Rudder, B. de (1937) *Über sogenannte « kosmische » Rhythmen beim Menschen*. Stuttgart, Georg Thieme.

Sargent, F. II, Shimkin, D.B. (1965) Biology, society and culture in human ecology. *BioScience*, **15**, 512-516.

Sargent, F. II (1974) *Human Ecology*. Amsterdam, North-Holland / New York, American Elsevier.

Sargent, F. II (1982) *Hippocratic Heritage – A History of Ideas About Weather and Human Health*. New York, Pergamon Press.

Sargent, F. II (1983) *Human Ecology - A Guide to Information Sources*. Detroit, Gale Research.

Sassaroli, E., Srivastava, Y., Swain, J. and Widom, A., (eds.) (1998) *Macroscopic Quantum Coherence*. Singapore, World Scientific.

Saunders, S., Brown, H.R., (eds.) (1991) *The Philosophy of Vacuum*. Oxford, Clarendon Press.

Savva, S. (1999) Biofield and a cybernetic model of the organism: sauggestion for empirical study. *National Geophysical Data Center Public Hypernews*, http://hypernews.ngdc.noaa.gov/hnxtra/savva_paper.html.

Schade, H. (1909) Kolloidchemie und Balneologie. *Medizinische Klinik*, **No.29**, 1070 and **No.30**, 1116.

Schade, H. (1912a) Physikochemische Beiträge zur Öedemfrage. *Verhandlungen des Deutschen Kongresses für Innere Medizin* (Wiesbaden), **No.29**, 526-535.

Schade, H. (1912b) Untersuchungen zur Organfunction des Bindegewebes. I.Mitteilung: Die Elasticitätsfunction des Bindegewebes und die intravitale Messung ihrer Störungen. *Zeitschrift für experimentelle Pathologie und Therapie*, **11 (3)**, 369-399.

Schade, H. (1913) Untersuchungen zur Organfunction des Bindegewebes. II.Mitteilung: Das Quellungsvermögen des Bindegewebes in der Mannigfaltigkeit seiner Erscheinungen. *Zeitschrift für experimentelle Pathologie und Therapie*, **14, (1)**, 1-29.

Schade, H. (1921a) *Die physikalische Chemie in der inneren Medizin. Die Anwendung und die Bedeutung physikochemischer Forschung in der Pathologie und Therapie*. Dresden, Theodor Steinkopff Verlag.

Schade, H., Neukirch, P., and Halpert, A. (1921b) Über lokale Acidosen des Gewebes und die Methodik ihrer intravitalen Messung, zugleich ein Beitrag zur Lehre der Entzündung. *Zeitschrift für experimentelle Medizin*, **24**, 11.

Schade, H. (1922) Die Physikochemie des Bindegewebes und ihre Bedeutung für die Lymph- und Ödembildung. *Verhandlungen der Deutschen Gesellschaft für Innere Medizin*, **34**, 283-287.

Schade, H. (1923a) Über die Gesetze der Gewebsquellung und ihre Bedeutung für klinische Fragen (Wasseraustausch im Gewebe, Lymphbildung und Ödementstehung). *Zeitschrift für Klinische Medizin*, **96**, 279-327.

Schade, H. (1923b) Von der klinischen Bedeutung des Bindegewebes. *Jahreskurse für Ärztliche Fortbildung*, **14, part III**, 10-17.

Schade, H. (1923c) Die Physikochemie der Entzündung. *Verhandlungen der Deutschen Pathologischen Gesellschaft*, 19th meeting, 69-80. Jena, Gustav Fischer Verlag.

Schade, H. (1926) Die Bedeutung der H-Ionenkonzentration in der Pathologie. *Kolloid-Zeitschrift*, **40**, 252-258.

Schilder, P. (1950) *The Image and Appearance of the Human Body*. London, Kegan, Paul, Trench, Trubner.

Schiller, J. (1967) *Claude Bernard et les Problèmes Scientifiques de son Temps*. Paris, Editions du Cèdre.

Schnitzler, H., Lieth, H., (eds.) (1983), *25 years activities of the International Society of Biometeorology and 25 years publication index, International Society of Biometeorology.* Lisse, Netherlands, Swets & Zeitlinger.

Schrödinger, E. (1926) Der stetige Übergang von der Mikro- zur Makromechanik. *Naturwissenschaften*, **28**, 665.

Schrödinger, E. (1944) *What is life ? The Physical Aspect of the Living Cell.* Cambridge, Cambridge University Press.

Schutz, A. (1975) *Collected Papers.* 3 vols. The Hague, Martinus Nijhoff.

Schwartz-Salant, N. (1988) *The Borderline Personality – Vision and Healing.* Wilmette, Illinois, Chiron Publications.

Sedlak, W. (1967) Elektrostaza i ewolucja organiczna (Electrostasis and organic evolution). *Roczniki Filozoficzne*, **3**, 31.

Sedlak, W. (1976) Bioplasma - a new state of matter. In *"Bioplasma". Proceedings of the 1st Bioplasma Conference, May 9, 1973, Catholic University, Lublin/Poland*, W.Sedlak, (ed.), Lublin (English translation 1978).

Sedlak, W. (1979) *Bioelektronika: 1967-1977.* Aleksandrowicz, Julian (ed.), Warsaw, Poland: Instytut Wydawniczy Pax.

Sedlak, W. (1980) *Homo Electronicus.* Warsaw, Poland, Instytut Wydawniczy Pax.

Sedlak, W. (ed.) (198x) *Bioelektronika - Materialy i Krajowego Sympozjum.* Catholic University, Lublin, Poland.

Sedlak, W. (1984) *Postepy fizyki zycia.* Warsaw, Poland: Instytut Wydawniczy Pax.

Segal, J. (1958) *Die Erregbarkeit der lebenden Materie.* Jena, Gustav Fischer.

Segal, J., Körner, U., Leiterer, K.P. (1983) *Die Entstehung des Lebens aus biophysikalischer Sicht.* Jena, Gustav Fischer.

Selleri, F. (ed.) (1988) *Quantum Mechanics versus Local Realism – The Einstein-Podolsky-Rosen Paradox.* New York, Plenum Press.

Selye, H. (1936) A syndrome produced by diverse nocuous agents. *Nature*, **148**, 84-85.

Selye, H. (1946) The general adaptation syndrome and the diseases of adaptation. *Journal of Clincial Endocrinology*, **6**, 117-196.

Selye, H. (1950) *Stress: The Physiology and Pathology of Exposure to Stress.* Montreal, Acta Medica.

Selye, H. (1967) *In Vivo – The Case for Supramolecular Biology.* New York, Liveright.

Shacklett, R.L. (1991) The physics behind the mind-matter link. In *"Subtle Energies and Energy Medicine: Emerging Theory and Practice",* conference proceedings of the First Annual Conference of the International Society for the Study of Subtle Energies and Energy Medicine. June 21-25, Boulder, Colorado, 92-95.

Sherrington, C.S. (1940) *Man on His Nature*. Harmondsworth, Middlesex, Penguin Books, 69-98.

Shimony, A. (1963) Role of the observer in quantum theory. *American Journal of Physics* **31 (10)**, 755-773.

Shipov, G.I. (1998) *A Theory of the Physical Vacuum – A New Paradigm*. Moscow, International Institute for Theoretical and Applied Physics RANS.

Shnol', S.E., Smirnov, B.R., Sadonsky, G.I. (1981) in Grechova, M.T. (ed.) *Autowave Processes in Systems with Diffusion* (in Russian). Gorky, 187.

Shnol', S.E., Smirnov, B.R., Sadonsky, G.I., Rovinsky, A.B. (1982) *Khimiya i Zhisn'*, 7, 68.

Shnol', S.E., (ed.) (1996) *Relations of Biological and Physico-Chemical Processes with Space and Helio-Geophysical Factors*. Abstracts of the 4th International Pushchino Symposium, dedicated to the centennial of helio-biology founder A.L.Chizhevsky (1897-1964), September 23-28. Pushchino, Academy of Sciences of Russia, Pushchino Research Centre.

Sigel, F. (1975) *Schuld ist die Sonne*. Thun-Frankfurt am Main, Harri Deutsch (Russian original ed. Moscow 1972).

Sinz, R. (1978) *Zeitstrukturen und organismische Regulationen*. Berlin, Akademie-Verlag.

Sinz, R. (1980) *Chronopsychophysiologie, Chronobiologie und Chronomedizin*. Berlin, Akademie Verlag.

Sit'ko, S.P., Gizhko, V.V. (1991) Towards a quantum physics of the living state. *Journal of Biological Physics*, **18 (1)**, 1-10.

Smith, C.W. (1994) Electromagnetic and magnetic vector potential bio-information and water. *Ultra-High Dilution – Physiology and Physics*, Endler, P.C. and Schulte, J. (eds.), Dordrecht, Kluwer Academic.

Smuts, J.C. (1926) *Holism and Evolution*. New York, Macmillan.

Sollberger, A. (1965) *Biological Rhythm Research*. Amsterdam, Elsevier.

Stapp, H.P. (1997) Nonlocal character of quantum theory. *American Journal of Physics*, **65 (4)**, 300-304.

Susskind, C., Bose, Jagadis Chunder (1970) in Gillispie, C.G. (ed.) *Dictionary of Scientific Biography*. New York, Scribner's, 325.

Szent-Györgyi, A. (1941) Energy levels in biochemistry. *Nature*, **148**, 157.

Szent-Györgyi, A. (1941) Towards a new biochemistry. *Science*, **93 (1941)**, 609.

Szent-Györgyi, A. (1957) *Bioenergetics*. New York, Academic Press.

Szent-Györgyi, A. (1960) *Introduction to a Submolecular Biology.* New York, Academic Press.

Szent-Györgyi, A. (1969) Charge-transfer and cellular activity. In *Theoretical Physics and Biology*, Proceedings Ist Int. Conf. on Theoretical Physics and Biology, Versailles, June 26-30, 1967, Marois, M. (ed.), 192-193, Amsterdam-London, North-Holland.

Szent-Györgyi, A. (1971) Biology and pathology of water. *Perspectives in Biology and Medicine*, **14 (2)** 239-249.

Takata, M., Murasugi, T. (1941) Flockungszahl-Störungen im gesunden menschlichen Serum, "kosmoterrestrischer Sympathismus". *Bioklimatische Beiblätter*, **8**, 17.

Takata, M. (1951) Über eine neue biologisch wichtige Komponente der Sonnenstrahlung. *Archiv für Meteorologie, Geophysik und Bioklimatologie*, **486**.

Tart, C. (1969) Psychedelic experiences associated with a novel hypnotic procedure, mutual hypnosis. *Altered States of Consciousness.* New York, Wiley, 291-308.

Taub-Bynum, E.B. (1984) *The Family Unconscious – An Invisible Bond.* Wheaton, Illinois, Quest Books.

Taylor, R. (1996) Scalar fields, subtle energy and the multidimensional universe (in English). *Soznanye i Fizicheskaya Realnost'* **1 (3)**, 79-

Taylor, R. (1998) A gentle introduction to quantum biology. *Consciousness and Physical Reality* **1(1)**, 64-72.

Tchijevsky, A.L. (1936) *Les Phénomènes Éléctrodynamiques dans le Sang.* Éditions Hippocrate, Paris.

Tchijevsky, A.L. (1940) Cosmobiologie et rhythme du milieu extérieur, in Holmgren, H.J. (ed.), *Verhandlungen der zweiten Konferenz der Internationalen Gesellschaft für biologische Rhythmus-Forschung am 25. und 26. August 1939 in Utrecht.* Stockholm, Acta Medica Scandinavica, 108, Suppl., 212-226.

Tettinger, L., Doljansky, F. (1992) On the generation of form by the continuous interactions between cells and their extracellular matrix. *Biological Reviews of the Cambridge Philosophical Society*, **67**, 459-489.

Thaheld, F.H. (1998) A proposed experiment concerning nonlocal correlations between a quantum observer and another person. *Physics Essays*, **11**, 422-425.

Thaheld, F.H. (2001a) A preliminary indication of controllable biological quantum nonlocality ? *Apeiron*, **8 (1)**.

Thaheld, F.H. (2001b) Proposed experiments to determine if there is a connection between biological nonlocality and consciousness. *Apeiron*, **8 (4)**, 53-66.

Thiele, H. (1947) Richtwirkung von Ionen auf anisotrope Kolloide - Ionotropie. *Naturwissenschaften*, **34**, 123.

Thiele, H. (1954) Ordered coagulation and gel formation. *Discussions of the Faraday Society*, 18, 294-301.

Thiele, H., Schacht, E. (1957) Über Ionenreaktionen bei Polyelektrolyten, Vorstufen bei der Bildung von Gelen. *Zeitschrift für physikalische Chemie*, 208, 42-58.

Thiele, H. (1967) *Histolyse und Histogenese. Gewebe und ionotrope Gele - Prinzip einer Strukturbildung.* Frankfurt am Main, Akademische Verlagsgeselllschaft.

Thompson, D'Arcy, W. (1917) *On Growth and Form.* Cambridge, Cambridge University Press.

Tigyi, J., Kellermayer, M and Hazlewood, C.F. (1991) *The Physical Aspect of the Living Cell.* Budapest, Akademiai Kiadó.

Tiller, W. (1977) Towards a future medicine based on controlled energy fields. *Phoenix*, 1 (1) 5-16.

Tiller, W. (1985) Energy fields and the human body, in White, J. (ed.), *Frontiers of Consciousness. The Meeting Ground between Inner and Outer Reality,* New York, Julian Press, 229-242.

Tiller, W. (1993) What are subtle energies ? *Journal for Scientific Exploration,* 7 (3), 293-304.

Tiller, W. (1995) Subtle energies in energy medicine. *Frontier Perspectives,* 4 (2), 17-21.

Tiller, W.A. (1999a) Towards a predictive model of subtle domain connections to the physical domain aspect of reality: The origins of wave-particle duality, electric-magnetic monopoles and the mirror principle. *Journal of Scientific Exploration,* 13 (1), 41-67.

Tiller, W.A., Dibble, W.E., and Kohane, M.J. (1999b) *Exploring robust interactions between human intention and inanimate/animate systems.* Excelsior, MN, USA, Ditron, LLC.

Tiller, W.A., Dibble, W.E., and Kohane, M.J. (2001) *Conscious Acts of Creation – The Emergence of A New Physics.* Walnut Creek, California, Pavior Publishing.

Tocquet, R. (1951) *Cycles et Rhythmes.* Paris.

Tomassen, G.J. (eds.) (1990) *Geo-Cosmic Relations; the Earth and its Macro-Environment.* Proceedings of the First International Congress on Geo-Cosmic Relations, Amsterdam, Netherlands, 19-22 April 1989. Wageningen, Pudoc.

Trincher, K.S. (1965) *Biology and Information: Elements of Biological Thermodynamics.* New York, Consultants Bureau.

Tromp, S.W. (1949) *Psychical Physics.* New York, Elsevier.

Tromp, S.W. (1961) Significance of border sciences for the future of mankind. In Boyko, H. (ed.), *Science and the Future of Mankind,* Den Haag, Uitgeverij Dr.W.Junk, 83-120.

Tromp, S.W. (19 63) *Medical Biometeorology*. New York, Elsevier.

Tromp, S.W. (1972) Possible effects of extra-terrestrial stimuli on colloidal systems and living organisms. In Biometeorology, 5, part II. Suppl. to *Intenational Journal of Biometeorology*, **16**, 239-248.

Troshin, A. S. (1966) *Problems of Cell Permeability*. Rev. & supplem. (ed.) Oxford, Pergamon Press. (Russian original ed.: Moscow: Academy of Sciences, 1956).

Usa, M., Kobayashi, M., Scott, R.Q., Maeda, T., Hiratsuka, R., Inaba, H. (1989) Simultaneous measurement of biophoton emission and biosurface electric potential from germinating soybean (*Glycine max*). *Protoplasma*, **149**, 64-66.

Uexküll, T. et al. (1997) *Subjektive Anatomie*. Stuttgart, Schattauer.

Vanable, J.W. (1991) A history of bioelectricity in development and regeneration, in Dinsmore, C.E. (ed.), *A History of Regeneration Research,* Cambridge, Cambridge University Press, 151-178.

Vernadsky, V.I. (1954-60) *Izbrannye Sochineniya* (Selected Works, in Russian). Vols.1-5. Moscow, Academy of Sciences.

Vernadsky, V.I. (1998) *The Biosphere*. With introduction and annotations by Mark A.S. McMenamin and Jacques Grinevald. Copernicus Books (first published in Russian in 1926).

Vincent, L.-C. (1954) Aperçu sur la bio-électronique. *Revue de Pathologie Générale et Comparée*, **54 (663)**, 121-130.

Vincent, L.-C. (1956a) Bio-Electronique. Définition des trois facteurs prhoniques pH, rH^2 et ρ. Bulletin de la Société de Pathologie Comparée, Séance du Mardi 14 Juin 1955, *Revue de Pathologie Générale et de Physiologie Clinique*, **56 (677)**, 643-656.

Vincent, L.-C. (1956b) Potentiel d'oxydo-réduction et rH^2. *Revue de Pathologie Générale et de Physiologie Clinique*, **56 (677)**, 657-678.

Vlès, F. (1927) *Cours de Physique Biologique*. Paris, Vigot Frères.

Vlès, F. (1929) Revue des notions actuelles sur un problème de physico-chimie pathologique: Les propriétés des points isolelectriques et du terrain physico-chimique dans l'organisme normal ou pathologique; leur application à l'étude des tumeurs. *Archives de Physique Biologique*, 7, fasc. suppl. **No.5**, 1-64.

Vlès, F., Reiss, P., and Deloyer, L. (1931) Le potentiel de platine et le rH du sang circulant des mammifères. *Comptes Rendus de la Société de Biologie*, **108**, 37.

Vlès, F. (1936a) Recherches sur les propriétés physico-chimiques des électrolytes. *Archives de Physique Biologique*, Vol.13, **No.1 (1936)**, 57-73 and 88-107.

Vlès, F. (1936b) Recherches sur les electrolytes. II. Les conditions de cohérence dans les solution. III: Remarques sur le rH de Clark. Comptes rendus des séances de la réunion de physique biologique de Strasbourg. Supplément aux *Archives de Physique Biologique*, **13**, 191-193.

Vlès, F., and Coulon, A. de (1936c) Recherches sur les propriétés physico-chimiques des tissus en relation avec l'état normal ou pathologique de l'organisme. 22e partie. Suite des expériences sur la cancérisation spontanée des souris au sol et isolées. *Archives de Physique Biologique*, **13 (2)**, 150-176.

Vlès, F. (1936d) L'eau colloidale. Comptes rendus des séances de la réunion de physique biologique de Strasbourg. *Supplément aux Archives de Physique Biologique*, **13**, 192.

Vlès, F. (1936e) Les données actuelles sur la constitution et les propriétés physico-chimiques de l'eau. *Archives de Physique Biologique*, **15 (1)**, 33-85.

Vlès, F. (1942) *Introduction à la photochimie biologique – radiations, analyse spectrale, photochimie*. Paris, Vigot Frères.

Vogelman, T.C. (1989) Penetration of light into plants. *Photochemistry and Photobiology*, **50 (6)**, 895-902.

Waddington, C.H. (1968-72) *Towards a Theoretical Biology*. Edinburgh, Edinburgh University Press.

Walker, E.H. (1970) The nature of consciousness. *Mathematical Biosciences* 7, 131.

Walker, E.H. (1974) Consciousness and quantum theory. In *Psychic Exploration*, White, J. (ed.), New York, Putnam's Sons, 544-568.

Walker, E.H. (2000) *The Physics of Consciousness*. Cambridge, Mass., Perseus Books.

Watson, A. (1997) Quantum spookiness wins, Einstein loses in photon test. *Science*, 277, 481.

Weiner, H (1992) *Perturbing the Organism – The Biology of Stressful Experience*. Chicago, University of Chicago Press.

Weiss, P., Garber, B. (1952) Shape and movement of mesenchyme cells as functions of the physical structure of the medium. Contributions to a quantitative morphology. *Proceedings of the National Academy of Sciences (USA)*, **38**, 264-280.

Welch, G. R. (1992) An analogical "field" construct in cellular biophysics: history and present status. *Progress in Biophysics and Molecular Biology*, **57**, 71-128.

Whipple, H.E., (ed.) (1965) Forms of Water in Biological Systems. *Annals of the New York Academy of Sciences*, **125**, Art.2, 249-772.

Wiener, N. (1948) *Cybernetics*. New York, Wiley.

Wiggins, P.M. (1971) Water structure as a determinant of ion distribution in living tissue. *Journal of Theoretical Biology*, 32, 131-146.

Wiggins, P.M. (1972) Intracellular pH and the structure of cell water. *Journal of Theoretical Biology*, **37 (2)**, 363-371.

Wiggins, P.M. (1990) Role of water in some biological processes. *Microbiological Reviews*, **54 (4)**, 432-449.

Wijk, R. van, Linnemans, W.A. (1993) The basic regulatory system - a system theoretical concept behind chronical diseases. In Lamoen, G.J. van (ed.), *Biologische Information und Regulation*, Heidelberg, Karl F.Haug Verlag, 36-62.

Williams, R.J. (1956) *Biochemical Individuality*. New York, Wiley.

Wnuk, M.J., Bernard, C.D. (2001) The electromagnetic nature of life – The contribution of W.Sedlak to the understanding of the essence of life. *Frontier Perspectives*, **10 (1)**, 32-35.

Wolf, E. (1963) Spatial coherence of resonant modes in a maser interferometer. *Physics Letters*, **3 (4)**, 166-168.

Wolkowski, Z.W., Sedlak, W. and Zon, J. (1983) The utility of bioelectronics and the bioplasma concept in the study of the biological terrain and its equilibrium. In *Proceedings of the International Symposium on Wave Therapeutics, Versailles 1979*, Wolkowski, Z.W. (ed.), 114-122, Créteil, Université de Paris-Val-de-Marne.

Wolkowski, Z.W. (1985) Le concept de champ: de la physique à la biologie. In *Synergie et Cohérence dans les Systèmes Biologiques*, Wolkowski, Z.W. (ed.), Paris, École Européenne d'Été d'Environnement.

Wolkowski, Z.W. (1988) Sur l'évolution de certain concepts physiques dans les sciences de la vie. *Revue Internationale de Systémique* **2 (4)**, 459-477.

Wolkowski, Z.W. (1995) Recent advances in the phoron concept: An attempt to decrease the incompleteness of scientific exploration of living systems. In *Biophotonics – Non-Equilibrium and Coherent Systems in Biology, Biophysics and Biotechnology*, Beloussov, L.V. and Popp, F.A. (eds.), Moscow, Bioinform, 33-51.

Wu, T.T. (1975) Concept of nonintegrable phase factors and global formulation of gauge fields. *Physical Review D* **12 (12)**, 3845-3857.

Wu, T.T. and Yang, C.N. (1975) Some remarks about unquantized non-Abelian gauge fields. *Physical Review D,* **12 (12)**, 3843-3844.

Yagodinskii, V.N. (1987) *Aleksandr Leonidovich Chizhevskii* (in Russian). Moscow, Nauka.

Zeiger, B.F. (1998) Photon emission of cereal seeds, "biophotons", as a measure of germinative ability and vigour. In Chang, J.J., Fisch, J., and Popp, F.A. (eds.), *Biophotons*. Dordrecht, Kluwer

Zeiger, B.F. and Bischof, M. (1998) The quantum vacuum and its significance in biology. Paper given at *the Third International Hombroich Symposium on Biophysics,* Neuss, Germany, August 20-24, 1998.

Zhang, C.L., Popp, F.A., and Bischof, M. (eds.) (1996) *Current Development of Biophysics.* Hangzhou, China, Hangzhou University Press.

Ziegler, E. (1960) *Messung und Bedeutung des Redoxpotentials im Blut in vivo und in vitro* (Habilitationsschrift, Universität Freiburg i.Br.). Arzneimittel-Forschung, 10.Beiheft. Aulendorf/Württemberg, Verlag Editio Cantor.

Zon, J.R. (1980) The living cell as a plasma physical system. *Physiological Chemistry and Physics,* **12 (4)**, 357-364.

Chapter 2

BIO-ELECTROMAGNETISM

Gerard J. Hyland
University of Warwick
Department of Physics
Coventry, CV4 7AL, England;

International Institute of Biophysics
Ehem. Raketenstation,
Kapellener Straße o.N.
D-41472 Neuss (Germany)

1. PROLOGUE

Bio-electromagnetism is the study (*a*) of the interaction of external electromagnetic (*EM*) fields and radiation – both natural and technogenic – with **living** organisms, and (*b*) of the electromagnetic fields and radiation **generated** by the alive organism itself. The aim of these studies is three-fold: firstly, to attempt to establish the bio-significance (if any) of different natural electromagnetic fields, such as the Earth's magnetic field and the Schumann resonances; secondly, to establish whether exposure to technogenic electromagnetic fields are harmful to the human organism; and thirdly, to learn more about the electromagnetic structure of the alive organism and the role of endogenous electromagnetic fields by studying (particularly with respect to their degree of coherence) the external fields and radiation associated with and emitted by the alive organism.

It is essential to appreciate at once that the kinds of influences that electric/ magnetic fields and electromagnetic radiation exert on an organism depends on whether it is alive or dead – a feature that in itself gives us some valuable clues concerning the organism's electromagnetic structure. The fact that there are any influences at all empirically tells us that the organism must contain *electromagnetic entities* to which the external fields can couple; on the other hand, the fact that *some* influences (generally the more interesting ones) occur *only* when the organism is alive indicates that *only then* are the relevant entities (bio-detectors) activated.

Electric fields couple to electric charges and electric dipoles, magnetic fields couple to electric currents and magnetic dipoles, whilst electromagnetic **radiation** couples to electric charges and endogenous electrical **oscillators**.

To understand the various couplings in more detail it is necessary to recall some basic electromagnetic and biological facts.

2. SOME RELEVANT ELECTROMAGNETIC THEORY

2.1 Basic Definitions and Units

In the *SI* system, there are 4 *vector* fields [1,2] – *E*, *D*, *B*, and *H*.

E is the *electric field*. This is defined in terms of the force exerted on an electric charge, q, irrespective of whether the charge is moving or not. The units of *E* are volts/metre (*V/m*, where $V \equiv joule/coulomb, J/C$).

B is the *magnetic induction*. This is also defined in terms of the force it exerts on an electric charge, q, but one that is *in motion*. The units of *B* are volt. second/square metre ($V.s/m^2 \equiv J.s/Cm^2$, or *tesla, T*).

The total force exerted on an electric charge, q, moving with velocity v (uniform or non-uniform) - the so-called Lorentz force, f_L - is given by:

$$f_L = q(E + v \times B) \tag{1}$$

D is the *displacement field* whose source is the density of electric charge, ρ (in coulombs/cubic metre, C/m^3), the units of *D* being ampere. second/square metre ($A.s/m^2 \equiv C/m^2$)

H is the *magnetic field* whose source is an electric *current density*, *J* (in A/m^2), the units of *H* being amperes/metre (*A/m*, where $A \equiv C/s$).

Thus, *E* and *B* are 'fields of force', whilst *D* and *H* are 'fields of source'.

2.2. The Maxwell Equations and the Quasi-Static Approximation

In general, E, B, D and H vary in space, r and time, t; thus, $E = E(r, t)$, for example - i.e. they are vector *fields* over space-time. Together they constitute the *electromagnetic* field, whose space and time evolution (subject to prescribed initial and boundary conditions) is given, *irrespective of the nature of the medium*, by the *Maxwell* Equations [1, 2]:

$$divD = \rho \qquad\qquad divB = 0 \qquad (2)$$
(Coulomb's law) \qquad\qquad (No magnetic monopoles)

$$curlE = -\partial B/\partial t \qquad\qquad curlH = J + \partial D/\partial t \qquad (3)$$
(Faraday's law) \qquad\qquad (Maxwell's extension of Ampère's law)

It is apparent that this set of equations comprises two scalar equations (those containing '*div*') - one inhomogeneous (Coulomb's law) and one homogeneous - and two vector equations (those containing '*curl*'), again one inhomogeneous (extended Ampère's law) and one homogeneous (Faraday's law). The differential operators '*div*' and '*curl*' are certain (linear) combinations of first order partial derivatives with respect to space, which in Cartesian coordinates have the form $\partial/\partial x$, $\partial/\partial y$, and $\partial/\partial z$. Thus, $divB \equiv \partial B_x/\partial x + \partial B_y/\partial y + \partial B_z/\partial z$, whilst $curlH \equiv i(\partial H_z/\partial y - \partial H_y/\partial z) + j(\partial H_x/\partial z - \partial H_z/\partial x) + k(\partial H_y/\partial x - \partial H_x/\partial y)$, where $\{i, j, k\}$ is an orthogonal triad of unit vectors in the directions of the x, y and z axes, respectively. In terms of ∇ ($\equiv$ grad $\equiv i\partial/\partial x + j\partial/\partial y + k\partial/\partial z$), $divB \equiv \nabla.B$, and $curlH \equiv \nabla \times H$.

In consequence of Maxwell's 'displacement' current, $\partial D/\partial t$, which extends Ampère's law ($curlH = J$, where J is the 'conduction' current density) to time dependent situations, ρ and J satisfy the so-called 'continuity equation'.

$$\partial \rho/\partial t + divJ = 0 \qquad (4)$$

This is the condition on the *densities* ρ and J that must hold *locally* at any instant in order to ensure that the **total** electric charge in a given system is conserved – i.e. $\partial (\iiint \rho \, d\tau)/\partial t = 0$, where the integral is taken over the total volume occupied by the system.

A further most important consequence of the $\partial \mathbf{D}/\partial t$ term is that in regions of space where ρ and $\mathbf{J}$ *vanish*, the fields satisfy wave equations – *see* Section 2.7.

It will be noted that because of the terms in Maxwell's equations containing the partial time derivatives, there is a *coupling* between the various fields. Thus, in the case of Faraday's law, for example, associated with a magnetic induction, $\mathbf{B}$, that changes with time, is an electric field, $\mathbf{E}$; as we shall see later, this is fundamental to the existence of electromagnetic waves and radiation. In the *static* limit, however - *i.e.* when there is no time dependence – there is no such coupling, and the $\mathbf{E}/\mathbf{D}$ and $\mathbf{B}/\mathbf{H}$ fields can be considered as *independent* entities, as is, of course, the situation in the case of electrostatics and magnetostatics. This decoupling of the electric and magnetic fields persists, however, *even in the time-dependent case* (the so-called 'quasi-static approximation'), provided the changes occur sufficiently slowly; this effectively means that the vacuum electromagnetic wavelength, λ ($=c/\nu$) corresponding to the frequency, ν, characterising the time dependence must be very much larger than the size of the system, so that the fields can be considered to be effectively uniform over the system.

2.3. Constitutive Relations and Non-Linear Properties of Living Biomatter

The properties of the **medium** (*including* free space) in which the fields exist are contained in the so-called '**Constitutive Relations**', *via* which $\mathbf{D}$ and $\mathbf{H}$ are related to $\mathbf{E}$ and $\mathbf{B}$ algebraically – *i.e.* in a way that does *not* involve any partial derivatives; thus:

$$\mathbf{D} = \varepsilon_0 \varepsilon_r \mathbf{E} \qquad\qquad \mathbf{B} = \mu_0 \mu_r \mathbf{H} \qquad (5)$$

The quantities ε_0 and ε_r are the permittivity of free space ($= 8.84 \times 10^{-12}$ *farad/m*) and the relative permittivity of the material dielectric medium, respectively, whilst μ_0($= 4\pi \times 10^{-7}$ *henry/m* $= 10^{-7}$ *newton/A²*) and μ_r are the corresponding permeabilities. Unlike ε_0 and μ_0, ε_r and μ_r are dimensionless, and are, in general, tensors. These four quantities govern the *response* of the free space/ medium to an applied *EM* field, the values of ε_r and μ_r, in general, varying with the space and time dependences of the applied fields, and with the temperature and density of the medium; in the case of free space, $\varepsilon_r = 1$, $\mu_r = 1$, of course.

The *farad* is the *SI* unit of capacitance, and is defined as the capacitance of a capacitor between whose plates there appears a potential difference of 1 volt when it is charged by 1 coulomb; a *henry* is the *SI* unit of inductance, which is defined as the inductance of a closed circuit in which an *emf* of 1 volt is produced when the current in the circuit varies uniformly at the rate of 1 amp per second.

It is to be noted that in the case of bio-matter, the values of ε_r and μ_r can depend on whether the system is alive or dead. In the case of living matter, the response can be *non-linear* – i.e. ε_r itself depends on the electric field intensity, **E**. Non-linear responses underlie a variety of remarkable electromagnetic effects, such as self-induced transparency of the medium (as illustrated, for example, by the superior transmission of light through *living* plant tissue), and the self-focusing of electromagnetic fields [3].

Finally, the following interrelations should be noted:

1. $(1/\varepsilon_o\mu_o)^{1/2} = c$, speed of light (electromagnetic radiation) *in vacuo*.

2. $(1/\varepsilon_r\mu_r)^{1/2} = v/c$, where v is speed (phase velocity) of light in a medium of *refractive index*, $n (= (\varepsilon_r\mu_r)^{1/2})$; in terms of n, $v = c/n$.

3. $(\mu_o/\varepsilon_o)^{1/2} = Z_o$, the so-called '*impedance of free space*' ($=377\Omega$).

2.4 Electromagnetic Field Energies and Fluxes

An electromagnetic field contains energy, the density, U, of which is given by:

$$U = \tfrac{1}{2}(\mathbf{E.D} + \mathbf{B.H}) \quad (J/m^3) \tag{6}$$

The rate, S, at which this energy crosses unit area per second is the (Poynting) flux (or *power density*), and is given by:

$$S = |\mathbf{E} \times \mathbf{H}| \quad (W/m^2) \qquad W \equiv watt (\equiv volt \times amp) = J/s \tag{7}$$

For an electromagnetic wave in free space, $S = E^2/Z_o = Z_o H^2$

2.5 Absorption of Electromagnetic Waves in Biological Media

Typically, biological media are 'lossy' dielectrics (*i.e.* have a non-zero electrical conductivity, $\sigma_r \neq 0$, at radio/microwave frequencies), within which absorption of energy *from* incident radio frequency (*RF*) or microwave (*MW*) radiation will occur. Accordingly, the amplitude of a wave progressively decreases as it penetrates the medium, resulting in its 'attenuation' [6].

The attenuation/unit length defines the attenuation constant, α, which is given in terms of the material parameters ε_r and σ, and the circular frequency $\omega\ (=2\pi\nu)$ of the radiation by:

$$\alpha = \omega[\mu_o \varepsilon_o \varepsilon_r/2\{(1 + \sigma/\omega\varepsilon_o\varepsilon_r)^{1/2} - 1\}]^{1/2} \qquad (8)$$

The ratio $(\sigma/\omega\varepsilon_o\varepsilon_r)$ is called the 'loss tangent' of the material, and is simply the magnitude of the ratio of the conduction current, $\mathbf{J}\ (=\sigma\mathbf{E})$ – representing an energy *loss* mechanism – to the 'displacement current', $\partial\mathbf{D}/\partial t\ (=i\omega\varepsilon_o\varepsilon_r\mathbf{E})$, which is 90° out of phase with $\mathbf{J}$, and represents energy *storage*. It should be appreciated, however, that in media containing permanent electric dipoles – such as *water*, for example, which is, of course, the dominant constituent of biosystems – the decrease of ε_r with increasing frequency increases the value of the loss tangent from that which it would otherwise have. The decrease in ε_r at RF/MW frequencies comes about because the permanent electric dipoles of the water molecules are here unable to fully follow the oscillations of field; consequently, the re-orientational polarisation lags behind the applied field, and eventually fails to follow it at all, resulting in the loss of this contribution to ε_r, which is accordingly reduced.

Materials in which the conduction current dominates the displacement current are conductors, whilst for dielectrics, the converse holds, the limiting cases of a 'good' conductor and a 'good' dielectric being characterised by $(\sigma/\omega\varepsilon_o\varepsilon_r) \gg 1$, and $(\sigma/\omega\varepsilon_o\varepsilon_r) \ll 1$, respectively; clearly, in the case of a *lossless* material (*i.e.* one with $\sigma = 0$), $\alpha = 0$. In the case of a lossless dielectric, it may also be noted that the phase velocity at which an electromagnetic wave propagates is lower than *in vacuo* (by a factor equal to the refractive index), whilst its wavelength is reduced by the same factor; accordingly, the frequency remains constant.

Associated with α in the case of a lossy material is the so-called **'penetration depth'**, d, given by:

$$d \sim 1/\alpha \qquad (9)$$

d is the distance inside the material at which the intensity has fallen to $1/e$ (*i.e.* to 37%) of its incident value at the surface, and within which 87% of the energy is dissipated; it *decreases* with increasing frequency. [In the case of metallic conductors, d becomes *independent* of $\varepsilon_o\varepsilon_r$ and is called the 'skin depth', $\delta = (2/\omega\mu_o\sigma)^{1/2}$, whilst in the case of a 'good' dielectric, $d \sim (2/\sigma)(\varepsilon_o\varepsilon_r/\mu_o)^{1/2}$, which is *implicitly* dependent on frequency *via* σ and ε_r.] In the case of biological materials, the value of d is strongly dependent on the water content, as shown in Table 1, which compares materials with relatively high and low values, such as muscle/skin and fat/bone, respectively.

Table 1:

Frequency (MHz)	ε_r		σ(S/m)		d (cm)	
	High	Low	High	Low	High	Low
300	54	5.7	1.4	(31-100)	4.0	31
900	51	5.6	1.6	(55-150)	3.0	18
2450	47	5.5	2.2	(100-200)	1.7	11
10,000	40	4.5	10.0	(300-500)	0.3	3.4

It is to be emphasised that the derivation of the above results assumes that the radiation has the form of a *plane* wave, and that the medium on which it is incident has a plane surface - conditions that are rarely realised, particularly in biosystems; furthermore, the cited values of ε_r and σ refer to *dead* biomatter.

2.6. Towards Electromagnetic Radiation

From the above we know that a moving electric charge (*i.e.* an electric current) gives rise to an electric field and a magnetic field.

If the charge is moving *uniformly* in a **vacuum**[1], the fields do **not** escape from it, but are confined to its vicinity, remaining *bound* to the charge.

To get the fields to *escape* as electromagnetic RADIATION, it is necessary (*in vacuo*) that the electric charges move in a **non**-uniform way — *i.e.* the charges must undergo an acceleration or deceleration [1, 2]. In the case of condensed systems this inevitably occurs as a result of thermal vibration of the constituent molecules (which contain electric charges), and is the origin of the incoherent (*see* Section 2.9) thermal emission from heated bodies, including that from the alive human body at physiological temperature (~36°C), which is maximal in the infra-red, at a wavelength of about 10μm.

To produce electromagnetic radiation technologically, a commonly employed strategy is to force electric charges to **oscillate** by applying an alternating voltage between two metal rods positioned end to end. This arrangement constitutes a prototype 'antenna' whose two limbs become alternatively positively and negatively charged. The associated electric field is in phase with the charge distribution, which itself has the form of an electric dipole, the *polarity* of which *alternates*. (It is to be stressed that it is in this sense that the dipole oscillates; the oscillation is *not in the separation* between two opposite electric charges that always maintain the *same* sign.) Associated with this time varying electric field is a magnetic field at right angles to it, and during one cycle of oscillation of the charges, one wavelength of radiation is produced.

2.7. Some Basic Results concerning the Electromagnetic Fields of an Oscillating Electric Dipole

The E and H fields emitted (into free space) by an electric dipole oscillating at frequency v, whose length is very much less than the wavelength, λ of the

---

In *matter*, however, even a *uniformly* moving charge produces radiation (*Cerenkov* radiation), provided its velocity exceeds the phase velocity of light in the medium (= c/n). It should be appreciated, however, that in this case the radiation comes not from the charged particle itself, but rather from the *medium*, which is asymmetrically excited by the passage of the rapidly moving charge. At $v > c/n$, this results in the formation of a *pulsating dipole* in its wake; it is this that constitutes the source of the radiation.

emitted radiation (a so-called Hertzian dipole) are mutually orthogonal, and have the following frequency and space dependences:

$$E \sim \nu^2(1/r), \ \nu(1/r^2), \ (1/r^3) \tag{10}$$

$$H \sim \nu^2(1/r), \ \nu(1/r^2) \tag{11}$$

It will be seen that in the static limit (*i.e.* $\nu = 0$), only the **electric** contribution $\sim 1/r^3$ survives - the familiar electro*static* dipole field.

At distances, $r \gg \lambda$, where the $1/r$ terms predominate, the so-called '**far**' field region is realised, where the fields 'escape' from the source as electromagnetic **radiation**. Here E and H assume the form of *plane* waves, in which the two fields are orthogonal to one another, and vibrate *transversely* to the common direction in which they both propagate with the speed of light, c. Since in the far field region, $E = c|B|$, where $|B| = \mu_0|H|$, it follows that here $|E|/|H| = Z_0$ – i.e. that $|E| = 377|H|$.

It is important to appreciate that whilst the presence of an electric charge/current is essential to produce E/H *in the first place*, once they have escaped from their source, the two fields mutually sustain one another – *i.e.* there is a kind of 'symbiosis' between them. This follows from the form of the two Maxwell equations containing time derivatives, which indicate that associated with a changing magnetic induction, $B (= \mu_0 H)$, is an E-field, and with a changing displacement field, $D (= \varepsilon_0 E)$ is a magnetic field, H.

The rate at which energy is carried away by the radiation across unit area defines the power density (or Poynting flux) S, which is given by:

$$S = |(E \times H)| \quad (W/m^2) \tag{12}$$

Since the electromagnetic (radiation) field is here so simply structured ($|E| = c|B| = Z_0 H$), measurement of *either* E or H is sufficient to calculate S.

Since, in the far field region, E and H are both proportional to $\nu^2(1/r)$, it follows that:

$$S \sim \nu^4 (1/r^2), \tag{13}$$

the energy flow being maximum in the equatorial plane, and zero along the axis of the dipole antenna.

To obtain the energy that is radiated away from the source, it is necessary to integrate S over a spherical surface enclosing the source. Because of the $1/r^2$ dependence, it follows that the system loses energy at a *constant* rate proportional to v^4.

At distances, $r << \lambda$, on the other hand, where the $1/r^2$ and $1/r^3$ terms predominate, the so-called '**near**' field region is realised, in which the field structure is *much more complicated*. Here, E and B are still orthogonal to each other, but are (*i*) *not* in phase with one another – rather, E is in phase with the electric charges in the antenna, and B with the currents - and (*ii*) are not always transverse to direction of propagation; in particular, there is a significant *longitudinal* E-field in the direction of the antenna axis. Accordingly, in some regions, the field can be mainly electric, and in others, mainly magnetic, so that here, $Z \neq |E|/|H|\ (=Z_o)$, and the fields must be considered to be effectively **decoupled**, it being here necessary to independently measure both E and H in order to calculate S. [Recall the 'quasi-static' case discussed earlier, which is realised under precisely the above condition, namely, $r << \lambda$]

In general, the value of Z mainly depends on the physical nature of the source antenna - one with a high voltage and low current (such as the oscillating electric dipole considered above) generating mainly an electric field (and $Z > Z_o$), whilst one with a high current and low voltage (such as a *magnetic* dipole source, constituted by a loop carrying an alternating electric current) produces mainly a magnetic field (and $Z < Z_o$). At sufficiently low frequencies, currents whose associated vacuum wavelength is long compared with the largest spatial dimension of the metallic circuitry effectively do *not* give rise to any radiation, since the range of the near field is here appreciable, becoming infinite in the static limit.

The $1/r^2$, $1/r^3$ components of the *electric* dipole field (which are sometimes called the 'inductive' and 'static' fields, respectively) also give rise to an energy flux, but one that *periodically flows back and forth from the antenna* (because the electric and magnetic fields are here 90° out of phase), never to escape as radiation, the energy remaining ***stored*** in the near field.

Thus, it is **only** *via* the field components $\sim 1/r$ (which in the near-field region are, of course, dominated by the $1/r^2$, $1/r^3$ components) that energy actually *escapes* from the antenna into the **far** field region as ***radiation***; accordingly, this radiation *necessarily* contains *incomplete* information about the source.

Bio-Electromagnetism

The extent of the near field region corresponding to different frequencies can easily be estimated using $\lambda = c/v$, and some values are given in Table 2 below:

Table 2:

v (Hz)	λ (m) ~ extent of the **near** field)
1	3.10^8
10	3.10^7
10^2	3.10^6
10^3 (kHz)	3.10^5
10^6 (MHz)	3.10^2
10^9 (GHz)	3.10^{-1}
10^{12} (THz)	3.10^{-4}

At the microwave carrier frequencies used in *GSM* telephony – namely, 900 and 1800MHz – $\lambda \approx$ 33cm and 17cm, respectively. Accordingly, during normal use of a mobile phone – *i.e.* when it is held close to the user's head - a rather localised region of the head is exposed to the *near* field of the phone's antenna. It is essential to appreciate, however, that, under these conditions, the near field is not just that emitted by the antenna, since the latter is partially scattered; this results in a *distortion* of the near field, *via* which energy *transfer* to the head occurs. Whether this occurs primarily capacitively, through the electric field of the microwaves, or inductively, through the time-dependent near field *magnetic* component (which, in accordance with Faraday's law, induces an internal *electric* field), depends on the kind of antenna deployed in the handset – *i.e.* whether it is in the nature of an electric or magnetic dipole.

It is important to appreciate, however, that the estimates of the penetration depth given in Table I are not applicable to this situation because (*i*) they not only assume that the electromagnetic field has the characteristics of a plane-wave – *i.e.* is in the nature of a *far* (radiation) field, rather than a near field, but also that the field is incident on an idealised plane surface, and (*ii*) the values of ε_r and σ on which the estimates of d are based refer to *dead* biomatter, whose dielectrical properties are most certainly quite different from those of living matter, which in certain essential aspects is *far* from thermal equilibrium, so that *non-linear* properties predominate (*see* Section 2.3).

It should also be appreciated that the designs of *practical* antennae differ in certain ways from the simple Hertzian dipole discussed above. In particular, to improve the efficiency with which it emits radiation, and in order to be able to incorporate the antenna into the necessary electrical circuitry in the most efficient way (*i.e.* so that its impedance matches that (73Ω) of co-axial cable), the size of the antenna is increased so as to make the length of each limb equal to ¼λ, the antenna being then a *half-wave dipole*. Like the Hertzian dipole, the half-wave dipole emits *omnidirectionally* in any plane perpendicular to its axis. Often, however, it is desirable, for various technical reasons, that the radiation pattern be focused in a *particular* direction. This can be achieved by the combined use of a *number* of omnidirectional antennae, and phasing the currents in each appropriately; by separating them sufficiently it is then possible (through constructive and destructive interference) to produce a beam that has some degree of directionality. It is, however, impossible to ensure that *all* the radiated power is unidirectional, owing to the existence of subsidiary 'side-lobes', whose directions differ from that of the main beam. It should be noted that whilst the structure of the far fields of most antennae approximate to that of the simple Hertzian dipole, the near fields are generally very much more complicated, with a quite different space dependence, which is generally weaker (approaching that of the far field).

The extent to which the radiated power is concentrated in a particular direction is called the 'gain' of the antenna.

At sufficiently long distances, r, from an omnidirectional antenna, the radiated power density, S, can be estimated using the following formula appropriate to an idealised, isotopic, point source antenna, to which a power, P is supplied:

$$S = P/4\pi r^2 \qquad (14)$$

This expression can be adapted to the case of a directional antenna by introducing the concept of the effective isotropic radiated power (or *EIRP*), which in terms of the linear (power) gain, G, is given by:

$$EIRP = G.P. \qquad (15)$$

The *EIRP* is simply the power that would have to be emitted by an isotropic antenna in order to realise at a given distance, the *same* power density as obtains *at the same distance* in the case of a **directional** antenna.

G is often expressed in decibels (db), according to:

$$G_{db} = 10 \log_{10} G \qquad (16)$$

For a Herztian dipole, $G_{db} = 1.76db$, or $G = 1.5$ (recall that there is *no* radiated field in the direction of the axis of the dipole, the emission being maximal in the equatorial plane), whilst for a half-wave dipole, $G_{db} = 2.15db$, or $G = 1.64$ – *i.e. neither* are particularly directional.

2.8 Classification of Electromagnetic Fields according to Frequency

It is usual to classify time-dependent fields according to frequency, as follows:

3Hz – 3kHz: Extremely Low Frequency (ELF)

3kHz – 30kHz: Very Low Frequency (VLF)

30kHz – 300MHz: Radio Frequency (RF)

300MHz – 300GHz: Microwave (MW)

Clearly at *ELF*s and *VLF*s, $\lambda \gg$ size of the human body, so that *near* field conditions always obtain, where E and B must be considered *separately*.

2.9 Coherence

Electromagnetic radiation consists of a succession of *wavepackets*, within which the fields undergo regular (sinusoidal) oscillations with a **fixed** phase and polarisation. Consecutive packets do not, in general, connect 'smoothly', however – *i.e.* they are **'out of phase'** with one another; in addition, the planes in which the E and H fields oscillate - *i.e.* their *polarisations* - are also different. This destroys the 'purity' of the macroscopic electromagnetic wave, in the sense that it is **not** perfectly monochromatic - *i.e.* it is *not* characterised by a *single* frequency, ν, despite the frequencies within each packet being the **same**. Instead, it covers a *range*, $\Delta\nu$, of frequencies [3, 4] given by $\Delta\nu = h/\Delta t$, where Δt is the time duration of an individual wavepacket, and h is Planck's constant (= 6.626×10^{-34} *J.s*).

Since within a packet there is **no** change in phase, Δt is a measure of the **'coherence time'**, τ_c, or, equivalently, the degree of 'temporal coherence'. Associated with τ_c is the (longitudinal) **coherence length**, l_L ($= c\tau_c$), which is a measure of the *distance* in the direction of propagation over which the radiation is phase stable - *i.e.* is **coherent**.

Some typical values of examples of τ_c and l_L are given below in Table 3:

Table 3:

	$\Delta v/v_o$	τ_c (s)	l_L
Sunlight	~ 1	10^{-15}	$0.1\mu m$
Gas discharge lamp	10^{-6}	10^{-9}	$0.3 m$
Radio transmitter	10^{-4}	10^{-6}	$300 m$
Laser	10^{-12}	10^{-3}	$3.10^5 m$

It is seen that *all* radiation is actually coherent, but to differing extents; Sunlight, however, can be considered to be effectively *temporally incoherent*, since $\Delta v/v_o \sim 1$. In the case of the laser, on the other hand, the long coherence time can be considered as a consequence of many ($10^{-3}/10^{-9} = 10^6$) successive wave-packets being in phase, whereby is effectively realised a macroscopic 'magnification' of a single 'ingredient' wave-packet, or equivalently, a macroscopic quantum state of the light field.

In addition to temporal coherence is the concept of *spatial* coherence. This reflects the **finite size**, a, of a *non-idealised* extended source *transverse* to the direction in which the radiation is emitted, and is responsible for the existence of *non-smooth* wavefronts. The extent to which they are smooth (*i.e.* phase stable) depends on distance, d, of a field point from the source in comparison with a and λ; the degree of smoothness is quantified in terms of the so-called **transverse** coherence length, l_T, which is given approximately by:

$$l_T = \lambda(d/a) \tag{17}$$

Clearly, unlike l_L, the value of l_T depends not only on the source, but also on the distance, d, of the field point from the source.

In the case of Sunlight at the Earth's surface, d is the Earth–Sun distance ($\sim 1.5 \times 10^{11} m$), and a is the diameter of the Sun ($\sim 1.4 \times 10^9$); thus for light in the middle of the visible spectrum ($\lambda \sim 500$nm),

$$l_T \sim 50\mu m \tag{18}$$

Associated with l_T and l_L, is the '**coherence volume**', V_c, defined by:

$$V_c \sim l^2_T l_L \tag{19}$$

Thus, in the case of Sunlight, for example, $V_c \sim 250(\mu m)^3$

In summary, the coherence of radiation gives both temporal and spatial information about the *source* of the radiation, although the latter depends on the field point.

3. SOME RELEVANT BIOLOGY

3.1 Cell Membrane Fields

Across the membrane of a *living* cell, ion pumping generates an electrical potential difference, ΔV, of magnitude $\Delta V \sim 100mV$.

Assuming a separation, a, of 10^{-8}m between the inner and outer layers of the bilipid membrane, the magnitude, E, of the associated electric field is given by:

$$E = \Delta V/a \ (= 10^7 V/m) \tag{20}$$

In **non**-living matter, an electric field of this magnitude is large enough to cause *dielectric breakdown*, suggesting that it plays a rather crucial role in the biophysics of *living* matter. One such role that can be immediately identified is that this large field protects the interior of the cell from external *DC* fields, but **not** technogenic *AC* (oscillatory) fields above a certain frequency - the value of which is determined by the electrical properties of the membrane [6].

Another physical consequence is that this enormous field results in a strong (possibly non-linear) polarisation of any biomolecules, such as proteins, that find themselves in the field, resulting in the formation of *induced* electric dipoles.

3.2 Electromagnetic Nature of the Alive Organism

Most familiar are the oscillatory electrical activities associated with heart beat and brain, which are measured by the electrocardiogram (*ECG*) and electroencephalogram (*EEG*). The associated frequencies lie between 1 and 1.5Hz, and between 1 and 35Hz, respectively, although much higher *EEG* frequencies have now been recorded using advanced techniques. Other electrical activities include those involved in nerve conduction and in the body's biochemistry, which is controlled and regulated electrically – often in an *oscillatory* way [4].

Less well-known, is an organised macroscopic electrical activity at cellular/sub-cellular levels, based on the **coherent** oscillation of electric dipolar units of a given kind, the frequencies of which range from the **microwave** band to values in the terahertz region – *see* Section 5.4.2

Further evidence of our electromagnetic nature *when alive*, is the fact that our bodies **radiate** electromagnetically – both incoherently (*thermal* radiation, which peaks in the infrared) and coherently, as indicated by both the ubiquitous phenomenon of biophoton emission, and by the recently reported emission of (non-thermal) microwave radiation – *see* Section 7.2.2.

Another indicator of the electromagnetic nature of living organisms is their extreme sensitivities to external electromagnetic fields, both natural and technogenic, whereby they can detect minute changes in their electromagnetic environment, even in the presence of much larger 'irrelevant' signals (noise) [5]. Indeed, their biosensors often operate near the quantum limit, as evidenced, for example, by the fact that the human eye can detect a single light quantum (photon), corresponding to a threshold intensity of the order of about 10^{-15}W/cm^2; interestingly, this value is close also to the thresholds of human hearing and *EEG* response.

In the animal kingdom, equally striking sensitivities can be found, such as in the case of birds and bees, which can detect changes in the Earth's magnetic field of a magnitude less than 1nT; some species of fish, on the other hand, can detect electric fields as low as 1µV/m [4].

4. THE INFLUENCE ON THE ALIVE ORGANISM OF ENVIRONMENTAL ELECTROMAGNETIC FIELDS – THE IMPORTANCE OF COHERENCE

4.1. The Natural Environment

Here, we are exposed to a variety of electric /magnetic fields and electromagnetic radiation – both ionising (*e.g.* cosmic rays), and non-ionising (both incoherent and spatially coherent): only the latter will be considered here, examples of which include:

a) The *Electrostatic* field of the Earth. This can range between (0.1 – 20) kV/m, depending on atmospheric conditions. (As a comparison, it may be noted that at 1m above ground beneath a 400kV power-line at 50Hz, the electric field is about 5kV/m, which will *light up a fluorescent tube*.)

b) The *Magnetostatic* field of the Earth. This is approximately 50µT, superimposed on which is a *variable* component of magnitude less than 2% of the steady value, which is synchronised with the solar and lunar rhythms. (Again, for comparison, the magnetic induction 1m above ground beneath a 400kV power-line at 50Hz is about 12µT.)

c) *Electromagnetic radiation* from the Sun. The blackbody spectrum of this essentially *temporally incoherent* radiation (which peaks in the **visible** part of the spectrum) results in an integrated intensity (power density) at Earth' surface of 1350W/cm^2 - the so-called *Solar constant*.

d) *Schumann radiation.* This spatially coherent, non-ionising, electromagnetic radiation represents the transverse *EM* normal modes of the Earth-ionosphere cavity, whose source is the totality of global lightening discharges. The structure of these resonant excitations at the Earth's surface principally consists of vertical **E** and horizontal **B**-fields, whose spectral fields strengths are $E_v \sim$ (100 – 200) µV/m /Hz$^{1/2}$, $B_h \sim$ 0.5pT/Hz$^{1/2}$, respectively, the associated spectral power density being of the order of $\sim 6 \times 10^{-15}$W/cm^2/Hz, and maximum at *local noon*.

In connection with the Schumann radiation, the following should be noted:

i) The Schumann power density is close to the threshold of the human *EEG* sensitivity.

ii) The lower frequencies correlate with those of the human *EEG*. Thus, 8Hz $\sim$ mid α_1 band (7-9.5Hz); 14Hz $\sim$ mid β_1 band (12.75-18.5Hz); 20Hz $\sim \beta_2$.
iii) Isolation from the Schumann field (and also the Earth's magnetic field) during space-travel *beyond* the ionosphere correlates with the onset of certain neurological disorders in astronauts. To counteract this, Schumann simulators are now fitted in spacecraft.

4.2 Some Influences of Exposure of the Alive Human Organism to Technological Electromagnetic Radiation

The human *EEG* and behaviour can be influenced in non-thermal, frequency-specific ways by exposure to external *EM* radiation of various kinds, when it is amplitude modulated/pulsed at *ELF*s close to those of the *EEG*.

Example 1 – **Photosensitive epilepsy:** Here, exposure to a light flashing at a rate between 15 – 20 times /sec can provoke seizures in people who suffer from photosensitive epilepsy. It is not the intensity of the light that triggers the seizure, but rather the fact that it flashes at the same frequency as characterises the electrical activity of the epileptic brain.

Example 2 – **Human behavioural modification:** It has been known for over 30 years that **microwave** radiation, pulsed at certain *ELF*s (near those now used in GSM telephony, incidentally) is psychoactive' – *i.e.* has the ability to exert a variety of behavioural and psychological effects on humans, a fact that is now exploited in '*non-lethal*' **weaponry.** Thus, for example, it is possible in this way to implant sounds and words into the brain, and to excite a range of moods by exposure to certain low frequencies modulations, such as depression (at 6.6Hz) and manic & riotous behaviour (at 11Hz).

4.3 The Importance of Coherence

Electromagnetic fields of technological origin are much more temporally **coherent** than those in the natural environment within which Man has evolved. Since human exposure to these highly **coherent** *EM* fields is relatively recent (~100 years only), we *cannot* have developed any evolutionary immunity to them. Two particular consequences of this high degree of coherence should be noted:

(*i*) It facilitates the discernment of technogenic fields by the living organism against the level of their own (incoherent) thermal emission at physiological temperature.

(*ii*) It assumes particular significance in view of the various *coherent EM activities* that the human organism supports when it is ALIVE. For there is then the possibility of *'oscillatory similitude'* between an external frequency and that of an endogenous bio-process, whereby an external *EM* field can influence the alive organism in certain frequency-specific, non-thermal ways.

5. COUPLING OF ELECTRIC/ MAGNETIC FIELDS AND ELECTROMAGNETIC RADIATION TO THE ALIVE BODY

5.1. Generalities

External E- fields couple – ***irrespective of aliveness*** to electric charges and electric dipoles [6], giving rise to:

a) An electric current density, *J* of *free* charges, in accordance with Ohm's law,

$$J = \sigma E, \quad (21)$$

where σ is the electrical conductivity of the bio-medium (in general a tensor quantity), the units of which are $(\Omega m)^{-1}$, or, equivalently, *S/m*, where *S* denotes the unit *siemen*.

b) An electric polarisation density *P*, of *bound* charges into (induced) electric dipoles, where,

$$P = \varepsilon_0 \chi E \quad (22)$$

Here, χ = electric susceptibility (tensor), which is related to the relative dielectric constant of the medium, ε_r, by $\chi = \varepsilon_r - 1$.

c) A reorientation of permanent electric dipoles.

External B-fields, on the other hand, couple [6] to electric currents and magnetic moments, both microscopic (atomic) and macroscopic, such as bio-magnetite, for example.

It should be appreciated, however, that because the electrical conductivity of the human body is very much greater than is that of the surrounding air, whilst its magnetic permittivity is approximately the same as that of air, the body is much more **electrical** than magnetic. Consequently, *E*-fields are *much more distorted* by the body than are *B*-fields [6]. We now consider in greater detail, the effects of electric, magnetic and electromagnetic fields – both static and time-dependent - on the human body.

5.2 Coupling to an External Electric Field

a) *Static fields*: Distortion of the field perpendicular to body surface results in increasing density of field lines, and hence a stronger *E*-field; for example, the intensity in the vicinity of the head is almost 10 times greater than that of the incident field. Since the electrical conductivity of the body is so much greater than that of air, the incident *E*-field is mainly confined to the *surface* of the body, there being very little penetration into the interior, where it is reduced by 10^{-6} [6]. In other words, the body acts rather like a Faraday cage. (a closed metallic box on which the redistribution of electric charge caused by exposure to an external *E*-field creates an *internal E*-field that exactly cancels the external field, so that there is ZERO net field inside the box). The total electric current flowing over/through the body is typically of the order of 15μA per *kV/m* of external field. The most conductive *internal* paths are blood and intercellular fluids.

b) *Time dependent ELF fields*: These induce surface charges, resulting in internal electric currents, which can be very non-uniform, depending on the shape and position of the exposed body relative to the field. For sinusoidal fields of frequencies below about 10MHz

$$|J| = vA\, E_{ext} \qquad (23)$$

A depends on σ, and size and shape; some values for different anatomical parts are given below in Table 4.

Table 4:

Part	A (S/Hz.m)
Head	2×10^{-9}
Body trunk (average)	6.7×10^{-9}
Neck	10×10^{-9}
Ankle	40×10^{-9}

5.3 Coupling to an External Magnetic Field

Since, as noted above, the magnetic permeability of the human body is approximately equal to that of free space; accordingly, there is *little distortion* of a *B*-field, which thus *easily* **penetrates the interior of the body**.

Bio-Electromagnetism

Once inside, it couples to any pre-existing electric current, magnetic moments and any magnetic material.

If B is *time*-dependent, it gives rise to an (induced) internal **ELECTRIC** field, E, via Faraday's law of electromagnetic induction - $curl E = - \partial B/\partial t$.

Thus, in the presence of free electric charges there is an associated *induced* electric current density, J_i, which circulates in a *closed loop*, the magnitude of J_i being given by:

$$|J_i| \approx \pi R \sigma \nu B_o, \qquad (24)$$

where:

R = radius (in m) of the inductive loop perpendicular to B
σ = electrical conductivity of the body (in S/m)
ν = the frequency of the B-field (in Hz)
B_o = the amplitude of the B-field (in T)

The appropriate inductive loop depends on the orientation of B relative to the body and on the value of the local electrical conductivity, σ.

J_i^{max} occurs in peripheral tissues and decreases linearly towards the centre, and is usually much less than those associated with the heart and brain.

It should be noted that (*i*) up to about 100kHz, time-dependent E/B-fields result in *negligible energy absorption* and no measurable temperature rise, and (*ii*) all the above processes occur **irrespective** of aliveness.

5.4 Coupling to External RF/Microwave Fields

5.4.1 By Energy Absorption from the Electric Field of the Microwaves

This results in **dielectric heating**, which again occurs *irrespective of whether the biosystem is alive or dead.*

The **mechanism** of absorption is the 'dielectric loss', discussed in Section 2.5, which results in the energy of the electromagnetic field being dissipated as heat in the medium, within the penetration depth, d.

Some calculated values of the penetration depth in water dominated (non-alive) systems are given in Table 5:

Table 5:

Frequency	d(cm)
100MHz	7
1GHz	3
10GHz	0.3

Energy absorption is quantified in terms of the **Specific Absorption Rate** (*SAR*), which is defined as the power absorbed/unit mass – *usually averaged over time (6mins)* and *mass (10gms)*:

$$SAR = \sigma E_i^2/\rho \quad (W/kg) \tag{25}$$

E_i is the magnitude of the time-averaged **internal** *E*-field ($= 1/\sqrt{2}\ E^{max}_i$) in V/m, σ is the electrical conductivity *at RF/MW frequencies* in $(\Omega m)^{-1}$, and ρ is the mass density in kg/m³.

The associated rate at which the temperature rises can be obtained using the specific heat, *C*, of the biomatter, which is given, in terms of the internal energy, *U* (in J/kgm) by:

$$C = dU/dT \tag{26}$$

Noting that *dU/dT* can be written as *(dU/dt) (dt/dT)*, and that *(dU/dt)* is nothing other than the *SAR*, it follows that:

$$dT/dt = SAR/C\ (= \sigma E_i^2/\rho C) \tag{27}$$

Thus, the higher the *SAR*, the more rapid the temperature rise; accordingly, the *SAR* is used as the dosimetric quantifier of thermal heating, although to obtain the magnitude of the actual temperature *rise* it is necessary to integrate over time.

To avoid health problems, Safety Guidelines governing human exposure restrict the *SAR* sufficiently to ensure that any temperature rise is kept below 1° C; this requires that the whole body average *SAR* be less than 0.4W/kg (when a safety factor of 10 is included).

It is important to appreciate that the absorption and distribution of *RF/MW* energy in the body are strongly dependent on its size and orientation relative to the wavelength and polarisation of the incident radiation, and is maximum when the long axis of the body is parallel to the direction of the electric field of the radiation, and when its length is roughly equal to the wavelength of radiation. For then the body acts like a dipole receiver, resulting in considerable (resonant) energy absorption. In the case of far field exposure under these conditions, whole body resonance of a grounded human of height 1.7m occurs at 32MHz, where the absorption is *15 times greater* than at 1GHz, for example. In the case of a human skull, on the other hand, the resonance occurs near 1GHz. It should also be noted that in the case of *cavity* structures, such as the skull, standing waves can occur, resulting in local 'hot spots'.

5.4.2 By Information Transfer from the Field to the Alive Organism

Unlike dielectric heating, which occurs irrespective of aliveness, the ability of an external electromagnetic field to exert an *informational influence* is strictly contingent on the exposed organism being **alive**. A possible mechanism is that based on an '**oscillatory similitude**' between the frequencies of the external electromagnetic field and those of the various *highly organised endogenous electrical activities* that the organism supports at different levels of complexity, when – and only when – it is **alive**, which play a role analogous to the tuned circuits in a radio.

Thus at the cellular /sub-cellular levels, for example, there are the '*coherent excitations*', with frequencies in the **microwave** band and above, mentioned in Section 3.2. These excitations are (longitudinal) collective electric modes (based on dipolar units of various kinds), which are suprathermally excited at the *lowest* collective mode frequency by a supercritical rate of energy supply from metabolism [7, 8]. Once the coherent excitation is established, the individual dipolar units oscillate together in synchrony, as a *single* '**giant' dipole**; in this way, the dipolar oscillations assume a *macroscopic significance* [9].

At the level of the whole alive organism, on the other hand, it is well known that the *EEG* can be influenced in non-thermal, frequency-specific ways by external *EM* radiation of various kinds, when the radiation is amplitude modulated/ /pulsed at *ELFs close* to those found in the *EEG*.

Three distinct possibilities of associated influences can readily be identified:

i) **Resonance** – perhaps to a physiologically *undesirably high* level.

ii) **Interference** – resulting in the degradation of some activity essential to homeostasis.

iii) **Entrainment** of an endogenous frequency to a value incompatible with homeostasis.

6. THE ALIVE ORGANISM AS A NON-EQUILIBRIUM MACROSCOPIC QUANTUM SYSTEM – NOVEL ELECTROMAGNETIC IMPLICATIONS

6.1 Generalities

The existence of coherent excitations (based on synchronised dipolar oscillations) is a rather general consequence of a living system being *an open, dissipative system, which is held far from thermal equilibrium (death) by the influx of metabolic energy associated with its vitality*.

The suprathermally excited nature of the (non-equilibrium) coherent mode means that its vibrational amplitude[2] far exceeds that realised in thermal equilibrium at physiological temperature. Quantum mechanically, this is equivalent to the number of quanta in this single mode being *far greater* than in thermal equilibrium. This single macroscopically occupied mode (sometimes called a 'condensate') effectively constitutes a *non-equilibrium*, **macroscopic quantum state** [8, 9].

6.2 Some Novel Implications

6.2.1 A Frequency-specific, Long Range Attraction:

Between two coherently oscillating systems of approximately equal frequency, whose size is much smaller than their separation, R, is a **long-range attraction** $\sim R^{-3}$ [7], which could form the basis of an 'intelligent' biochemistry based on **frequency recognition**, [9] and which, moreover, can be switched on/off according as whether the necessary coherent oscillations are excited or not. Through this selective attraction, particular systems (*i.e.* those with approximately equal frequencies) can be brought into sufficiently close proximity for them to be able to react *chemically*, at a certain stage of

---

[2] It should be noted that being in the nature of a **non**-linear limit cycle oscillation, the amplitude (and also frequency) of the coherent excitation is always the same, *irrespective* of the initial conditions.

their evolution. If in turn, as a result, the frequency of one or both of the coherent modes is altered to a value close to that of *another* molecular species, then this species will be attracted into the vicinity, thereby facilitating the *next* stage of some complex multi-stage biochemical process. Evidence in support of this interaction is provided by the observation of *rouleaux* formation by red blood cells [7-9], where, below a certain separation of the order of microns (which is orders of magnitude *greater* than the range of chemical forces) the cells rush towards each other to form coin-like stacks (*rouleaux*) at a rate *much faster* than is expected on the basis of random Brownian motion. Upon removal of the membrane potential, however, the rate reduces to the Brownian value (as it also does when metabolism is inhibited), but increases again when the potential is restored. Furthermore, in the case of a *mixture* of mammalian cells, it is found that cells of the *same* species tend to aggregate predominantly, again in contrast to the prediction based on Brownian motion.

A particularly interesting application of this R^{-3} interaction is a model in which the *EEG* frequency can be understood in terms of an electrochemical limit cycle associated with 'auto-catalytic' excitation/de-excitation of an enzyme-substrate system in the Greater Membrane of the brain, the supply of reactants being maintained by the long-range interaction [7-9].

In consequence of the possibility of inducing a collapse of this electric limit cycle by exposure to a weak *external* oscillatory electric field, sufficient energy can be made available to initiate certain physiological processes (such as nerve conduction, for example), the threshold of which *exceeds* the energy available in the incoming signal alone. In effect, the limit cycle stores the energy of the incoming signal until its (finite) storage capacity is reached, after which it collapses - *i.e.* the role of the external field is simply to "trigger" the collapse of the limit cycle, the stored energy thereby released effectively amplifying the (small) incoming energy, possibly to a physiologically significant level.

The interaction of external electric fields with the internal limit cycle is the subject of continuing theoretical research. One of the most interesting findings is that in the presence of an oscillatory external electric field the model can exhibit deterministic chaos.

6.2.2 Absorption and Emission of Electromagnetic Radiation

Being in the nature of a 'giant', oscillating electric dipole, a coherent excitation should be **electromagnetically active** – *despite its longitudinality* *(in the bulk)* – in consequence of the existence of internal bounding surfaces that effectively endow it with a degree of **tranversality** [9]. Two particular consequences should be noted:

(*a*) The excitation can be **accessed** electromagnetically, *via* resonance, interference or entrainment, and can even be **turned-on** by external radiation, should the rate of endogenous energy supply be insufficient, the duration of irradiation depending on the power deficit and the proximity of the external frequency to that of the excitation eventually established. It is possible that the efficacy of Microwave Resonance Therapy – a modality in which ultra-weak microwave irradiation of highly specific frequencies applied to relevant acupuncture points is able to re-establish homeostasis in a wide range of human pathological conditions – can be understood in this way: namely, that the applied hyper weak external microwave radiation acts to *switch back on* some particular coherent excitation essential to homeostasis (which had for some reason ceased to be excited), thereby restoring homeostasis [9].

(*b*) The coherent excitation must **radiate** electromagnetic radiation of approximately the *same* frequency. It should be noted that the *monochromaticity* of this radiation empirically distinguishes it from *biophoton emission* [10].

6.2.3. Bio-Communication *via* the (magnetic) Vector Potential

The realisation of (non-equilibrium) **macroscopic quantum** states in living systems means that they should be able to respond (in a way *similar to a superconductor*) to an external **magnetic vector potential** [7]. This would entail the possibility of a quite novel form of electromagnetic bio-communication that would escape detection by conventional devices that register only electric and magnetic fields.

The existence of a vector potential, A (as well as a scalar potential, ϕ) follows [2] from the two homogeneous Maxwell equations, $curl E = -\partial B/\partial t$ and $div\ B = 0$, in consequence of the vector identities and $curl(grad S) \equiv 0$, for *any* scalar field, S, and $div(curl V) \equiv 0$, for *any* vector field V. In terms of the potentials A and ϕ, the E and B fields are given by:

$$E = - grad\phi - \partial A/\partial t, \quad B = curl A \qquad (28)$$

Thus the electric field, *E*, is seen to be related to the way in which the vector potential, *A*, changes in *time*, whilst the magnetic induction, *B*, reflects its variation in *space*; thus only in the time-independent case can *A* strictly be called a *magnetic* vector potential.

The existence of an asymmetry in the above expressions for *E* and *B* should be noted – namely, that the expression for *B* holds under *both* time dependent and time-independent conditions, which is not the case with *E*, which in the time-independent limit reduces to the familiar electrostatic form $E = -grad\phi$, where ϕ is the familiar time-independent electrostatic potential.

Because of the first of the above vector identities, *B* is invariant under the addition to *A* of the *spatial* gradient of a scalar field, $\chi(r, t)$; to simultaneously preserve the invariance of *E* it is then clearly necessary that under $A \rightarrow A + grad\chi$, ϕ transforms according to $\phi \rightarrow \phi - \partial\chi/\partial t$. The scalar field $\chi(r, t)$ is called a 'gauge' field, and the transformations of *A* and ϕ under which *E* and *B* remain invariant are called 'gauge transformations'. This gauge freedom means that it is always possible to define the potentials (*A*, ϕ) such that the following condition (the so-called Lorentz condition) holds, the associated (relativistically covariant) gauge being called the Lorentz gauge:

$$divA + (1/c^2)\partial\phi/\partial t = 0 \qquad (29)$$

In the absence of sources, the Lorentz condition ensures that *A* and ϕ separately satisfy wave equations.

In the absence of sources it is also possible, however, to 'gauge away' the scalar potential completely [2], so that:

$$E = -\partial A/\partial t, \quad B = curl A \qquad (30)$$

In this case, the gauge is called the radiation or transverse gauge, satisfying $divA = 0$, and is a special case of the so-called Coulomb gauge [2], which can, in general, be defined in the presence of sources, the origin of the name being that the scalar potential is here just the instantaneous Coulomb potential, which contributes only to the near field. By contrast, in the far field region, only the transverse radiation field survives, which can be expressed *solely* by the vector potential.

It is to be noted that it is through the 'underlying' vector (and scalar) electromagnetic potentials that electrically charged particles in a quantum system (described by a wavefunction ψ) couple to an external electromagnetic field, and that it is the radiation or transverse gauge that underlies the quantum mechanical description of the radiation field in quantum electrodynamics, in which photons are the quanta of the *transverse* **vector potential** field, satisfying $divA = 0$.

The above mentioned possibility of a novel form of electromagnetic biocommunication arises from the fact that there is a non-trivial solution for the potentials, corresponding to $E = 0$ and $B = 0$, namely:

$$\phi = \partial\chi/\partial t, \text{ and } A = grad\chi \tag{31}$$

i.e. the vector potential is *longitudinal* ($curlA = 0$) – *i.e.* is derivable purely from the *scalar* field $\chi(r, t)$, whose time derivative is the associated scalar potential ϕ. In turn, it follows from the Lorentz condition that $\chi(r, t)$ satisfies the wave equation:

$$\nabla^2\chi - (1/c^2) \partial^2\chi/\partial t^2 = 0 \tag{32}$$

In the presence of the above longitudinal vector potential, the quantum mechanical wavefunction, ψ, of the system takes the following form, where q is the electric charge to which A couples:

$$\psi = \psi_o \exp(iq\chi/h) \tag{33}$$

where ψ_o describes the system in the absence of A. Although in the case of a single electric charge, χ is unobservable (since $|\psi|^2 = |\psi_o|^2$), this no longer holds in the case of the *relative* phase factor $(\chi_1 - \chi_2)$ of two interfering charges; thus **information** can be transferred by the scalar field, *even though* (because $E = 0$, and $B = 0$) there is **no transport of electromagnetic energy**. If, however, the scalar field interacts with a highly non-linear medium, it is possible for A to develop a *transverse* contribution (with nonzero *curl*), associated with which is a magnetic field, $B(= curlA)$ that is detectable in the conventional way.

7. ELECTRIC AND ELECTROMAGNETIC FIELDS GENERATED BY LIVING SYSTEMS

7.1 Electric Fields

7.1.1 MHz-Fields

Time-dependent electric fields with frequencies near 1, 7-8, 50-80MHz (with $\Delta v \sim 50Hz$) have been detected [9] in the *immediate* vicinity of synchronised yeast cells near the time of cell division. The close proximity to the living system at which the fields have been detected, and the values of the frequencies are consistent with the fields being in the nature of **near** fields, and thus **non**-radiative. It may be noted that the detection of short-range electric fields near 6-7MHz has been independently corroborated [9] in the same system.

7.1.2 kHz-Fields

Micro-dielectrophoresis and cellular spin resonance experiments on a variety of **dividing** cells (bacteria, algae, yeast, avian and mammalian) indicate [9] presence of a *very short-range* electric fields at somewhat lower frequencies, in the range ~ 1–30kHz; again, it is likely that these are in the nature of **non**-radiative, **near** fields. [It should be noted that this frequency range is consistent with the requirement that the frequency of the oscillating source dipole must be much greater than the masking rate (due to dielectric relaxation) of the aqueous medium if it is to be detectable.]

In connection with frequencies in the kHz region, it is interesting to note that it is possible to influence (in a way characteristic of a **non**-linear oscillator) the growth rate of yeast cells by **irradiation** with low intensity ($5pW/cm^2$) microwaves, which have been square wave amplitude modulated at 8kHz [11].

7.2 Electromagnetic Fields

7.2.1 Thermal

As already noted in Section 2.6, living systems emit incoherent thermal radiation appropriate to their physiological temperature, which for humans is about 36°C, at which the spectral emission is a maximum in the infra-red part of the spectrum near 10microns.

7.2.2 Non-Thermal

Living systems containing *DNA* are also observed to spontaneously emit photons called 'biophotons' approximately uniformly across the visible, with a (total) intensity typically between 100 and 1000 photons per second per square centimetre [10]. In contrast to thermal radiation, however, biophoton emission is *coherent*.

Recently, another variety of non-thermal, coherent emission has been reported in the *mm* microwave region, which is considered to be an intrinsic hall-mark of the alive [12], possibly related to the coherent excitations mentioned above in Section 5.4.2.

8. COHERENCE IN WATER

The possibility of coherence in water (whether biological or otherwise) was predicted some time ago [13], based on the introduction of the ideas of quantum electrodynamics into condensed matter physics. This indicated the possibility of realising coherence, *without* pumping, as a result of an instability in the usual ground-state, which occurs above a certain critical density $(= 0.31 \text{g/cm}^3$, which is close to the thermodynamic critical density) when account is taken of a coupling between certain electronic excitations of the water molecules and the (zero-point) **fluctuations** of the electromagnetic field - a coupling that is usually neglected on the grounds that (on average) the vacuum electromagnetic field vanishes.

The predicted coherence, which is here *self-confined* to domains whose spatial extent is of the order of $0.1 \mu m$, involves synchronised electronic transitions between the ground-state and a certain (almost unbound) excited state of each of the 10^7 water molecules contained within a given domain. Coupled to this is an oscillatory, strong (evanescent) electric field $(\sim 10^{11}$ V/m) that is *in phase* with these molecular transitions, the common frequency being close to $60 THz$. With increasing temperature, an increasing number of water molecules break away from these coherent domains, resulting in the formation of a *two-fluid* system - a temperature-dependent mixture, consisting of those water molecules that continue to behave coherently and those (the incoherent fraction) that do not, the coherence disappearing completely at the critical temperature, $T_c (= 500K)$.

The implications of water coherence could be far-reaching. One area that is currently attracting attention relates to the *memory* of water, where it has been proposed [14] – based on the non-Maxwellian propagation of an external electromagnetic fields through a coherent medium (wherein the photons acquire a finite rest mass, an associated *longitudinal* polarisation, and a frequency that is proportional their velocity) that the (time-dependent) magnetic **vector potential** plays a fundamental role in carrying the frequency information (*bio*-information, in the case of a living system), the water itself being 'formatted' either by an additional magnetic field (possibly that of the Earth) or, alternatively, by succussion.

In the case of biosystems, the existence of such coherent domains (and associated inter-domain Josephson phenomena) would elevate water from its traditional bio-role as a passive space-filling solvent to a position of singular importance, the full significance of which is yet to be fully elucidated.

SUGGESTED READING

Electromagnetic Theory:

1. Feynman, R.P., Leighton, R.B. and Sands, M. (1963) *The Feynman Lectures in Physics*. Vol. II, Addison-Wesley Publishing Co. Reading, Mass..

2. Jackson, J.D. (1998) *Classical Electrodynamics*. Wiley & Sons, New York.

3. Popp, F.A. (1994) *Electromagnetism and Living Systems*. Chapter 2 of Bioelectrodynamics and Communication', eds. Ho, M-W, Popp, F.A. and Warnke, U., World Scientific, Singapore

Biophysics

4. Smith, C.W. and Best, S. (1989) *Electromagnetic Man*. J.M. Dent & Sons Ltd., London.

5. Popp, F.A. *et. al* (1989) *Electromagnetic Bio information*. 2nd Edition, Urban & Schwarzenberg, Munich.

6. Pollock, C. & Polk, E. eds. (1996) *Handbook of Biological Effects* 2nd Edition, CRC Press, New York.

7. Fröhlich, H. ed. (1988) *Biological Aspects of Coherence and Response to External Stimuli*. Springer-Verlag, Berlin.

8. Hyland, G.J. (1998) 'Quantum Coherence and the Understanding of Life'. *Biophotons* eds. Chang, J.J., *et al.*, Kluwer Academic Publishers, Dordrecht.

9. Hyland, G.J. (2002) 'Coherent Excitations in Living Biosystems, and their Implications'. *What is Life*, World Scientific. In Press

More specific references

10. Popp, F.A.: *Frontier Perspectives* 11(1) (2002), 16-28.

11. Grundler, W. & Kaiser, F.: *Nanobiology* 1 (1992), 163-176.

12. Sit'ko, S.: *Physics of the Alive* **6** (1998), 6-10; see also other contributions to this issue.
13. Arani, F *et al.*: *J. of Mod. Phys. B.* **9** (1995), 1813-1841.
14. Smith, C.W. (1994) 'Electromagnetic and magnetic vector potential bio-information and water'. *Ultra High Dilution –Physiology and Physics*, eds. Endler, P.C. & Schulte, J., Kluwer Academic Publishers, Dordrecht.

Chapter 3

DEVELOPMENTAL BIOLOGY AND PHYSICS OF TODAY

Lev V. Beloussov
Department of Embryology,
Moscow State University
Moscow 119899, Russia
E-mail: lbelous@soil.msu.ru

Abstract: We start by analyzing two alternative approaches which can be used for explaining natural events: a physicalistic one, directed towards formulation of the overall laws which should be as non-specific as possible, and an "instructivistic" one, which aim is to create a list of as specific as possible "instructions". We discuss whose one of them is mostly relevant for describing the development of organisms. We point to some deep discrepancies between classical physics ("physics of yesterday") and the fundamental holistic properties of the developing organisms. It was this controversial situation which inclined researchers to use instructivistic approaches. Meanwhile, the "physics of today" (and, first of all, a theory of a self-organization) is much more adequate for interpreting holistic properties of biological development. We give several examples illustrating how modern physical ideas and approaches may be used for clarifying some fundamental problems of development. We consider, in particular, the field of an "embrymechanics", which is studying the feedback relations between passive and active mechanical stresses. In the last section we represent briefly some results of a new field of a "developmental biophotonics".

1. INTRODUCTION

The development of a multicellular organism from an egg cell up to the adult state is undoubtedly the most ordered and complicated process ever taking place in all of Nature without a human's intervention. to become aware of this, one may look at the development of one single species, for example a sea-urchin, a beloved object for several generations of biologists (Fig. 1). Its egg is firstly divided into a definite number of precisely located daughter cells, the blastomeres. When this initial period of the development, called the cleavage, is completed, the embryonic cells start to change their mutual positions in a highly coordinated fashion, making invaginations, bendings, pockets and so on. At the same time or later the individual cells start to differentiate, that is, to change in a definite way their internal structure and chemical contents. So, in the first approximation the development of a multicellular organism can be regarded as a precisely regulated space-temporal succession of cell divisions, cell movements and cell

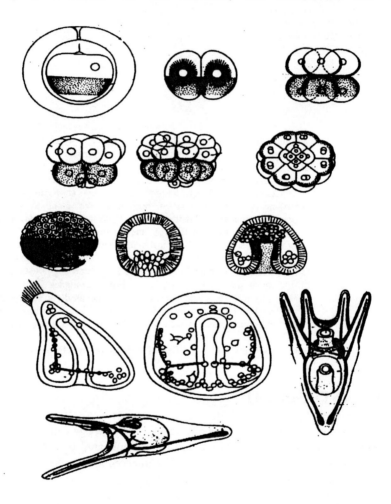

Figure 1: The development of a sea-urchin embryo from an egg cell up to a free-swimming larvae (from top to bottom). Two upper rows illustrate egg cleavage, next row shows transformation of a blastula to a gastrula. Two last pictures give a view of a free-swimming larva.

differentiation. How does all of this become possible? In what way can any small part of an embryo "know" what it should do at this very time moment and at this spatial location, and why is this enormously complicated succession of events do precisely reproduced during each next generation? Why is different species develop in different ways sharing at the same time some common features? In this or other formulations, such questions were a matter of trouble for the naturalists since long ago. Aristotle, the famous ancient Greek philosopher was the first who introduced scientific methods in attempting to resolve these questions. He believed that all the natural bodies, both living and non-living ones, have some intrinsic formative powers ("causae formalis") bringing them towards specific goals. Much later meanwhile, in the Renaissance time, the sciences of a living and a non-living world have been drastically separated from each other, and the "causae formalis" were resolutely expelled out of the second ones. Such a procedure provided an enormous progress in physics and in the related sciences dealing with non-living bodies. But, on the other hand, the gap between biology and non-biology increased enormously. "So let it not look strange – wrote a famous German philosopher Immanuel Kant in the year of 1755 in his "Natural History of a Sky" – if I claim that it is much easier to explain the movement of the giant celestial bodies than to interpret in mechanical terms the origination of just a single caterpillar or a tiny grass" [1]. Only recently such a situation of a deep discrepancy between the physical approach to the natural world and the living events started to diminish, and a way for a new fundamental and fruitful alliance between physics and biology opened up. The main aim of this paper is to trace this way and to show how some modern trends in physics and in the related sciences can promote much better understanding of the developmental events than we had in a recent past. Our first task will be in characterizing briefly the logical structure of the physical and non-physical explanations and to decide which of them can be better used for interpretation of the developmental events.

2. CLASSICAL PHYSICS AND CLASSICAL EMBRYOLOGY: POINTS OF MISUNDER-STANDING

One of the best definitions of physics belongs to James Clarke Maxwell, a founder of a theory of an electromagnetic field. By his words, "Physical Science is that department of knowledge which relates to the order of nature, or, in other words, to the regular succession of events" [2]. And the main way by which physics is treats "the order of nature" is in formulating the embracing laws which contain both the variables and the constant values (parameters) and which can be applied to as wide range of events as possible (according to a trivial text-book example, from celestial bodies up to Newton's apple).

On first glance, the developmental processes, being highly ordered and repetitive, fit perfectly Maxwell's definition. Why shouldn't one try to formulate the embracing laws of cell divisions, movements and cell differentiation by regarding them, according to physicalistic methodology, as some invariable functions of space and time? However, with few remarkable exceptions, such an approach remained alien for classical developmental biology. The most obvious reason was that the developmental processes are too various and complicated to be described by some simple mathematical laws. Not only is a set of the required variables is in most cases extremely large but, quite unusually from the view of a routine physics, this very set is drastically changed while passing from one developmental period to another (and also from one biological species to another). But there are also some deeper reasons, to be discussed below.

In any case, in spite of numerous declarations that modern biology strictly follows the line of physical-chemical sciences, it is actually enslaved by an alternative approach which is borrowed mainly from the realm of the complicated behavioral and/or social events and may be called "instructivistic". We will illustrate it by the following allegory.

Let us watch a trajectory of a postman bringing letters to this or that destinations, or the motions of the soldiers making their exercises, or, for becoming a bit more spiritual, the movements of a performer's fingers while playing a piece. Certainly, all of these processes obey some common physical laws (first of all, mechanical). But who can be brave enough to claim that these modes of behavior (the trajectories of the bodies, or of the fingers in space and time) can be ultimately *derived* from these laws? Obviously, this is not the case, and the explanatory power of physics is here very much restricted. Instead of physical laws, the observed kinds of behavior can only be derived from some kinds of *instructions*, let them be given in the form of postal addresses, military commands or musical scores. In many respects the instructions are something quite opposite to the physical laws. While the latter should be, as mentioned before, as wide as possible (applicable to as large as possible number of events), the instructions should be always specific, properly addressed and separated from each other. *Ex definitio,* one instruction cannot oblige a subject to do two different things. To "explore" instructions means nothing more than to enumerate, or to codify them one after another. This is an approach of a lawyer, not of a physicist.

However, an instructivistic approach is not something completely alien to physics. The print is that even if a fundamental physical law is known, the behavior of the natural bodies remains in some cases and to some extent unpredictable. Yes, we may calculate a position of a celestial body at any time moment, but what about the time of Newton's apple's falling to the ground? Such a situation of a practical non-predictability is avoided in physics by introducing a notion of the initial and/or boundary conditions which are allowed to be established in quite an arbitrary way. Hence, they are in some sense similar to what we have defined as instructions. However, a "good" physical description of the process demands the number of arbitrarily and independently introduced initial and/or boundary conditions to be as small as possible.

Now we are ready to formulate our main question: is it possible to give to the developmental events a "good enough" physical description, not requiring too large a number of arbitrarily introduced initial/border conditions (=instructions), or can this not be done? Firstly we will discuss this question from the point of view of a classical physics. This will bring us to a somehow paradoxical conclusion: we will see, that a classical physics gives some serious arguments just in favor of instructivistic approaches. But beforehand we have to make some brief excursion into the theory of symmetry, which is very useful for understanding developmental events.

One of the basic concepts of this theory is that of a *symmetry order* of a given physical body (or a process). This is a number of transformations which combine a body with itself. The transformations treated in the elementary symmetry theory are the rotations around a certain axis, the reflections from a certain axis (or a plane) and the linear (translational) shifts. For example, an equilinear triangle can be combined with itself after rotating to 120^0, 240^0 and 360^0 around its center. Correspondingly, the order of its rotational symmetry is 3, and it has also 3 axes of reflection. For a quadrangle the orders of a rotational and reflection symmetry are 4. A mostly symmetrical body in the whole Universe is a sphere, which has an indefinite number of the rotational axes and reflection planes. Hence, its symmetry order is indefinitely large. Now, if we deform a sphere by flattening its small part, or by putting a colored mark onto any of its surface points, we'll reduce the number of the symmetry axes up to one (which will pass through the flattened part, or through a marked point) while the number of the reflection planes will be reduced up to a bundle passing via this axis. By making the same procedure(s) for the second time, we reduce the order of a rotational symmetry up to 1 (only 360^0 rotation is now permitted) but still preserve the reflection plane. By repeating it for the third time we'll devolve a "former sphere" from a reflection symmetry as well. Such a procedure is called dissymmetrization, or a symmetry break. An example of a translational dissymmetrization is a transformation of a uniformly colored tape into that bearing a numerously repeated design.

Dissymmetrization lies in the very heart of a biological development. Each of its important steps is associated with the reduction of a symmetry order. Very early in development, a more or less homogeneously structured (sphere-like) egg cell acquires a single axis of the rotational symmetry which is called the animal-vegetal axis. Somewhat later in most cases an egg cell or an embryo loses completely the rotational symmetry but preserves (if neglecting the left-hand asymmetry, which is quite a specific property not to be discussed here) a single definite reflection plane (so-called saggital). Further development of an embryo can be regarded as a chain of successive reductions of a translational symmetry. For example, in many animals two symmetrical rows of mesoderm are split into a definite number of segments (somites) and in Vertebrate embryos an anterior part of a neural tube is subdivided into several brain vesicles.

Symmetric transformations are taking place in non-living nature as well and belong hence to the realm of physics. As early as in the year 1894 a French physicist Pierre Curie formulated a law describing these processes from the classical point of view [3]. One of his remarks looks as written especially for the embryologists (although he hardly knew this science): "C'est la dissymmetrie qui crée l'event" (This is a dissymmetry which creates an event). It is a very deep idea, because there are just the successive symmetry breaks which transform an egg cell into an adult organism. However, the most influential but at the same time controversial idea of the Curie law is the forbidding of a "spontaneous" dissymmetrization of any physical body: by his idea any step of a dissymmetrization requires a certain "cause" (either coming from outside, or residing in a hidden form within a given body) bearing a required element of a dissymmetry. Intuitively this is obvious: returning again to the dissymmetrization of a sphere, those are our hands which flatten it or put a point onto it. And if we see that the flattening or a color painting takes place without any obvious intervention from outside, our first idea would be that the body had already these structural heterogeneities in a hidden (invisible) form (like an already exposed but as yet undeveloped film).

The Curie principle is not a solitary statement. Rather, it is deeply rooted in the classical concept of physical causation, known as Laplacian determinism. It was a French mathematician Pierre Simon Laplace (a contemporary of Immanuel Kant, sharing with him the honor to create a first "natural history of a sky") who claimed that the main goal of science is to split Nature into as detailed as possible one-to-one cause-effect relations, such that one cause should bring into being no more than one event. If (when) this will be done, all the events of a whatever remote past could be, by Laplace's idea, completely reconstructed, and the future ones predicted. The Curie principle can be regarded as a concretization of the Laplacian approach for the case of symmetry breaks: for apprehending such processes, we must discover and enumerate, one after another, all the "dissymmetrizers". As one can see, this is not so far from an "instructivistic" approach.

Now, what can we learn from embryological experiments about the causes of the symmetry breaks during development? Can we really point to discrete "dissymmetrizers", providing each next symmetry break? On the first glance, it looks so. For example, the direction of a polar axis of an egg cell is determined in some cases by an external agent, such as a gradient of oxygenation or (in brown algae) by the direction of external illumination. Similarly, a setting down of a saggital plane in some species' eggs coincides with the entrance point of a spermatozoon. All of these factors are typical external dissymmetrizators. Further exploration showed however, that in no way is this necessary. Eggs of brown algae can acquire polarity under as uniform illumination as possible [4], and frog eggs can establish the plane of bilateral symmetry either without spermatozoon or with that penetrated precisely into the animal pole [5], so that it cannot act as a dissymmetrizer. Those developmental events that take place at the later stages, depend even less on external factors.

May it be then that it is any kind of an internal (maybe invisible) heterogeneity of an egg or an early embryo which provides its visible dissymmetrization? This question brings us to one of the most prolonged and important discussions in developmental biology, known as a counterstanding of a so-called preformism and epigenesis. The adepts of the preformism believed in the existence of such an invisible heterogeneity, while the supporter of epigenesis denied it. The discussion lasted several centuries until a German embryologist Hans Driesch at the end of the XIX century did not get its final experimental solution. By a remarkable coincidence, he made his first experiments roughly at the same time when Curie formulated his principle, although the two scientists never knew each other and could not suppose that their investigations are somehow interrelated (there is also a third temporal coincidence with Driesch and Curie works which we'll mention below).

The main idea of Driesch's experiments was very simple: suggest that a given stage embryo is internally heterogeneous in such a way that each of the developing adult structures has already at the given stage its individual unique material precursor (say, a cell, or a group of cells, or even a part of a cell). Then, and only then the deterministic causation (including Curie's principle), as well as the "instructivistic" approach can be saved (because any causes should be applied to, or any instructions should be received by a definite material element). The opposite situation, when any developing structural element does not have such a precursor but can be developed, instead, from quite various parts of an embryo, means that this latter does not possess at a given stage of development any "informative" heterogeneity. In order to determine which one of these statements is true, it is enough to dissect eggs/embryos into parts and to see what will be developed out of them. If each material element of a dissected whole can produce only that adult structure which it gives normally (and, correspondingly, a part of an embryo produces just a definite part of an adult) the preformistic idea of the "informative heterogeneity" may be true. If however under certain conditions some parts of an embryo produce other structures than normally (or not *only* those structures), such an idea should be definitely rejected.

In the year 1892 Driesch succeeded to separate from each other two first blastomeres of a sea-urchin egg. If the preformists were correct, each one of the blastomeres would produce just a half of a sea-urchin larva. The result was meanwhile quite the opposite: each one of the blastomeres produced an *entire*, well-structured whole larva, although of a diminished size. This phenomenon was called "embryonic regulation". A careful analysis made by Driesch' successors showed that during embryonic regulations virtually all of the cells-descendants of each one blastomere have changed their developmental fates. Similar experiments have been repeated by dozens of investigators on quite different species and stages of development. In spite of an unavoidable diversity of concrete results, the following general conclusions valid for all the cases could be made:

- Within each species' developmental history there is a certain initial period (sometimes more and sometimes less extended) when a part of an embryo can produce a whole organism;
- Even at those latest stages when the capacity to produce a whole organism out of its part is ceases, those embryonic parts which produce different organs (eye, limb etc) still preserve regulatory capacities within themselves: a complete eye, or a limb can develop also from a small part of an entire region, or even from a random mixture of cells taken from this part. So, never does an embryo become a complete "mosaic".

In any case, what follows from these experiments is that the developing embryos do not obey any kind of a deterministic causation, including Curie's principle: a developmental fate of each one material structure is not determined from the very beginning and the symmetry breaks can proceed without discrete dissymmetrizers.

Although Driesch was not familiar with the details of a contemporary physics, he felt intuitively this fundamental contradiction. He claimed therefore that the developing embryos obey some specifically biological, or vitalistic laws irreducible to those of physics and chemistry. If taking into mind the physics and the chemistry of his days (a "physics of yesterday" [6]), Driesch was absolutely correct, although most of his contemporaries did not forgive him for this. On the other hand, he could not know that already in this very time physics started to evolve in such a way that it began, after several decades, to include his results into a general physical vision of nature. Even less could he predict that contrary to his own vitalistic prejudices he himself put the first stone in the basis of this new approach bringing physics and biology much closer to each other than ever before. What we mean here is his famous law interpreting a capacity of an embryonic part to produce a whole.

We have already mentioned that during embryonic regulations all the cells are changing their developmental fates (that is, all their future way of development). Can one find an overall law which could describe from the common positions all these fate changes as well as the normal developmental fate of a given part of an embryo? By searching for such a law, Driesch noticed a simple thing: if we, in our imagination, impose a coordinate grid (linked, for example, with the animal-vegetal egg axis) both onto a normal intact early embryo and onto that one developing from its part (during embryonic regulation), the cells having the same coordinates will have also the same fates (irrespective of their origination). This led to a laconic formulation known as Driesch' law:

The fate of a part of an embryo is the function of its position within a whole.

One cannot find within embryology another law which would embrace such a huge amount of empirical data as this one! And nevertheless its own fate could be regarded as a happy one. Although nobody could claim it to be incorrect, many people did not know what to do with it, and some interpreted it in a wrong way. Driesch himself did not develop further his law and did not notice (or pay attention to) a similarity of his statement with the idea of a *field* already known in physics. This was done somewhat later by the Russian biologist Alexander Gurwitsch [7].

Here we come to another, and probably the most important point. The question is, whether the fate-determining positions of any embryonic element can be referred to some other distinct ("privileged") material elements of the same embryo, or this is not the case, so that the fate-determining positions can only be referred to some topological or geometrical structures of an embryo irrespectively of what particular cells they are constructed. Most embryologists accepted the first alternative because it was more compatible with the deterministic ideas. It created a basis of a very popular concept claiming that development is nothing more than a chain of cells' reactions to so-called embryonic inductors, the active "centers" which drive the development of the surrounding tissues in this or that direction in the positional-dependent way. Many people felt themselves comfortable with the inductors because they could be regarded as sources of specific instructions and revive thus the instructivistic approach.

The inductive centers do really exist. For example, in the development of amphibian embryos one can trace, successively, a so-called Nieuwkoop center affecting the development of the mesodermal layer, then Spemann's center, providing the formation of a central nervous system, and so on [8]. But can these or other inductive centers be regarded as some containers of unambiguous instructions, dissymmetrizers and/or the reference points for the fate-determining positions? None of these suggestions is really confirmed. What we know today after 80 years of studying the inductive processes can be summarized as follows:

- There is nothing like "one inductor - one organ" rule. On one hand, a same organ can be induced by quite different agents, some of them being definitely non-specific. For example, an amphibian limb, during its normal development hardly has any discrete inductor at all; but it can be easily induced by a rudiment of an internal ear (otic vesicle), nasal placode and even a piece of a cellophane transplanted to a flank region of an embryo although none of the above mentioned rudiments (not to say about a piece of cellophane) have anything in common with the limb and cannot induce it in their normal locations.

- At no size scale are there any indications of one-to-one correspondence between the spatial structure of an inductor and of an induced organ: just a small piece of an inductor, or an inductor dissociated into single cells and then re-associated, etc produces practically the same result as an intact whole inductor.

- A symmetry order of an induced organ or a set of organs is as a rule lower than that of the inductor. Therefore, a causal link between an inductor and an induced organ should imply a symmetry break somewhere in between.

- There is a lot of developmental processes going on without any discrete inductors.

- By the modern views [9], some important inductive agents play an inhibitory, rather than a promoting role, so that the capacity to differentiate in different directions should be better regarded as an intrinsic property of embryonic tissues.

- After microsurgical operations followed by embryonic regulations, any one of the material structures of an embryo (including the inductive centers) will be shifted to the positions geometrically non-homologous to those occupied by these very points in normal embryos (Fig. 2). And even more: these new positions will be different if the regulations started from ½, ¼ or 1/8 of an embryo. Consequently, if referring the fate-determining positions in a .point-to-point fashion, the fates of the geometrically homologous parts will be each time different, which contradicts the facts of embryonic regulations.

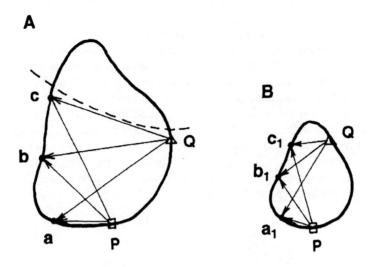

Figure 2: Embryonic regulations are incompatible with the assumption of any pre-localized "privileged" material elements (say, P and Q, frame A), regarded as the sources of a "positional information" (PI). After the dissection of a part situated above the dotted line shown in A and closure of the wound (B) the positions of the elements P and Q (as well as all the others) will become geometrically non-homologous to the same elements' positions in A. As a result, any points of an embryo which occupy in A and B homologous positions (say, a and a_1, b and b_1, c and c_1) will perceive quite different PI signals, which is incompatible with embryonic regulations.

All of the above said means, that neither inductive centers, nor any other "privileged" material elements of an embryo can serve as reference points for the fate-determining positions. And where, then, should we look for the reference points of a real fate-determining coordinate system? The only possible answer will be that: *the fate-determining positions of embryonic parts should be referred to some symmetrical, and/or topological, and/or geometrical features of a whole embryo which are preserved during embryonic regulations (axes or planes of symmetry, curvatures, some singularities etc) irrespectively of what material structures are located in these very areas.* We shall call these features topo-geometrical.

Such a topo-geometrical character of the positional dependence was unusual for Driesch's time physics: even the contemporary concepts of physical fields were closely related to distinct sources (for example, electrical charges). Rieman's geometrical approach did not yet affect considerably the physical picture of the world: this was done later. So, the perspectives to bind physics and developmental biology looked in Driesch's times and even later still gloomy.

Meanwhile, a number of arguments favoring the holistic approaches to developmental processes increased progressively. Not just the experiments by dissecting embryos, but even the careful observations of their normal development demonstrated that the trajectories and the final positions of its small constituent parts (cells or groups of cells) may be quite variable: the whole turned out to be much more precise than its parts. For example, by labelling a definite cell of an amphibian embryo at a 32-cell stage, one can see a wide variability in the arrangement and, hence, developmental fates of its descendants [10] (Fig. 3A). Similarly, the formation of precisely symmetrical leaves is compatible with wide variations of the growth rates of their constituent parts (Fig. 3B).

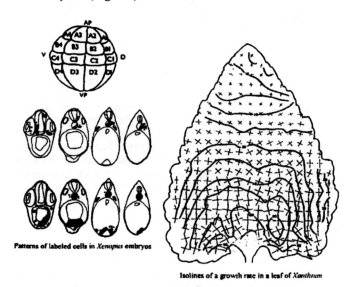

Patterns of labeled cells in *Xenopus* embryos

Isolines of a growth rate in a leaf of *Xanthium*

Figure 3: A whole is more precise than its parts. To the left: the descendants of a C1 blastomere (upper frame) occupy quite different and disconnected positions in the advanced embryos (two lower rows). To the right: growth isolines on a leaf surface (curves joining the areas with the same growth rate) are largely non-symmetrical while an entire leaf shape is perfectly symmetrical.

That brings us towards a peculiar analogy with an uncertainty principle in modern physics: in a wide enough range, the more precisely do we localise an embryonic element, the less precisely can its developmental fate be predicted. Namely, if we take a whole egg, we are 100% sure that it will produce a copy of its parents. If we label a large enough embryonic region, containing about several hundred cells, its normal fate is also in most cases (although still not in 100%) more or less the same. Meanwhile, the smaller is the labelled element, the less precise will be our predictions.

So, developmental biology shows remarkable parallels with physics of our time, but not with that of Driesch's time. The only precise non-biological shapes known at that time (those of macrocrystals) were based upon a precise molecular order, and when the molecules were allowed to move in random (as in gases and liquids) no regular shapes were registered at that time. More than that: according to classical thermodynamics a spontaneous evolution of a closed system can lead to nothing more than its chaotization and homogenisation (a "thermal death"). But the more we revealed the properties of the developing organisms, the less they fit the postulates of a classical physics.

3. A SELF-ORGANIZATION THEORY HELPS TO CREATE A BRIDGE BETWEEN PHYSICS AND DEVELOPMENTAL BIOLOGY

Already in Driesch's time, and even earlier, a movement towards new physics started. We have already mentioned Riemann's approach directed towards geometrization of physics. Much later it became a basis of Einstein's theory of general relativity. In those very years when Driesch made his experiments on sea urchin eggs, a Russian mathematician Alexander Lyapunov presented his dissertation named "On the stability of movement". Again as in the case of the Curie principle, nobody would suggest that there can be something in common between the works of a German biologist and a Russian mathematician. But long ago it became clear that Lyapunov's work, together with outstanding studies of Henry Poincaré paved a way for a new branch of science, lying in between physics and pure mathematics and being later transformed to what is called now the self-organisation theory (SOT).

It is impossible, within the framework of this paper, to describe SOT in a comprehensive way. There are many books of different levels on this subject (e.g., [11-13]) and, by my view, SOT should be carefully learned by any person interested to understand properly the fundamental laws of an entire Nature, including, in the first instance, the living systems. Within this paper's framework it is possible to formulate some of its conclusions only in mostly general outlines.

Firstly, SOT claims overtly that there is a large class of the natural system (including non-living ones) which are able to move "spontaneously" (or, better to say, under negligibly small perturbations) towards a more complicated state (that of a decreased symmetry order), rather than to a homogenisation and chaotization. For acquiring such properties, a system should be considerably removed from a thermodynamic equilibrium (to be opened). So far as living systems fulfil this requirement, their morphogenetic capacities (including "spontaneous" symmetry breaks) should look now as quite natural.

Correspondingly, SOT uses a causal structure, much different from that postulated by classical determinism. The unidirectional and discrete relations between the "causes" and the "effects" of a same value order are regarded, from the SOT position, just as particular situations. Instead, the self-organising systems are interlinked by a network of multi-level non-linear feedback, the non-linearity meaning here that the "signals" and the "responses" may differ from each other in several orders of magnitude: weak signals may produce extensive responses, and vice versa. In the first case the systems are called unstable while in the opposite case stable. Systems, described by SOT, obey dualistic causation: they are regulated both dynamically and parametrically. This means the following.

Dynamic regulation has some similarities with classical determinism. This kind of regulations is mediated by the changes of so called dynamic variables, which are just the manifestations of a system's activity. They are definitely located both in space and in time. The parameters, on the other hand, are as a rule smoothed-out throughout an entire system (both in space and in time), have much greater linear dimensions and smaller rates. The parametric regulation is achieved by changing the parameters' values. It provides as a rule much more profound (often threshold-like) changes in a system's behaviour than the dynamic regulation can do.

Now we would like to show, that even if being employed in a mostly abstract way, SOT may clarify some fundamental developmental problems.

The first of them is: why does any developing system possess quite a restricted number (instead of a smooth continuum) of discrete developmental pathways (DP), among which it selects, within certain space-temporal points, a single one? Any attempts to resolve this question within the framework of an "instructivist" paradigm can only be done by using some non-verifiable *ad hoc* assumptions. Meanwhile, SOT offers quite another, much more general and heuristically valuable interpretation. Let us compare the characteristic time T_d (a reverse rate) of a development *per se* with those of the dynamic variables just producing the motive forces of development: transformations of a cytoskeleton, cell membranes and cell-cell contacts etc) Let us denote the latter as T_f. It is known that T_d ranges from dozens of minutes to hours, while T_f from seconds to dozens of seconds. Hence, in SOT terms we may regard $1/T_d$ as the rate of the parameters (let us denote them as K) while $1/T_f$ as the rate of the dynamic variables (denoted as X). Since T_f is negligibly small as compared with T_d we may write for any DP

$$dX/dt \approx 0 \qquad (1)$$

$$dK/dt \neq 0 \qquad (2)$$

In addition, for making DP stable to finite disturbances, we should add

$$d^2X/dt^2 < 0 \qquad (3)$$

This gives us a full explanation of why a set of DPs is so restricted and why they are so discrete: this is because *the DPs correspond to the stable roots (solutions) of a dynamic equation describing a feedback network within a given system.* Consequently, the DPs are determined by a holistic structure of a network and only slightly (in a very degenerative way) depend upon the dynamic variables's values: preciseness of a system's behaviour in no way implies a precise setting of all its elements! Since the lowest order of a differential equation permitting production of two alternative stable solutions (and one unstable in between) is that of a third order, its mostly general form is:

$$dX/dt = F_1(X^3) + F_2(X^2) + F_3(X) + A \qquad (4)$$

An entire development can be regarded as a succession of selections (bifurcations) between such pairs of stable solutions. SOT offers two principles of such a selection.

First, at the moment of selection both stable roots may occur. Under these conditions the selection will be made in a more or less stochastic way. If both roots have equal attraction basins on the dynamic variable axis (Fig. 4A) each of them may be selected with an equal probability, so that the final outcome may be random. Sometimes this takes place during development. If, however, their attraction basins are unequal (Fig. 4 B), that root which has a largest one has more chances to be selected. Such a situation is a most usual one, and in many important cases the inequality of the attractors' basins is really enormous, so that one of the solutions is distorted even by small fluctuations (it is non-robust) while another is non-affected at all by much greater ones (it is robust). A tendency towards selecting the robust solutions may be regarded as a powerful driving force of the individual development and of evolution.

A second mode of selection is in eliminating, at this moment, one of the stable roots, simply by changing the value of a free member A in eq (4) (Fig. 4 C, D).

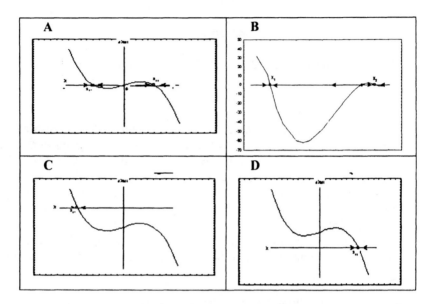

Figure 4: Discrete developmental pathways may be regarded as the stable roots of 3rd order differential equations. Horizontal axes: values of a dynamic variable X. Vertical axes: dX/dt. A: graph of an eq. $dX/dt = kX - k_1X^3$. Both stable roots have equal attraction basins. B: graph of an eq. $dX/dt = -x^3 - 9x^2 - 8x$. The left stable root has much greater attraction basin (is much more robust) than the right one. B, C: graphs of an eq $dX/dt = kX - k_1X^3 \pm A$. Due to the vertical shifts of the equation graph each time only one stable root is preserved.

This is very effective, "non-invasive" and informationally "cheap" way of regulation because it provides a uniform, fully deterministic solution without violating an entire feedback network of a given system, which is described by non-free members. It is also largely non-specific: any factor which considerably modulates the rates of the dynamic variables, may play such a role. Therefore, it is reasonable to suggest that just this way is used for the mostly important developmental decisions, including those mediated by inductors. By taking such a view, many above mentioned contradictions and paradoxes associated with the action of inductors can be ruled out.

Another fundamental question is how some spatially heterogeneous structures can be initiated, or affected by much less heterogeneous factors. Not only the inductors, but also the genes fit this category. Why, indeed, does a mutation of a certain gene which is equally represented in the nuclei of all the somatic cells change the number of the floral petals from 4 to 5 [14], or modify the number of the segments in *Drosophila* appendages, and so on? How can something equally smoothed throughout an entire organism (and thus being unable to play a role of a dissimetrizator) produce what is called the patterns? For about the entire 20th century this problem was a kind of nightmare for many biologists. Meanwhile, from the SOT point of view, the principal answer is very simple: this is just the parametric regulation which may produce perfect patterns with the use of completely homogeneous or rather smoothed factors. So, if we regard the genes (or, better, the genome as a whole) as one of the developmental parameters, these effects become in principle explicable, although concrete models on how this might go still should be created.

Does it mean, as is often claimed by modern biologists, that it is enough to clone all the genes, one after another, and we'll resolve all the problems of development and become its full masters? In no way. The print is, that *the parameters themselves have no definite meaning when taken as themselves, out of the context of an equation (or a model, or a feedback circuit) in which they participate.* Their meaning is absolutely context-dependent. Suggest for example that we know the exact value of a gravitational constant without knowing how the gravity equation looks like – what will we do with it? Certainly, for physics such a situation is completely absurd, because the parameters are, so to say, born out of an equation and cannot live independently. But in the present day biology the situation is quite another: to clone single genes is a fashionable and a well supported task, while very few persons are interested in revealing a structure of the dynamic systems into which they may be involved. Meanwhile by regarding the genes as something belonging to the parameters, some kind of knowledge about the feedback circuits within a system becomes a matter of primary importance.

In any case, SOT fully rehabilitates a holistic approach. Most of the self-organising effects, including so-called auto-waves, are essentially macroscopic: "Auto-waves exemplify a new type of dynamical processes generating a macroscopic linear scale due to the local interactions, each of the latter possessing no linear scale at all" [15]. In other words, the self-organisational macroscopic effects do not have as a rule any microscopic residues, they are essentially emergent and cannot be understood by simply looking upon a system under increased magnifications.

4. RETURNING TO DEVELOPMENTAL BIOLOGY: A SELF-ORGANIZING MECHANICS INSTEAD OF A SET OF ISOLATED INSTRUCTIONS

In spite of all the above mentioned SOT advantages, it is not so easy to pave a way for its valuable application to the developmental problems. What we need here are not so much any isolated models but a possibility to find, with the help of an SOT, those embracing laws for prolonged developmental successions which could replace a chain of isolated "instructions". Some few biologists followed this way long before the emergence of an SOT. The first of them was Gurwitsch who devoted his whole life to discovering such holistic laws of development (rather than any microscopic instructions independent from a whole) which could derive any next stage embryonic configuration from that of a preceding stage ("shape out of a shape"). By his own words, "...out of an adequate description of any momentary state of a living system its transition to the next one should inevitably follow" [8].

Now we can see that this point of view anticipated an SOT approach insofar as it implied the existence of feedback relations between the successive shapes. Making just a step further, we can see easily that, for introducing a dynamics instead of a pure geometry, the "shapes" themselves should be exchanged by the patterns of mechanical forces, moulding these very shapes: "A shape should be regarded as a diagram of forces" [16]. To be more precise, the notion of forces should be exchanged by that of mechanical stresses, that is, by balanced forces related to the squares to which they are applied. Later on we will show that the view of an embryo as a stressed solid body perfectly correlates with a notion of coherent states which may be regarded as a modern physicalistic version of holistic approaches.

The notion of stresses adequate for developmental purposes increase the more that the stresses are extensively geometric-dependent. Indeed, the patterns of tensile stresses that emerge on the surface of an internally pressurised body largely depend upon its geometry. For example, if a cylindrical body comes under internal pressure, the circular tensions on its surface will be two times greater than the longitudinal ones. In several cases a uniform pressure will produce a definite number of discrete folds on the body surface [17]. This is a real self-organising event diminishing the symmetry order of a body without using any external dissymmetrizator. A presence and the regular changes of mechanical stresses in embryonic tissue has been demonstrated elsewhere [18,19]. Relaxation of stresses distorts developmental order and the proper relations between the main embryonic rudiments [20].

Two kinds of mechanical stresses should be distinguished, the *passive* stresses, imposed to a given embryonic part from outside (as a rule, from the neighbouring part) at a certain time of the past, and the *active* ones generated in this very part during that time period which we consider as present. Now the above mentioned Gurwitsch's claim upon deriving each next step of development from the preceded one can be re-formulated as a suggestion *to discover the rules governing the feedback between the passive and active mechanical stresses*: if we know these rules we can really derive each new embryonic configuration from a preceding one.

Within two last decades or so several models of such a kind have been suggested. In [21] it was postulated that the passive stretching of a part of a cell layer induces its active tangential contraction. This one stretches the next part of a layer, and so on. By the authors' idea, such a "+, +" stretch-contraction feedback can interpret several important morphogenetic processes.

In [22,23] more complicated "+, - " feedback relations were assumed. The first of these works exploited the authors' discovery that the isolated cells seeded onto a substrate which they can themselves stretch become organized into regular clusters connected by elongated cell files. The postulated feedback circuit implied: (1) a stretching of a substrate by the cells aggregated due to short-range adhesive forces; (2) a long-range inhibition of further aggregation due to decrease of cell density onto a stretched substrate. Therefore, the positive feedback link turned out to be short-ranged while the inhibitory one long-ranged.

In [23] similar ideas have been applied to the formation of polarized cell domains out of initially homogeneous epithelial layer. As a short-range positive link a so-called contact cell polarization, namely, a transmission of a polarized state from one adjacent cell to another has been taken. If the layers' edges are fixed, such a polarization inevitably increases tangential tensions, which inhibit further polarization. Not only does this model predict the segregation of an epithelial layer into the domains of polarized and flattened cells but it shows that the relations of both domains' lengths are scale invariant. Thus, the model reproduces the phenomenon of Driesch's embryonic regulations, which is not an easy task.

Although each one of the described models can imitate some specific morphogenetic situations, it would be desirable to suggest some kind of an overall law embracing all of them together, even if in a non-quantitative manner. This was the aim of a so-called hyper-restoration (HR) hypothesis suggested in [18,19]. By its idea, *a cell or a piece of embryonic tissue, after being shifted by any external force* (either imposed from outside artificially, or coming from another part of the same embryo) *from its initial stress value* (is relaxed, stretched or compressed) *develops an active mechanical response which is directed towards restoring this stress value but as a rule overshoots it in the opposite direction.*

Correspondingly, if a cell or a tissue piece is relaxed, it tends to increase its tension at least up to the initial value. Similarly, after being stretched, it tends to (hyper) relax itself, up to producing the internal pressure force. A concrete way for producing any of these responses depends largely upon the sample's nature and the initial (border) conditions. So, if a stretched sample is able to overcome the stretching force (its edges are not firmly fixed) it may contract itself up to the initial length, and even further. This is just what corresponds to "+,+" stretch-contraction feedback. Meanwhile, if the edges are firmly fixed the relaxation may be achieved only owing to the mutual cell-cell insertion in the direction perpendicular to that of a stretching force. Such an insertion is a well-known cell intercalation [24]. It may be oriented either in the plane of a cell sheet, or perpendicular to it. In any case, such an insertion may go with an overshoot and create a real pressure force which provides either isotropic spreading of a cell sheet, or the active elongation of a rudiment in the direction of the initial stretching force.

On the other hand, the only way for a relaxed tissue piece with its edges firmly fixed to (hyper) restore the initial tensions is to contract tangentially at least a part of its surface. This is inevitably associated with a columnarization of at least some of its cells. If a piece has free edges, some other ways for doing the same are opened (for example, the rolling around free edges).

The above described reactions bring into being several *standard morphomechanical situations* (SMS) which provide the advancement of a development. Let us describe two of them.

SMS-1: active contraction-extension feedback. We start from a cell layer of a uniform (or zero) curvature being exposed to a more or less uniform stretching force. In a particular (but wide spread) case this may be a spherical sheet enclosing a cavity with an increasing turgor pressure inside. According to the HR idea, the layer's stretching will trigger cell intercalation which will now extend a layer in an *active* way (and make it thinner). Meanwhile, a completely uniform active extension is non-robust: under slight deviations from a homogeneity, or (what is mostly universal) under any resistance to further extension at least one or even several zones of a layer will be at least relaxed, or even compressed. Assume that there is one such zone. For (hyper) restoring the preceded stress value, it must actively contract itself in the tangential direction. Such a contraction will lead to further stretching of the adjacent parts and hence to further intercalation-like processes in these latter, and so on. Different versions of such a feedback are illustrated by Fig. 5 A, B_1-B_3, D_1, D_2. They dominate at the initial stages of gastrulation, neurulation and other important developmental processes.

SMS-2: increase of a curvature. Now we start from a slightly tensed cell layer which is bent (curved) by an external force. As a result, the concave part of the layer will be compressed (or at least relaxed), while the convex one stretched (Fig. 5 C_1). In which way will the HR tendencies now be expressed? If the layer cannot be straightened back in a spring-like manner, the following should be expected: the concave side relaxed/compressed cells, in order to (hyper)restore the initial tension should either contract tangentially their free surfaces, or/and migrate towards the convex side (centrifugally). Correspondingly, the convex side cells should enlarge their outer surfaces, either by the insertion of new portions of plasma membrane, or by continuing the centrifugal migrations. In any case, both processes will enhance each other and be directed towards the *active* increase of the *passively* initiated curvature (Fig. 5 C_2).

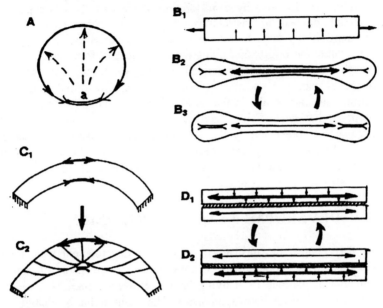

Figure 5: Different examples of an active contraction-extension feedback. Double arrows indicate the active while single arrows the passive contraction or extension. A: an internally pressurized sphere segregated (due to demand of robustness) into the actively contracted zone *a* and the surrounding extending zone. Dotted lines show a possible emigration of cells from a contracted zone. $B_1 - B_3$: interplay of the active-passive contraction-extension in a cell layer in which cell intercalation has been triggered by an external stretching force (B_1). $D_1 - D_2$: similar situation in two mechanically bound cell layers: the active extension of an upper layer induces the passive extension of a lower one which is transformed into the active extension of a lower layer, etc. $C_1 - C_2$: transformation of a passive bending into the active one.

This will automatically produce the areas of an opposite curvature on the flanks of a given curvature maximum. These latter will also pass from the passive to the active state, producing the next curvature maxima of the opposite sign on their periphery, and so on. In such a way, any mechanically active epithelial layer which is under not too great tension tends to be folded. The number, shape and location of the folds depends certainly upon a number of the geometrical and mechanical properties of a layer (and, correspondingly, of a lot of epigenetic and genetic factors affecting these properties).

Already now we have strong enough evidence for suggesting that within prolonged enough and crucial periods of development each next step can be derived from a preceding one by using the same feedback construction, namely, the idea of a hyper-restoration of stresses. This is a physicalistic approach, opposite to "instructivistic" one. By using the first approach we do not deny in any way the role of inductors or genes expression but like to concentrate our efforts on finding the physical laws from which an observed morphology can be directly derived and into which framework the inductors or genes can play a role of parameters and/or initial conditions. By formulating our results in the terms of the morphogenetic fields, one may propose that such fields may be, at least in the first approximation, adequately represented by the patterns of the mechanical stresses and the active hyper-restoring cell reactions to the latter. In such a view, the morphogenetic fields are self-organized space-temporal entities.

5. DEVELOPMENTAL BIOPHOTONICS

At first glance, the mechanical aspects of development have nothing in common with biophotonics, dealing with the emission of some small amounts of photons from living bodies. However, there is a fundamental similarity between both approaches. The main point is, that it is just a mechanically stressed state of a solid body which provides conditions for creating collective excited states of a matter (exitons) which may be extended over macroscopic distances. Within exitons thermal energy can be concentrated up to high potentials and be deposited onto some few degrees of freedom which are highly improbable under equilibrium conditions. Or, in the other words, in the mechanically stressed bodies a creation of large coherent domains is highly probable. Therefore, the above mentioned correlation between the existence of mechanical stresses and a developmental order obtains a profound physical interpretation in the terms of coherent states. And, as a reader can see from many papers of this volume, one of the main aims of biophotonic studies is just to explore the coherent states of living matter.

Actually, developmental biophotonics is still itself in its embryonic state. However, it may be a time to summarize, maybe tentatively, some of its results, which are as follows:

Developmental Biology and Physics of Today

1. As a development proceeds, an average intensity of a spontaneous biophoton emission decreases while the spectral (including Fourier) bands of a biophoton emission become wider.

This was shown firstly on the skeletal muscles of newborn rabbits [25; see also paper by Voeikov, this volume). At the same time, the intensity of so-called degradational radiation (initiated by slight reversible damaging influences and suggested to destroy just the collective excited states) increased. Consequently, such states are progressively generated and enlarged within development. It is highly probably, that they have properties of a "photon sucking" (see below and also in other papers from this volume).

A progressive decrease of a spontaneous biophoton emission have been shown also in *Drosophila* eggs [26], hen's embryos (the intensity went down to a background level in about 3-4 days embryos) and fish (*Misgurnus fossilis*) eggs and embryos [27]. In the latter case we have used so-called zero-statistics (measuring the probability of zero counts under 0.1 s dwell time). As seen in Fig. 6, this number increased gradually during a cleavage period, approaching a background level. Its dynamics perfectly correlates with the increase of DNA/cytoplasm ratio, confirming that DNA can play a role of a photonic storage [28].

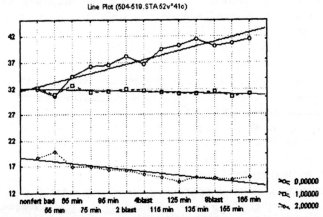

Figure 6: Dynamics of biophoton emission (in terms of "0,1"-statistics) for the initial period of the development of *M. fossilis* eggs (from non-fertilized eggs up to 165 min after fertilization). Vertical axis: percents of 0, 1 and 2 counts per 0.1 s. Note a gradual increase of zero counts.

From a mostly general point of view, a high intensity of the biophoton emission in early development may be associated with an outflow of entropy from an embryo, which is required for increasing an ordered complexity of a developing organism. By another formulation, a decrease of a biophoton emission as development proceeds may be associated with the formation of exitons with "photon sucking" properties.

Another interesting point is the change of the width of the spectral bands of biophoton emission during development. This was also firstly shown in [25] while studying the spectra of newborn rabbits's muscles in UV range: narrow emission bands typical for early newborns fused together into broader bands as development proceeds. We could observe the same tendency by studying Fourier spectra of *M. fossilis* eggs and embryos (Fig. 7). In most general terms, this may point to an increased interaction of the coupled oscillators or, in other words, to a progressive enlargement of the collective excitations areas.

2. Developing eggs and their outer membranes (shells) create together a common non-additive biophotonic whole which store external light and increase its coherency [27, 29].

A hen's egg, irrespectively whether it contains a developing embryo, is non- fertilized or even boiled, emits, after being taken from daylight, within many hours a considerable amount of photons, mostly in the red spectral range. Its only significant emitter is an egg shell because yolk+embryo emits an order less photons and do this only during the first 3-4 days of development. Egg whites do not emit photons at all. However, only in the dead and non-fertilized eggs the rate of photon emission of a whole intact egg is roughly equal to that of its isolated shell. Whole developing, 2nd day eggs emit, in most cases, many more photons than their isolated shells, while for 9 days developing eggs the reverse is true. Hence, at the relatively early stages an eggs' interior stimulates, while at the later stages somehow suppresses its shell's emission. This latter effect may be due to an active absorption ("sucking") of photonic energy from a shell by an embryo, possibly via chorio-allantoic blood vessels which closely approach the interior shell surface just at this stage.

A non-additivity of a photon emission can be also demonstrated with the use of an artificial "egg model", consisting of a 3-days yolk+embryo, gently poured into a quartz cuvette and covered by a piece of a shell taken from the same egg . The rate of emission of an entire system turned out to be much greater than the sum of the emission rates of its components while taken individually [29].

Developmental Biology and Physics of Today

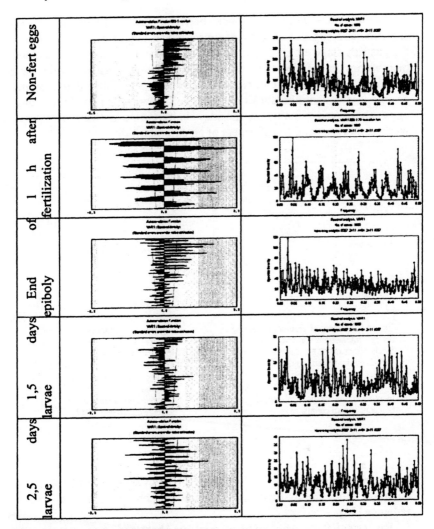

Figure 7: Autocorrelation patterns (left column) and Fourier spectra (right column) for successive developmental stages of *M. fossilis* eggs and embryos (shown at left). Note disordered patterns in non-fertilized eggs, quite regular and uniform short-wave patterns in fertilized eggs and the emergence of a long-range order in more advanced stages.

By a hyperbolic decay criteria [30], the emission of an isolated shell is highly coherent, obviously due to its crystalline structure (see Musumeci et al. in this volume). Same is true for a whole developing egg, but not for a boiled and a fresh non-fertilized egg. Consequently, only the developing eggs can support coherence of a shell-emitted light.

Similar non-additive relations between the emission of developing embryos and egg membranes are observed in amphibians [26]. Namely, in the gastrula stage embryos the amount of photon emission of the isolated egg membranes is smaller than of the embryos+membranes, while for more advanced neurula stage embryos the reverse is true. This again may point to an increase of "photon sucking", as development goes on. In any case, the eggs' shells and membranes, besides their routine protective role seem to work as storages and transformers of an external light.

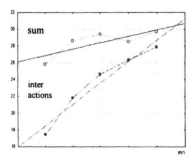
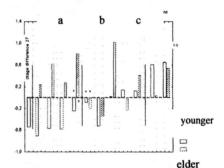

Figure 8: Increase of zero-counts during optical interactions of two embryonic batches goes on with a much greater rate (filled circles) than while these same batches "do not see" each other ("sum"). Horizontal axis: minutes of common incubation, in 10 units. Vertical axis: percents of zero counts.

Figure 9: Changes in kurtosis (vertical axis) during optical interactions of two different age embryonic batches. The greater is the age difference, the greater is the increase of kurtosis in a more advanced embryonic batch
a: stage difference 15-17,
b: stage difference 9-10,
c: stage difference 4-5

3. A population of optically contacting developing embryos create a common biophotonic field showing the properties of a sub- and super-radiance [31].

By establishing the optical contacts between two quartz cuvettes containing fish (*Misgurnus fossilis*) eggs or embryos, the rate of their common biophotonic signal gradually goes down, while if the same batches "do not see" each other, the signal remains constant (Fig. 8). This may point to a sub-radiance (destructive interference) between optically contacting batches. On the other hand, the amount of single bursts of a biophoton emission (as revealed by kurtosis, Fig. 9) increases during optical interactions [32]. This may be the evidence of a super-radiance (constructive interference). Both events can only take place in a coherent biophotonic field.

6. A FEW WORDS IN CONCLUSION

In this paper we suggest that the best thing which can be borrowed by developmental biology from the "physics of to-day" is the concept of a field endowed by holistic and coherent properties irrespectively of whether this field is manifested in the mechanical stresses, biophotons or something else. This concept should replace to a great extent a still dominating paradigm of multiple specific micro-events and their interactions. Does it mean, however, that such a replacement must be a total one, going on up to a complete exclusion of the latter paradigm? One should be very cautious in this respect. My personal view is that at any time a certain part of biology, including the developmental one, will be occupied by some "specific" entities which have to be, so to say, remembered by heart and irreducible to any physicalistic laws. But another part may be treated in a much more physicalistic way than it is done in the present time. It will be a matter of a scientific skill to keep a proper balance between these two.

ACKNOWLEDGMENTS

This work was partly supported by the Russian Fund for Fundamental Investigations (RFFI), grant 02-04-49004.

REFERENCES

1. Kant, I. (1755) (1910) *Allgemeine Naturgeschichte und Theorie des Himmels.* Königsberg – Leipzig (Gesammelte Schriften Bd. 1 Berlin).

2. Maxwell, J.C. (1871) (1991) *Matter and Motion.* Dover, London.

3. Curie, P. (1894) De symmetrie dans les phenomenes physique: symmetrie des champs electrique et magnetiques. *J. De Physique,* 3, 393-427.

4. Jaffe, L.F. (1968) Localization in the developing *Fucus* eggs and the general role of localizing currents. *Advances in Morphogenesis,* 7, 295-328.

5. Nieuwkoop, P.D. (1977) Origin and establishment of an embryonic polar axis in amphibian development. *Curr. Top. Devel Biol.,* 11, 115-117.

6. Dürr, H.-P. (2000) Unbelebte und Belebte Materie: Ordnungsstructuren immaterialler Beziehungen. Physicalische Wurzeln des Lebens. In *Elemente des Lebens,* Dürr, H.P., Popp, F.A. and Schommers, W. (eds.), Die Graue Edition, 179-208

7. Gurwitsch, A.G. (1991) *Principles of Analytical Biology and a Theory of a Cellular Field.* Nauka, Moskva (in Russian).

8. Gilbert, S.F. (1988) *Developmental Biology*. Sinauer Ass., Inc. Sunderland, Mass.

9. Dosch, R., Gawantka, V., Delius, H., Blumenstock, C., Niehrs, Chr. (1997) BMP-4 acts as a morphogen in dorsoventral mesoderm patterning in Xenopus. *Development,* **124**, 2325-2334.

10. Dale, L., Slack J.M.W. (1987) Fate map for the 32-cell stage of Xenopus laevis. *Development,* **99**, 527-551.

11. Nicolis, G., and Prigogine, I. (1977) *Self-organization in Non-equilibrium Systems*. Wiley, N.Y.

12. Prigogine, I. (1980) *From Being to Becoming: Time and Complexity in the Physical Sciences*. W.H. Freeman and Co., N.Y.

13. Loskutov, A. Ju., Michailov, A.S.(1990) *Introduction to Synergetics*. Nauka, Moskva (in Russian).

14. Goodwin, B.C. (1994) *How the Leopard Changed Its Spots. The Evolution of Complexity*. Weidenfeld and Nicolson, London.

15. Krinsky, V.I., Zhabotinsky, A.M. (1981) Autowave structures and perspectives of their investigations. In *Autowave Processes in Diffusional Systems*, Grechova, M.T. (ed.), Gorky, Inst. Appl. Physics Acad. Sci. USSR, 6-32 (in Russian).

16. Thompson, D'Arcy (1942) *On Growth and Form*. Cambridge University Press, Cambridge.

17. Martynov, L.A. (1982) The role of macroscopic processes in morphogenesis. In *Mathematical Biology of Development*, Zotin, A.I., Presnov, E.V. (eds.), Nauka, Moskva, 135-154 (in Russian).

18. Beloussov, L.V., Saveliev, S.V., Naumidi, I.I., Novoselov V.V. (1994) Mechanical stresses in embryonic tissues: patterns, morphogenetic role and involvement in regulatory feedback. *Int Rev Cytol.,* **150**, 1-34.

19. Beloussov, L.V. (1998) *The Dynamic Architecture of a Developing Organism*. Kluwer Academic Publishers, Dordrecht/Boston/London.

20. Ermakov, A.S., Beloussov, L.V. (1998) Morphogenetic and differentiation consequences of the relaxations of mechanical tensions in *Xenopus laevis* blastula. *Ontogenez (Russ J Devel Biol),* **29**, 450-458.

21. Odell, G.M., Oster, G., Alberch, P., Burnside, B. (1981) The mechanical basis of morphogenesis. I. Epithelial folding and invagination. *Devel Biol.,* **85**, 446-462.

22. Harris, A.K., Stopak, D, Warner, P. (1984) Generation of spatially periodic patterns by a mechanical instability: a mechanical alternative to the Turing model. *J. Embryol. Exp. Morphol.,* **80**, 1-20.

23. Belintzev, B.N., Beloussov, L.V. and Zaraisky, A.G. (1987) Model of pattern formation in epithelial morphogenesis. *J. Theor. Biol.* **129**, 369-394.

24. Keller, R.E. (1987) Cell rearrangements in morphogenesis. *Zool. Sci.*, **4**, 763-779.

25. Gu, Qiao (1992) Quantum theory of biophoton emission. In *Recent Advances in Biophoton Research and its Applications,* Popp, F.-A., Li, K.H. and Gu, Q. (eds.), World Scientific, 59-112.

26. Ho, M.W., X.Xu, S.Ross and P.T. Saunders (1992) Light emission and rescattering in synchronously developing populations of early *Drosophila* embryos. In *Recent Advances in Biophoton Research and its Applications,* Popp, F.-A., Li, K.H. and Gu, Q. (eds.), World Scientific, 287-306.

27. Gurvich, A.A. (1992) Mitogenetic radiation as an evidence of nonequilibrium properties of living matter. In *Recent Advances in Biophoton Research and its Applications,* F.-A. Popp, K.H. Li and Q. Gu (eds.), World Scientific, 457-468.

28. Beloussov L.V., Louchinskaia N.N. (1998) Biophoton emission from developing eggs and embryos: non-linearity, holistic properties and indications of energy transfer. In *Biophotons*, Chang, Jiin-Ju, Fisch, J. and Popp, F.-A. (eds.), Kluwer Academic Publishers. Dordrecht, Boston, London, 121-142.

29. Beloussov L.V., Popp, F.-A., Kazakova, N.I. (1997) Ultraweak photon emission of hen eggs and embryos: non-additive interaction of two emitters and stable non-equilibricity. *Ontogenez (Russ. J. Devel Biol.),* **28**, 377-388.

30. Popp, F.-A. And K.H. Li (1992) Hyperbolic relaxation as a sufficient condition of a fully coherent ergodic field. In *Recent Advances in Biophoton Research and its Applications,* Popp, F.-A., Li, K.H. and Gu, Q. (eds.), World Scientific, 47-58.

31. Li, Ke-hsueh (1992) Coherence in physics and biology. In *Recent Advances in Biophoton Research and its Applications,* Popp, F.-A., Li, K.H. and Gu, Q. (eds.), World Scientific, 113-156.

32 Beloussov, L. V., Louchinskaia, N.N. and Burlakov, A.B. (2000) Biophoton emission in fish (*Misgurnus fossilis L.*):developmental dynamics and optical interactions. http://www.lifescientists.de

Chapter 4

CELLULAR AND MOLECULAR ASPECTS OF INTEGRATIVE BIOPHYSICS

Roeland Van Wijk
International Institute of Biophysics,
Neuss, Germany; and
Department of Molecular Cell Biology,
Utrecht University, Utrecht, The Netherlands
e-mail: *meluna.wijk@wxs.nl*

Keywords: coherent excitations, cytoskeleton, fluid movement, membranes, microtubules, microfilaments, water.

1. INTRODUCTION

The past several decades the scientific literature has grown so rapidly that it is almost a full-time job to keep abreast of the major developments relating to cellular organization and behavior. As a consequence of this enormous profusion of information we have witnessed an explosive growth in our understanding of the properties and functions of living cells. In particular, this has led to more emphasis on the interplay of DNA, RNA and proteins as, what has often been called the central molecules of life. Thus, recombinant DNA technology has profoundly transformed biochemistry of the cell. The genome is now an open book. Cloning and sequencing of millions of bases of DNA have greatly increased our insight in how the genome is organized and its expression is controlled. The reading of the genome has also provided a wealth of amino acid sequence information that has resulted in knowledge on how proteins fold, catalyze reactions, transport ions, and transduce signals. It is interesting to see that as information accumulates, disconnected facts give way to rational explanations, and simplicity emerges from chaos. However, gradually the essential large areas of ignorance in cell biology and the many facts that cannot be explained come into focus. It is evident that molecular biology by itself is not enough to understand the "time" and "space" aspects of the internal working of cells. The temporal and the spatial organization of a cell or an organism are essentially determined by its dynamic structure which in turn is based on metabolic organization.

A eucaryotic cell contains a billion or so protein molecules, and there are thought to be about 10,000 different types of protein in an individual vertebrate cell. Most of them are highly organized spatially. In all cells proteins are arranged into functional complexes, most consisting of perhaps 5 to 10 proteins. But other complexes are as large or larger than ribosomes. A further level of organization involves the confinement of functionally related proteins within the same membrane or aqueous compartment of a membrane-bound organelle, such as the nucleus, mitochondria, or Golgi apparatus. An even higher level of organization is created and maintained by the cytoskeleton. It enables the living cell, like a city, to have many specialized services concentrated in different areas but extensively connected by path of communication.

In order to explain this internal dynamic organization and the adaptive behavior of biological systems one has to deal with the dynamic non-equilibrium energy-driven processes and the constraints based on quantum coherent phenomena of the biological organization. In general, biologists seem to pay little attention to physical theories that attempt to explain living systems without detailed references to the elaborate structure that we know exists, notably at the cellular and subcellular levels. This explains, at least to some extent, why Fröhlich's theory of coherent excitations has almost gone unnoticed by the biological community. He explained the dynamics of enzymes and high-order structures by coherent phenomena.

The quantum coherent state allows for information transfer with very high efficiency. On macroscale a coherence represents another scale of coherent phenomena. It is a physical representation of the integrative principle coordinating individual events in the organized state.

The goal of this chapter is to describe certain features of the structure and function of eucaryotic cells, trace coherent events from the macromolecular to the cellular level, and define possible aspects that seem relevant for a description of the organization of cells and biosystems in general.

In preparing this chapter we have focused in the first instance on the fundamental principle of domain dynamics that guide cellular organization and function of eucaryotic cells. To accomplish this we have necessarily been selective both in the types of examples chosen to illustrate key concepts and in the quantity of scientific evidence included. The examples chosen include the dynamics of the membrane and the cytoskeleton structure. In addition we discuss the domain that integrates these two major structures by including the structural dynamics of cellular water. We think it is important to understand some of the critical scientific evidence that has led to the formulation of the central concept of domain dynamics without going into experimental detail. Moreover, we have ventured the hypothesis of coherence for the business of organizing long-range order in water and the structure and function of the eucaryotic cell.

2. MEMBRANE STRUCTURE

2.1 Composition of Biological Membranes

Membranes are crucial to the life of the cell. The plasma membrane encloses the cell, and maintains the essential differences between the cytosol and the extracellular environment. Inside the cell many other membrane structures are present. They belong to a variety of organelles with different composition, structure and function. The organelles have been named differently, depending on their function, and we distinguish in eucaryotic cells the endoplasmic reticulum, Golgi apparatus, mitochondria, chloroplasts and other membrane-bounded organelles. The membranes maintain the characteristic differences between the contents of each organelle and the cytosol. The universal basis for cell-membrane structure is the lipid bilayer (Jain, 1988; Alberts et al, 1994). All of the lipid molecules in cell membranes are amphipathic (or amphiphilic) – that is, they have a hydrophilic ("water-loving", or polar) end and a hydrophobic ("water-hating", or nonpolar) end. They constitute about half of the mass of most animal cell membranes, nearly all of the remainder being protein. It is the shape and amphipathic nature of the lipid molecules that cause them to form bilayers. Thus, when lipid molecules are surrounded by water, they tend to aggregate so that their hydrophobic tails are buried in the interior and their hydrophilic heads are exposed to water.

The lipid bilayer of cell membranes is composed of three major classes of membrane lipid molecules – phospholipids, cholesterol, and glycolipids. The lipid compositions of the two faces of a cell membrane, the inner and outer monolayers, are different. Different mixtures of lipids are also found in the various organelle membranes of a single eucaryotic cell, as well as in the membranes of cells of different types. Why eucaryotic membranes contain such a variety of lipids, with head groups that differ in size, shape, and charge, can only be understood if one thinks of the biological membrane lipids as constituting a two-dimensional solvent of lipid molecules in which membrane proteins are embedded. Thus, some membrane proteins require specific lipid head groups in order to function. Just as water constitutes a three-dimensional solvent for proteins, where many enzymes in this solution require a particular ion for activity.

Although the basic structure of biological membranes is provided by the lipid bilayer, proteins carry out most of the specific functions. Accordingly, the amounts and types of proteins in a membrane are highly variable. In the membrane, which serves as electrical insulation for nerve cell axons, less than 25% of the membrane mass is protein, whereas in membranes involved in energy transduction (such as the internal membranes of mitochondria and chloroplast, approximately 75% is protein. The usual plasma membrane contains about 50% of its mass as protein. Because lipid molecules are small in comparison to protein molecules, there are always many more lipid molecules than protein molecules in membranes – about 50 lipid molecules for each protein molecule in a membrane that is 50 % proteins by mass.

The membrane proteins have all types of function. The plasma membrane contains proteins that act as sensors of external signals, allowing the cell to change its behavior in response to environmental cues; these protein sensors, or receptors, transfer information. Other membrane proteins establish ion gradients across membranes; these gradients can be used to synthesize ATP, to drive the transmembrane movement of selected solutes, or, in nerve and muscle cells, to produce and transmit electrical signals. Despite their differing functions, all biological membranes have a common general structure; each is a very thin film of lipid and protein molecules.

It was in the early 1970s that it was recognized that individual lipid molecules are able to diffuse freely within their own lipid bilayers (Frye and Edidin,1970; Petris and Raff,1973; Jacobson et al, 1987). Most types of lipid molecules, however, very rarely tumble spontaneously from one monolayer to the other. Like membrane lipids, membrane proteins do not tumble across the bilayer, but they are able to move laterally within the membrane. The precise fluidity of cell membranes is biologically important. Certain membrane transport processes and enzyme activities, for example, can be shown to cease when the bilayer viscosity is experimentally increased beyond a threshold level. The fluidity of a lipid bilayer depends on both its composition and temperature. Thus, the characteristic freezing point temperature at which a bilayer changes from a liquid state to a rigid crystalline (or gel) state, called a phase transition, is lower if the hydrocarbon chains are short or have double bonds.

2.2 Membrane Domains and Dynamics

The recognition that biological membranes are two-dimensional fluids was a major advance in understanding membrane structure and function. It has become clear, however, that the picture of a membrane as a lipid sea in which all proteins float freely is greatly oversimplified. Many cells have ways of confining lipids and membrane proteins to specific domains in a continuous lipid bilayer. A few beautiful examples are commonly mentioned in this respect (Gumbiner and Lourard, 1985; Myles and Primakoff, 1985; Jacobson and Vaz, 1992; Sheetz, 1993). In epithelial cells, such as those that line the gut, for example, certain plasma membrane enzymes and transport proteins are confined to the apical surface of the cells, whereas others are confined to the basal and lateral surfaces. This segregation permits a vectorial transfer of nutrients across the epithelial sheet from the gut lumen to the blood. The lipid composition of these two membrane domains is also different. Other membrane proteins serve as selective permeability barriers, separating fluids on each side that have different compositions. These so-called tight junctions are thought to confine the transport proteins to their appropriate membrane domains by acting as diffusion barriers within the lipid bilayer of the plasma. The occurrence of these domains demonstrates that in epithelial cells the diffusion of lipid as well as protein molecules between the domains can be prevented. Another example of a specific membrane domain is seen in the mammalian spermatozoon. This single cell that consists of several structurally and functionally distinct parts is covered by a continuous plasma membrane. In this examples again, the diffusion of protein and lipid molecules is confined to specialized domains within a continuous plasma membrane. A third illustration is the cell-cell adherens junction. In epithelial sheets they often form a continuous adhesion belt around each of the interacting cells in the sheet, located near the apex of each cell just below the tight junction. The adhesion belts in adjacent epithelial cells are directly opposed and the interacting plasma membranes are held together at this point by linked proteins. Apparently, cells seem to have drastic ways of immobilizing certain membrane proteins. In the search for an explanation it has been argued that a way of restricting lateral mobility of specific membrane proteins is to tether them to macromolecular assemblies either inside or outside the cell. It has been seen that in most cellular types plasma membrane proteins can be anchored to the cytoskeleton, or to the extracellular matrix or to both.

2.3 Regulation of Domain Traffic

The regulation of membrane protein's mobility and immobilization and domain formation is illustrated by of the processes called patching and capping. When ligands, such as antibodies, that have more than one binding site (so called multivalent ligands) bind to specific proteins on the surface of the cells, the proteins tend to become aggregated, through cross-linking, into large clusters or patches (Edidin, 1992). This indicates that the protein molecules are able to diffuse laterally in the lipid bilayer. Once such clusters have formed on the surface of a cell capable of locomotion, such as a white blood cell, they are actively moved to one pole of the cell to form a cap.

The dynamics of membrane domains becomes even more impressive when we consider the elaborate internal membrane system of eucaryotic cells (Gruenberg and Clague, 1992; Watts and Marsh, 1992). This system allows cells to take up macromolecules by a process called endocytosis. Localized regions of the plasma membrane invaginate and pinch off to form endocytic vesicles. The vesicles are then delivered to digestive enzymes that are stored intracellularly in lysosomes. Endocytosis takes place both constitutively and as a stimulated response to extracellular signals. It is so extensive in many cells that a large fraction of the plasma membrane is internalized every hour.

Metabolites generated from the macromolecules by digestion are delivered from the lysosomes directly to the cytosol as they are produced. The plasma components (proteins and lipids) are continually returned to the cell surface in a large-scale endocytic-exocytic cycle. Besides providing for regulated digestion of macromolecules by the endocytic pathway, the internal membrane system provides a means whereby eucaryotic cells can regulate the delivery of newly synthesized proteins and carbohydrates to the exterior. The biosynthetic-secretory pathway leads from the endoplasmatic reticulum towards the Golgi apparatus and cell surface, with a side route leading to lysosomes, while the endocytic pathway leads inward toward endosomes and lysosomes from the plasma membrane (Hong and Tang, 1993; Kelly, 1991). This traffic is highly organized. In fact, the lumen of each compartment along the biosynthetic-secretory and endocytic pathways is topologically equivalent to the exterior of the cell. The compartments are all in constant communication with one another, at least partly by means of numerous transport vesicles. These vesicles continually bud off from one membrane and fuse with another. To perform its function, each transport vesicle that buds from a compartment must take up only the appropriate proteins and must fuse only with the appropriate target membrane. A vesicle carrying cargo from the Golgi apparatus to the plasma membrane, for example, must exclude proteins that are to stay in the Golgi apparatus, and it must only fuse with the plasma membrane and not with any other organelle. While participating in this constant flow of membrane components, each organelle must maintain its own distinct identity. Another interesting point is that each molecule during its travel through the multiple compartments of the biosynthetic-secretory pathway can be modified in a series of controlled steps, stored until needed, and then delivered to a specific cell-surface domain by a process called exocytosis.

The cell contains 10 or more chemically distinct membrane-bounded compartments and vesicular transport mediates a continual exchange of components among them. In the presence of this massive exchange, how does each compartment maintain its specialized character? To answer this question, it has been considered what defines the character of a compartment (Kreis, 1992; Waters et al, 1991). Above all, it is the nature of the enclosing membrane markers displayed on the cytosolic surface of the membrane that guides the targeting of vesicles, ensuring that they fuse only with the correct compartment and so dictating the pattern of traffic between one compartment and another. Given the presence of distinct membrane markers for each compartment, the problem is to explain how specific membrane components are kept at high concentration in one compartment and at low concentration in another. The answer depends most fundamentally on the mechanisms of transport vesicle formation and fusion, by which patches of membrane, enriched or depleted in specific components, are transferred from one compartment to another. The creation of a transport vesicle involves the assembly of a special coat on the cytosolic face of the budding membrane. Such coats serve as devices to suck membrane that is enriched in certain membrane proteins and depleted in others out of one compartment, so that specific proteins can be delivered to another compartment. Thus, transport vesicles bud from special coated regions of the donor membrane. The assembly of the coat allows the transport of the vesicle.

Other well-defined proteins provide a molecular link between these coats and specific membrane receptors and thereby mediate the selective uptake of cargo molecules into the respective coated vesicles. Coated vesicles have to lose their coat to fuse with their appropriate target membrane in the cell. There are two types of coated vesicles, one type of vesicle mediates selective vesicular transport from the plasma membrane, while the other type mediates non-selective vesicular transport from the endoplasmatic reticulum. It is evident that one type of coat is lost soon after the vesicle pinches off from the donor membrane whereas the other type of coat is lost after the vesicle has docked on the target membrane. The differences between the membranous compartments of a cell are maintained by an input of free energy, driving directional, selective transport of particular membrane components from one compartment to another. Several classes of proteins help regulate various steps in vesicular transport, including vesicle budding, docking, and fusion. These proteins are incorporated during budding into the transport vesicles and help ensure that the vesicles deliver their contents only to the appropriate membrane-bounded compartment. Other specialized cytosolic proteins then catalyze the membrane fusion. They assemble into a fusion complex at the docking site. The involvement of the cytoskeleton in the transport processes will be discussed below.

3. CYTOSKELETON

3.1 Compositon of Cytoskeleton

The ability of eucaryotic cells to adopt a variety of shapes and to carry out coordinated and directed movement depends on a complex network of protein filaments that extends throughout the cytoplasm. This network is called the cytoskeleton (Alberts et al, 1994). The cytoskeleton spatially arranges a eucaryotic cell. This is a remarkable phenomenon because a eucaryotic cell has a diameter of 10 micrometers or more and the cytoskeletal protein molecules are a few thousands times smaller in linear dimensions. This structure occurs by polymerization. In fact, any protein with an appropriately oriented pair of complementary self-binding sides on its surface can form a long helical filament.

The cytoskeleton is a highly dynamic structure that reorganizes continuously as the cell changes shape, divides, and responds to its environment. The cytoskeleton has sometimes been called the cell's musculature because it is directly responsible for crawling and many other changes in shape of a developing vertebrate embryo. It also provides the machinery for intracellular movements, such as the transport of organelles from one place to another in the cytoplasm and the segregation of chromosomes at mitosis.

The diverse activities of the cytoskeleton depend on three types of protein filaments – actin filaments, microtubules, and intermediate filaments. Each type of filament is formed from a different protein subunit: actin for actin filaments, tubulin for microtubules, and a family of related fibrous proteins, such as vimentin or lamin, for intermediate filaments. For each of the three major types of cytoskeletal protein, thousands of identical protein molecules assemble into linear filaments. These filaments can be long enough to stretch from one side of the cell to the other. Each of the three principal types of protein filaments that make up the cytoskeleton has its specific arrangement in the cell and a distinct function. By themselves, however, the three types of filaments cannot be considered as the sole constituents that could provide shape or strength to the cell. Their functions depend on accessory proteins that link the filaments to one another and to other cell components, such as the plasma membrane. Although the main subunits of the three classes of cytoskeletal polymers, as well as many of the hundreds of accessory proteins that associate with them, have been isolated and their amino acid sequence determined, it has been frustratingly difficult to establish how these proteins function in the cell. The difficulty arises from the observation that the function of the cytoskeleton depends on the cooperative groups of proteins, which bind to the cytoskeletal filaments. It is relatively straightforward to examine the effect on a filament of a single accessory protein but very much more difficult to analyze the effects of a mixture of many different proteins. Another difficulty in understanding the functions of the cytoskeleton is that the movements of the polymeric structure are without any major chemical change.

3.2 Cytoskeleton Domains and Dynamics

We shall now elaborate on the dynamics of the cytoskeleton filaments (Gelfand and Bershadsky, 1991; Lee et al, 1993). Microtubules are polar structures: one end is capable of rapid growth (plus end), while the other end tends to lose subunits if not stabilized (minus end). This stabilization is commonly done in a structure called the centrosome. The rapidly growing ends are then free to add tubulin molecules. At any one time, several hundreds of microtubules are growing outward from a centrosome, so that their plus end is at the edge of the cell. Each of these microtubules is a highly dynamic structure that can shorten as well as lengthen: after growing outward for many minutes by adding subunits, its plus end may undergo a sudden transition that causes it to lose subunits, so that the microtubule shrinks rapidly inward and may disappear. Thus, the microtubule network that emanates starlike from the centrosome is constantly sending out new microtubules to replace the old ones that have depolymerized.

In general, microtubules in the cytoplasm function as individuals, whereas actin filaments work in networks or bundles. Actin filaments lying just beneath the plasma membrane, for example, are cross-linked into a network by various actin-binding proteins to form the cell cortex. The network is highly dynamic and functions with various myosins to control cell-surface movements. The actin filament, like a microtubule, is a polar structure, with a relative inert and slowly-growing minus end and a faster-growing plus end. There is a fundamental difference in the way the actin filaments and the microtubules are organized and function in cells. Actin filaments are thinner and more flexible, and usually much shorter than microtubules. Total length of the actin filament in a cell is at least 30 times greater than the total length of the microtubules. However, the actin filaments rarely occur in isolation in the cell but rather in cross-linked aggregates and bundles. Actin filaments, unlike microtubules, show the dynamic behavior called tread-milling, which occurs when actin molecules are added continually to the plus end of the filament and are lost continually from the minus end, with no net change in filament length. This tread-milling is a non-equilibrium behavior that requires an input of energy. This energy is provided by the ATP hydrolysis that accompanies polymerization of actin. In tubulin polymerization the energy is provided by GTP hydrolysis.

3.3 Regulation of Domain Traffic

Domain characteristics become evident when we observe the essential role of accessory proteins in the controlled assembly of the protein filaments in particular locations (Lee, 1993; Olmsted, 1986). Accessory proteins control where and when actin filaments and microtubules are assembled in the cell by regulating the rate and extent of their polymerization. Other accessory protein function as motor proteins, which hydrolyze ATP to produce force and either move organelles along the filaments or move the filaments themselves (Bloom, 1992; Janmey, 1991). Thus, nucleation sites in the plasma membrane control the location and orientation of the cortical actin filaments. Moreover, different regions of the membrane direct the formation of distinct actin-filament-based structures. Localized extracellular signals that impinge on a portion of the cell surface can induce a local restructuring of the actin cortex beneath the corresponding part of the plasma membrane(Theriot and Mitchison, 1992). In a reciprocal way the organization of the actin cortex has a major influence on the behavior of the overlying plasma membrane. Mechanisms based on cortical actin filaments, for example, can push the plasma membrane outward to form long, thin microspikes or sheetlike lamellipodia (Condeelis, 1993). Or, they can pull the plasma membrane inward to divide the cell in two. The latter process of cell division consists of nuclear division followed by cytoplasmic division.

The nuclear division is mediated by a microtubule-based mitotic spindle, which separates the chromosomes (Wadsworth, 1993). The cytoplasmic division is mediated by an actin-filament- based contractile ring (Cao and Wang, 1990). The mitosis is largely organized by the microtubule asters that form around each of the two centrosomes produced when the centrosome duplicates. This duplication of the centrosome begins during the DNA replication phase of the cell cycle and the duplicated centers move to opposite sides of the nucleus at the onset of mitosis to form the two poles of the mitotic spindle. As a part of this process large membrane-bounded organelles, such as the Golgi apparatus and the endoplasmatic reticulum, break up into many smaller fragments, which ensures their even distribution into daughter cells during cytokinesis.

The role of cytoskeleton in mitosis demonstrates that cytoskeletal filaments serve not only as structural supports but also serve as lines of transport. In general, the cytoskeleton plays also an important part in positioning membrane-bounded organelles within an eukaryotic cell under steady state conditions. The membrane tubules of the endoplasmic reticulum, for example, align with microtubules and extend almost to the edge of the cell, whereas the Golgi apparatus is localized near the centrosome. When cells are treated with a drug that depolymerizes microtubules, both of these organelles change their location: the endoplasmic reticulum collapses to the center of the cell, while the Golgi apparatus fragments into small vesicles that disperse throughout the cytoplasm. When the drug is removed, the organelles return to their original positions. This demonstrates the involvement of the cytoskeleton in the movement and location of organelles.

Another impressive situation in which microtubules and actin filaments act in a coordinated way to polarize the whole cell is with the killing of specific target cells by cytotoxic T lymphocytes (Kupfer and Singer, 1989). Cytotoxic T cells kill other cells that carry foreign antigens on their surface. This is an important part of vertebrate's immune response to infection. When receptors on the surface of the T cell recognize antigen on the surface of a target cell, the receptor signals to the underlying cortex of the T cell, altering the cytoskeleton in several ways. First, proteins associated with actin filaments in the T cell reorganize under the zone of contact between the two cells. The centrosome then reorients, moving with its microtubules to the zone of T-cell-target contact. The microtubules, in turn, position the Golgi apparatus right under the contact zone, focussing the killing machinery – which is associated with secretion from the Golgi – on the target cell.

In general, a cell becoming polarized shows the following general pattern of changes. The plasma membrane has sensed some difference on one side of the cell and a transmembrane signal is generated. The actin cortex reorganizes in a local area beneath the affected membrane. The centrosome moves to that part of the cell, presumably by pulling on its microtubules. The centrosome in turn positions the internal membrane systems in a polarized way. The net result is a cell with a strong directional focus.

The intracellular movements are generated in two ways. The first manner is that organelles are dragged by motor proteins and move along the cytoskeleton. In case of vesicles of the endoplasmatic reticulum and the Golgi apparatus the normal position of each of these organelles is thought to be determined by a receptor protein on the cytosolic surface of the membrane that binds to specific microtubule-dependent motor proteins. They use the energy derived from repeated cycles of ATP hydrolysis to move steadily along it. The dozens of different motor proteins that have been identified, differ in the type of filament they can bind to, the direction in which they move along the filament, and the cargo they carry. As with microtubules motor proteins are also important in actin filaments. All of the actin filament motor proteins belong to the myosin family. The second manner for intracellular movement is based on the polymerization-depolymerization cycle of the cytoskeleton structures. In general terms it can be concluded that the latter mechanism results in gross changes, while motor proteins drive organelles along the microfilaments or microtubules and might be responsible for more specific changes.

4. WATER AND THE COUPLING OF STRUCTURES

4.1 Structure of Cell Water

Now that the oversimplifications and artificiality of most biochemical analyses are becoming better appreciated, the internal operations of a cell are less likely to be compared with a jumble of chemical interactions taking place in bulk aqueous phase in which laws of mass action apply. While a number of reactions undoubtedly take place in the bulk aqueous phase, the question is whether they represent the majority or a very small minority of the total. Already Sherrington in 1938 stated "Although it is fluid and watery, most of the cell is not a true solution. A drop of true solution, of homogenous liquid, could not "live". It is remote from "organization". In the cell there are heterogeneous solutions. The great molecules of proteins and aggregated particles are suspended not dissolved. A surface is a field for chemical and physical action. The interior of a pure solution has no surface. But the aggregate of surface in these foamy colloids which are in the cell mounts up to something large. The "internal surface" of the cell is enormous. The cell gives chemical results which in the laboratory are to be obtained only by temperatures and pressures far in excess of those of the living body. Part of the secret of life is the immense internal surface of the cell" (Sherrington, 1940). An important step forwards in the recognition of the role of water in cellular organization was the conference on "Forms of water in biological systems", organized in 1965 by the New York Academy of Sciences (Whipple, 1965). At that time most people still considered water as a homogeneous fluid with definite physical properties subject to variation or alteration as a function of temperature and pressure. At this conference the alternative look upon the cell as a drop of polyphasic colloid was officially presented as opposite to viewing the cell as a membrane bag filled with dilute solution. The drop of polyphasic colloid is a mixture of different environs; some areas are rich in particles, others are not. Water serves as the dispersion medium or continuous phase and protein, nucleic acids, carbohydrates, lipids and ions are the dispersed particles. Since then, many researchers have speculated on the association of enzymes with the vicinal water and with other proteins, including the filaments. Thus, Ling (1962; 1969), Drost-Hansen (1965; 1971), Clegg (Clegg, 1983;1984; Clegg and Drost-Hansen, 1991), and Wiggins (1990; 1995) have suggested that the water vicinal to the protein molecules was structured differently than normal water. This is due to the unique properties of water. Water is singular as a liquid because of its ability to form three-dimensional networks of molecules, mutually hydrogen bonded. It results from the fact that each water molecule has four fractional charges directed in three-dimensional space toward the corners of a regular tetrahedron. Two of these are positive (the hydrogen atoms), and two are negative (the lone pairs of electrons on the oxygen atom). The consequence of this distribution of positive and negative poles is that each water molecule can make up to four bonds (hydrogen bonds) with neighboring water molecules, a positive pole of one molecule being attracted to a negative pole of another. The macromolecular network of the protoplasm and the enclosed water interact in a mutual manner. Thus, bio-polymers have an effect on the structure of water, while the water structure is significant for the protoplasmic organization. Bio-polymers have an indirect effect on water. Bio-polymers characteristically present to water both charged and hydrophobic surfaces. Globular proteins in solution, for example, bury much of their hydrophobic surface area internally. Nearly all charged and hydrophilic groups, which endow the large molecules with solubility in water, are on the external surface. Inevitably, therefore, those patchy surfaces generate patchy interfacial water structures: zones of dense water grow round the charged groups, while zones of low-density water grow round the hydrophobic regions. A variety of recent publications deal in detail with the structure and properties of water at biological interfaces (Ide et al, 1997; Laaksonen et al, 1997; Nadig et al, 1998; Sharp and Madan, 1997; Svergun et al, 1998)

These zones of water have consequences for the solubility of small molecules as well as for the macromolecular organization. In fact, water molecules near the weakly hydrogen-bonding surfaces are in a state of higher enthalpy than more distant molecules that form hydrogen bonds with one another. Since hydrogen-bonding is co-operative several layers of water molecules have weaker than normal hydrogen-bonding which increases the local chemical potential.

4.2 Water Structure and Protoplasmic Organization

The significance of water structure for the protoplasmic organization is then also evident. If water exists in different hydrogen-bonded states in different regions of the protoplasm, both protein conformations and aggregation/de-aggregation states are likely to be affected. Low-density water selectively accumulates chaotropic ions (like Cl^-, K^+, charged amino acids, and glucose. It selectively excludes highly hydrated ions (Mg^{2+}, Ca^{2+}, H^+, Na^+), and predominantly hydrophobic molecules. It follows that the buried hydrophobic proteins can more readily emerge, and the proteins unfold, when the surrounding water is weakly bonded. Conversely, stabilization of the native folded conformation and aggregation of monomer proteins by hydrophobic interactions are promoted by strongly bonded water. Preferential association of proteins with the lattice is to be expected because both their hydrophobic patches and their charged patches of surfaces partition into weakly bonded water. In a cell with only ions as osmolytes, extreme forms of water that are deleterious to enzyme stability and function coexist. Water adjacent to the filament is probably so weakly bonded that internal hydrophobic moieties emerge and enzymes unfold and denature. However, in case the macromolecular network of the protoplasm contains not only ions, but also compatible solutes as osmolytes, the excess osmolality is offset, to a degree, by accumulation of compatible solutes into the strongly bonded regions adjacent to hydrophobic patches of surface and in between filaments. Water near the charges on the filaments can equilibrate with a less drastic increase in density and reactivity, while the rest of the water maintains a rather open structure, because of the hydrophobic portions of its dissolved solutes. In this case, enzymes still accumulate into the more reactive water associated with the filaments, but they do not denature. The view of a cell as a highly integrated system of multiple components interacting with one another leads to the question on the quantity of water in the vicinity of surfaces. It is evident that moving away from the surfaces and charged groups, the cell water begins to act more like free water. This is a continuum theory; it does not imply that there are two kinds of water but rather that water changes. The early evidence already indicated that the great majority of intracellular macromolecules, including enzymes are probably not free in solution, but are instead loosely associated with formed structure. The view is based on studies on metabolic flux measurements, calculations of metabolite-transit times, and kinetic tracer studies of intracellular pools. The concept of water in the vicinity of surfaces is furthermore based on cell disruption studies. From the known physical properties of extruded protoplasm, which maintains its physical integrity, it is likely to suggest that cell water occupies a large portion of the cell volume in the vicinity of surfaces, where it may have a longer residence time and moves more slowly. With respect to a more exact determination of the water composition of all the compartments it is necessary to realize that this water is exposed to a truly vast surface area presented by cellular architecture. An early example has been mentioned by Clegg (1983). Using image analysis methods the surface area and volume of the cytoskeleton and associated cytoskeletal components were estimated for a single cell of 16 µm diameter; the surface area of these cytoplasmic structures is between 47.000 and 95.000 µm^2. On the basis of the surface to volume ratio of the cytoplasm of about 30 to 60 it is expected that a large fraction of cell water should be in reasonably close proximity to ultrastructural surfaces. In this respect it is significant that several studies have indicated that water within 25-30 Angstrom of biological surfaces is known to exhibit physical properties that differ significantly from those of pure water. This influence can extend even over 500 Angstrom from such surfaces. Most experimental evidence for long-range ordering comes from the thermodynamics of dilute solutions of "effectively hydrophobic" compounds in which solute-solute interactions, transmitted through their ordering effect on water, are detectable at surprisingly low concentrations. It is even possible that none of cell water has the properties of pure water, a view that has been advocated since 1960 by Ling (1962, 1969). All these non-traditional views on cell water invoke the influence of intracellular surfaces. Thus, there is a sound basis for the perturbation of a major fraction of the water in living cells.

4.3 Water and Functional Coupling of Domains

The significance of altered cellular water, compared with the pure liquid, can now be considered. With respect to the functional state of the domain(s) in our example, i.e. between plasma membrane and the actin networks, it is of great importance that extremely extensive networks are made. Keeping in mind that many extracellular signals, a number of hormones for example, exert their influence on the cytoskeletal organization, it is likely that the very weak stimuli impinging on the cell surface switch the organized water domain consisting of membrane and cytoskeletal structures interacting with one another from one state to another, creating conditions to switch the rate of actin addition to higher levels chiefly at nucleating sites to bring about filament growth near to specific receptors in the plasma membrane. The feature of relevance is, that generate many other chemical reactions might, also influence this domain by switching it on or off. If the view that virtually all metabolism occurs at or near the surfaces of cellular structures is accepted then, it is evident that the enzymes that perform this role find themselves in a microenvironment that is influenced by the changes taking place in the membrane-water-cytoskeleton domain. It automatically results in a shift in the composition of the domain that could lead to a subsequent switch in the domain leading to changes in protein interactions and a de-sensitization of the previously activated receptors. Significant metabolic effects could then result from very low levels of energy input. However, an important feature to be kept in mind is that the domain discussed has not a fixed boundary in space. Two general features are responsible for the domain boundaries. First, the domain depends on the location of the receptor under consideration. Secondly, any particular domain is influenced by a multitude of metabolic processes that are taking place in this domain as well. Each enzyme can be considered as a basic unit of biological organization. Recognition of substrates by enzymes is connected with the correspondence of the active site of the enzyme to certain molecular coordinates of a substrate. The energy released when a substrate is recognized by the enzyme molecule turns the latter into the different alternate conformation which results in the realization of the actual substrate conversion into the product. The enzyme-substrate interaction can be represented as a cyclic structure (enzyme molecule undergoes cyclic conformational transition), whereas substrate conversion into a product is a linear process. An early example studied experimentally is the activation of myosin adenosinetriphosphatase (ATPase) by actin (Knight and Wiggins, 1979). The ATPase operates by undergoing a conformational change which generates a zone of interfacial water of greatly increased hydrogen-bonded structure. The resulting cyclic changes in free energy of hydration of all solutes or functional groups within the modified aqueous zone could then drive the association-dissociation cycles that are needed for other enzymatic processes (in this domain).

5. TOWARDS COHERENCE AND A METABOLON

5.1 Coherence and Metabolic Domains

Long-range mobile protonic states are the most essential coherent events in organized cellular processes via which interactions are realized. Coherence within cycles is maintained by the supramolecular domain organization of enzymes; the enzymes of a domain form cycles that are assembled in multienzyme metabolic domains (Igamberdiev 1993; 1998; 1999). Such a metabolon is possible because of the specific recognition of molecules of different enzymes. The enzymes constituting metabolons are organized in such a way that the enzymes of the direct and reverse routes are regulated reciprocally in a metabolon structure. Such elementary cycle is observed in the metabolic reactions that involve ATP utilization and that can be coupled with the reactions of ATP generation. This principle is also demonstrated in the reciprocal regulation of the enzymes forming NADH that are connected with the reactions utilizing NADH. Although these cyclic structures seem to be disadvantageous in terms of energy waste, their transition to a higher level of organization with strongly coupled reaction networks converts them into a powerful mechanism of reciprocal regulation of metabolic pathways. Then, they are considered as serving as metabolic switchers. A particular feature is their action as allosteric regulators for the reciprocal regulation of the two enzymes that act in a reciprocal manner. It results in the effective regulation of both – direct and reverse – routes in such a way that these reactions become separated in time and energy loss is minimized. The separation between the direct and reverse reactions seems to be the initial precondition for the appearance of networks of increasing lifetime and stability. It is a major characteristic for defining the coherent volume of a domain. The cost of regulation has been estimated to be very high and it can reach 20% of energy available in the biosystem. However, maintenance of homeostasis is the gain of this cost (Koshland, 1987). The increase of number and diversity of enzymatic and metabolic links within a domain results in a biosystem with lesser but larger domains of long-lived and stable networks. Any realization of a new cycle is therefore a realization of a new dynamical coherent structure. The coherent volume of the domain is then extended. The generation of oscillations is theoretically connected with the two depots between which the substances can flow with a certain period. Along this view transport between depot compartments within a domain is necessary for providing the oscillations. Although the importance of transport is evident, this property has been hardly included in the consideration of coherence in cells. It is clear that diffusion is not a very likely explanation for the regulation of this metabolic structure. Cells therefore are perceived to need some new strategy for an intracellular directional flow of nutrients. Interestingly, a fluid movement in relative large cells as unicellular Acetabularia cells has received broad attention with respect to the question how metabolic events within these giant cells might operate in a controlled manner (Koop and Kiermayer, 1980; Van Wijk et al, 1999). In earlier studies it was generally accepted that there is an upper limit to the size a cell can attain before diffusion becomes too slow a process at the molecular level to satisfy metabolic demand. Big cells, therefore were perceived to need some strategy for dispersal of nutrients, and came up with the process of protoplasm streaming. However, streaming is seen in large cells for no other reason than that these cells lend themselves to such observation. In fact, similar fluid movement necessarily occurs in all cells. It is evident that if movement occurs in a concerted manner, and all the particles move in a certain direction, these particles can move from one location to another much more quickly than could ever be achieved by diffusion random movements of molecules in a non-stirred environment. Wheatley has pointed up the fact that directionality of the flow of molecules provides information to maintain vital activity (Wheatley 1985; 1999; Wheatley and Clegg, 1994). The intracellular directional flow increases the complexity of the organization of the cell. With such view we are beginning to get a more complete image of the structural and functional organization of the protoplasm. A significant parameter therefore in relation to metabolic activity is the flow. The speed of flow, its control, and its directionality have been long overlooked in favor of concentration at the cell level. At basal metabolic rate, a low flow rate might suffice to keep life going, but for cells to work under extreme conditions it demands that fuelling is up-regulated almost immediately. This brings in another factor, dealing with the relationship between structure and function of the protoplasm. If the protoplasmic mass anticipates work, some restructuring has taken place before it actually begins its exertion. And as soon as it comes into operation the flow automatically is faster in the protoplasm structure. This element of purpose is most appropriately introduced into classical cell physiology by assuming that the cell is directed and effective in particular circumstances through innate reflexes. According to the model proposed in the present chapter, the proper cooperation between cellular processes can be understood if it is assumed that the concerted movement of particles is an essential part of the coherent structure. The directed movement can be seen as reflecting the dynamic coherent domain (Van Wijk et al, 1999).

5.2 Concept of Stored Mobilizable Energy

The view that cells are large and complex dynamically structured ensembles of thousands of different molecules, in which a cascade of increasingly macroscopic changes occur, motivates the concept of stored mobilizable energy in a space-time structure. According to this view molecules do the work in a cell by a direct transfer of stored energy (Popp, 1986; Popp, 1989). In principle, thermalized energy might be converted into stored energy over all space-times, so that it is mobilizable all over the system. The key to understanding the thermodynamics of a living system should therefore be energy storage under energy flow. The energy storage under energy flow is effective if cycles are coupled together in a coherent domain. The more cycles there are, the larger the coherent domain is, the more energy is stored, and the longer it takes for the energy to dissipate. The average residence time of energy is therefore a measure of the organized complexity of the system. An important step forward in the description of this view was the model constructed by Fröhlich (1968; 1983). From his point of view biological materials function in a most systematic manner even though they form extremely complicated chemical systems. And it was the search for a physical characteristic common to most biological materials that led Fröhlich to examine their dielectric properties. He represented the biological system by two highly polarisable dielectrics capable of giant dipole oscillations that were separated by a distance greater than their dimensions. The relevance of dielectrics to biological cells and tissues will be clear, since there is a very high electric field across live biological membranes. Fröhlich modeled this situation mathematically in terms of oscillating dielectric dipoles and concluded that non-linear, coherent excitations of the dielectric dipoles were theoretically possible leading to long-range interactions on a frequency-selective basis.

6. FINAL REMARK

An intriguing question is ultimately the role of genetic information in biological information and morphogenetic processes. The genetic information defines the parameters of morphogenetic processes but it cannot give much for understanding morphogenesis. According to Popp the specific recognition and molecular complementarity might be the basis of biological information processing and can be based on condensed bosonic fields (Popp, 1989). DNA itself can then be a vehicle for the self-assembly model of computing, since here the conformational dynamics can change in distinctly different ways in response to different input signals. This may be realized via coherent interactions in DNA-sequence-specific biophoton transfer. The coherence of biophotons may form a Gestalt – information essential for morphogenesis. An important role in morphogenetic processes belongs to the heterogeneities which participate in the process of self-organization and realize coherent events as energy-driven self-consistent dynamics. Biomembranes provide the deposition of charges and determine the transport of ions and metabolites which results in the intracellular autoelectrophoresis. This polarization of the cell is an important preconditon of morphogenetic processes. In this respect morphogenesis is not only a chemical, but it is a mechanochemical process in which heterogeneities, phase boundaries and mechanical tensions play an important role (Beloussov, 1998). The water-membrane interface is considered to be a substrate for a proton superflow based on the pairing of mobile hydrogen bonds by propagating electronic oscillations in polar side groups of the membrane. Cytoskeleton is an important milieu for providing coherent events. The position of many multienzyme complexes on cytoskeleton is an important precondition for their self-according operation. Thus, morphogenesis can be understood only in the frame of limiting conditions that define self-generating processes. Functional activity of biosystems emerges during the process of morphogenesis via non-programmable generation of computable events through the coherent phenomena. Coherent events and computing phenomena percolate from elementary submolecular level to organismic entities by generating the dynamic informational field that operates within the metabolic network. We believe that the arguments will strongly favor the further substantiation of the concept of coherence in biology. Many of the ideas are hard to test experimentally at this point in time. However, in our opinion further progress in this area should significantly contribute to our understanding of the processes underlying the dynamic structuring and functioning of living cells and organisms.

REFERENCES

Alberts,B, Bray,D, Lewis,J, Raff,M, Roberts,K, Watson,J.D. (1994) *Molecular biology of the cell.* New York, Garland Publ..

Beloussov,L.V. (1998) *The dynamic architecture of a developing organism.* Kluwer, Dordrecht.

Bloom,G.S. (1992) Motor proteins for cytoplasmic microtubules. *Curr.Opin.Cell Biol.*, **4**, 66-73.

Cao.L.G., Wang,Y.L. (1990) Mechanism of the formation of contractile ring in dividing cultured animal cells. II. Cortical movement of microinjected actin filaments. *J.Cell Biol.***111**, 1905-1911.

Clegg,J.S. (1983) Intracellular water, metabolism and cell architecture. In *Coherent Excitations in Biological Systems*, Fröhlich, H. and Kremer, F. (eds.), Springer Verlag, Berlin, 162-177.

Clegg, J.S. (1984) Properties and metabolism of the aqueous cytoplasm and its boundaries. *Am. J. Physiol*, **246**, 133-151.

Clegg, J.S. and Drost-Hansen, W. (1991) On the biochemistry and cell physiology of water. In: *Biochemistry and Molecular Biology of Fishes*, vol.1., Hochachka, P.W., Elsevier, Amsterdam 1-23.

Condeelis,J. (1993) The formation of cell protrusions. *Annu.Res.Cell Biol.*, **9**, 441-444.

Drost-Hansen,W. (1965) The effects on biological systems of higher-order phase transitions in water. *Annals of New York Acad. of Sciences,* **125**, 471-501.

Drost-Hansen,W. (1971) Structure and properties of water at biological interfaces. In *Chemistry of the Cell Interface, B.*, Brown, H.D. (ed.) Academic Press, New York, 1-184.

Edidin,M. (1992) Patches, posts and fences: proteins and plasma membrane domains. *Trends Cell Biol.,* **2**, 376-380.

Fröhlich, H. (1968) Long range coherence and energy storage in biological systems. *Int.J.Quant.Chem.*, **2**, 641-649.

Fröhlich, H. (1983) Evidence for coherent excitations in biological systems. *Int.J.Quant.Chem.*, **23**, 1589-1595.

Frye, L.D., Edidin,M. (1970) The rapid intermixing of cell surface antigens after formation of mouse-human heterokaryons. *J.Cell Sci.* **7**, 319-335.

Gelfand, V.I., Bershadsky,A.D. (1991) Microtubule dynamics: mechanism, regulation and function. *Annu.Rev.Cell Biol.*, **7**, 93-116.

Gruenberg, J., Clague, M.J. (1992) Regulation of intracellular membrane transport. *Curr.Opin.Cell Biol.*, **4**, 593-599.

Gumbinor, B., Louvard, D. (1985) Localized barriers in the plasma membrane: a common way to form domains. *Trends Biochem.Sci.*, **10**, 435-438.

Hong, W., Tang, B.L. (1993) Protein trafficking along the exocytotic pathway. *Bioessays*, **15**, 231-238.

Ide, M., Maeda, Y., Kitano,H. (1997) Effect of hydrophobicity of amino acids on the structure of water. *J.Phys.Chem.B.*, **101**, 7022-7026.

Igamberdiev, A.U. (1993) Quantum mechanical properties of biosystems: a framework for complexity, structural stability and transformation. *Biosystems*, **31**, 65-73.

Igamberdiev, A.U. (1998) Time, reflectivity and information processing in living systems: a sketch for the unified informational paradigm in biology. *Biosystems*, **46**, 95-101.

Igamberdiev, A.U. (1999) Foundations of metabolic organization: coherence as a basis of computational properties in metabolic networks. *Biosystems*, **50**, 1-16.

Jacobson, K., Ishihara, A., Inman, R. (1987) Lateral diffusion of proteins in membranes. *Annu.Rev.Physiol.*, **49**, 163-175.

Jacobson, K, Vaz, W.L.C. (1992) Domains in biological membranes. *Comm.Mol.Cell Biophys.*, **8**, 1-114.

Jain, M.K. (1988) *Introduction to biological membranes*. New York, Wiley.

Janmey, P.A. (1991) Mechanical properties of cytoskeletal proteins. *Curr.Opin.Cell Biol.*, **3**, 4-11.

Kelly, R.B. (1991) Secretory granule and synaptic vesicle formation. *Curr.Opin.Cell Biol.*, **3**, 654-660.

Knight, V.A., Wiggins, P.M. (1979) A possible role for water in the performance of cellular work. II. Measurements of scattering of light by actomyosin. *Bioelectrochem.Bioenergetics*, **6**, 135-146.

Koop, H.U., Kiermayer, O. (1980) Protoplasmic streaming in giant unicellular green alga Acetabularia mediterranea. I. Formation of intracellular transport systems in the course of cell differentiation. *Protoplasma*, **102**, 147-165.

Koshland, D.E. (1987) Switches, thresholds and ultrasensitivity. *Trends Biochem.Sci.*, **12**, 225-229.

Kreis, T.E. (1992) Regulation of vesicular and tubular membrane traffic of the Golgi complex by coat proteins. *Curr.Opin.Cell Biol.*, **4**, 609-615.

Kupfer, A., Singer, S.J. (1989) Cell biology of cytotoxic and helper T cell functions: immunofluorescence microscopic studies of single cells and cell couples. *Annu.Rev.Immunol.*, **7**, 309-337.

Laaksonen, A. Kusalik, P.G., Svishchev, ,I.M. (1997) Three-dimensional structure in water-methanol mixtures. *J.Phys.Chem., A*, **101**, 5910-5918.

Lee, G. (1993) Non-motor microtubule-associated proteins. *Curr.Opin.Cell Biol.*, **5**, 88-94.

Lee, J., Ishihara, A., Therior, J.A., Jacobson, K. (1993) Principles of locomotion for simple-shaped cells. *Nature*, **362**, 167-171.

Ling, G.N. (1962) *A physical theory of the living state: The Association-Induction Hypothesis.* Blaisdell, New York.

Ling, G.N. (1969) A new model for the living cell: a summary of the theory and recent experimental evidence in its support. *Int.Rev.Cytol.*, **26**, 1-61.

Myles, D.G., Primakoff, P. (1985) Sperm surface domains. In *Hybridoma Technology in the Biosciences and Medicine*, Springer, T.A. (ed.) 239-250, New York, Plenum.

Nadig, G., Van Zant, L.C., Dixon, S.L., Merz, K.M. (1998) Charge-transfer interactions in macromolecular systems: a new view of the protein/water interface. *J.Am.Chem.Soc.*, **120**, 5593-5594.

Olmsted, J.B. (1986) Microtubule-associated proteins. *Annu.Rev.Cell Biol.*, **2**, 421-457.

de Petris, S., Raff, M.C. (1973) Normal distribution, patching and capping of lymphocyte surface immunoglobulin studied by electron microscopy. *Nature*, **241**, 257-259.

Popp, F.A. (1986) *In Disequilibrium and Self-Organization.* Kilmister, C.W. (ed.), Reidel, Dordrecht, 207-221.

Popp, F.A. (1989) Coherent photon storage of biological systems. In *Electromagnetic Bio-Information.* Popp,F.A. (ed.), Urban & Schwarzenberg, München, 144-167.

Sheetz, M.P. (1993) Glucoprotein mobility and dynamic domains in fluid plasma membrane. *Annu.Rev.Biophys.Biomol.Struct.*, **22**, 417-431.

Sharp, K.A., Madan, B. (1997) Hydrophobic effect, water structure, and heat capacity changes. *J.Phys.Chem.B*, **101**, 4343-4348.

Sherrington, C.S. (1940) *Man on his Nature.* Gifford Lectures 1937/8, Penguin Books, Harmondsworth, Middlesex, 69-98.

Svergun, D.I., Richard,S, Koch, M.H.J., Sayers, Z., Kuprin, S, Zaccai,G (1998) Protein hydration in solution: experimental observations by X-ray and neutron scattering. *Proc.Natl.Acad.Sci.*, USA, **95**, 2267-2272.

Theriot, J.A., Mitchison, T.J. (1992) The nucleation-release model of actin filament dynamics in cell motillity. *Trends Cell Biol.*, **2**, 219-222.

Van Wijk, R., Scordino, A., Triglia, A., Musumeci, F., (1999) Simultaneous measurements of delayed luminescence and chloroplast organization in Acetabularium acetabulum. *J.Photochem.Photobiol.*, **49**, 142-149.

Wadsworth, P. (1993) Mitosis, spindle assembly and chromosome motion. *Curr.Opin.Cell Biol.*, **5**, 123-128.

Waters, M.G., Griff, I.C., Rothman J.E. (1991) Proteins involved in vesicular transport and membrane fusion. *Curr.Opin.Cell Biol.*,**3**, 615-620.

Watts, C., Marsh, M.,(1992) Endocytosis: what goes in and how? *J.Cell Sci,* **103**,1-8.

Wheatley, D.N. (1985) On the possible importance of an intracellular circulation. *Life Sci.*, **36**, 299-307.

Wheatley, D.N. (1994) On the vital role of fluid movement in organisms and cells: a brief historical account from Harvey to Coulson, extending the hypothesis of circulation. *Medical Hypothesis*, **52**, 275-2284.

Wheatley, D.N., Clegg, J.S. (1994) What determines the basal metabolic rate of vertebrate cells in vivo? *Biosystems*, **32**, 83-92.

Whipple, H.E. (ed.) (1965) Forms of Water in Biological Systems. *Annals of the New York Academy of Sciences,* vol.125,1965.

Wiggins, P.M. (1990) Role of water in some biological processes. *Microbiol.Rev.*, **54**, 432-449.

Wiggins, P.M. (1995) Micro-osmosis in gels, cells and enzymes. *Cell Biochem.Function*, **13**, 165-172.

Chapter 5

PHYSICAL BASIS AND APPLICATIONS OF DELAYED LUMINESCENCE

Francesco Musumeci
*Dipartimento di Metodologie Fisiche e Chimiche per l'Ingegneria,
Università di Catania, INFM Unità di Catania,
Viale A. Doria 6,
I-95125 Catania (Italy)*

1. INTRODUCTION

The ultra weak luminescence of biological systems is a phenomenon which was first measured about 50 years ago [L. Colli and U. Facchini, 1954] and since then has received increasing interest. Notwithstanding the fact that this phenomenon has a very low intensity, variable from about ten to several thousand photons per second per square centimetre, it seems important because experiments show that it depends on the biological state of systems.

Even today the mechanisms that cause this phenomenon are not well understood, and it has not been possible to establish whether the luminescence is due to chemical or metabolic reactions [D.Slawinska and J.Slawinski, 1983, B. Deveraj et al., 1997] or if it is due to a coherent electromagnetic field, inside biological systems, that play a fundamental role in the promotion and control of vital processes [F.A. Popp, 1986]; the last mentioned author calls the quanta of the above mentioned field "biophotons".

To understand the origin of ultra weak luminescence is necessary both to evaluate the importance of this phenomenon from a biological point of view, and to extend the applicative possibilities.

The intensity of the phenomenon is however so low as not to allow a direct measurement of the coherence. To study the phenomenon of ultra weak luminescence one must try to increase the intensity disturbing the system with physical or chemical methods. One method, considered non-invasive, is that of illuminating the sample. This provokes a transitory increase of the intensity of the ultra weak luminescence, which then returns to its normal value.

Photo induced ultra weak luminescence, often called, in literature, Delayed Luminescence (DL), can be easily characterised from a spectral point of view and from the time trend because its intensity is much greater than spontaneous ultra weak luminescence. This fact has made possible the development of a large number of experimental works on different biological systems [A. Ezzahir et al., 1992, R. Van Wijk et al., 1993, H. J. Niggli, 1993, M. W. Ho et al., 1998, A. Triglia et al., 1998] that show how there is a close connection between the state of biological systems themselves and their photo induced ultra weak luminescence. In some cases it has been possible to express this connection through an analytical relationship between a biological parameter and one of DL [F. Musumeci et al., 1994, A. Scordino et al., 1996].

This fact has produced notable interest for the applicative potential of the DL in different sectors, for example the control in real time of pollution in water, the checking of the quality of food, or clinical diagnosis. However until it will be possible to understand properly the mechanisms that underlie the phenomenon it will be difficult to obtain further applicative developments. It is therefore necessary, so that Delayed Luminescence can become a widely used analysis technique, at first further to study the problem of its origins.

In the literature there are at the moment two hypotheses. The first [J. Lavorel et al., 1986] considers the DL as the visible evidence of an imperfection in the mechanism of separation and transport of charge induced by the light or as the inevitable consequence of the presence of a multiplicity of excitable states inside biological systems. The second, by F.A. Popp [F.A. Popp, 1986, F.A. Popp, 1989] sees this phenomenon as strictly connected to spontaneous ultra weak luminescence, rising from the excitation and the successive decay of a coherent field, present inside biological systems.

Notwithstanding the close link that we have mentioned between the state of biological systems and their DL, the second hypothesis could seem more correct; both theories identify, unfortunately, the emission mechanism in such a generic way as to make impossible a strict theoretical prediction, on which to construct a control experiment. [A. Scordino et al., 2000].

The identification of such a mechanism is difficult: biological systems are made up of relatively high density material and present a series of ordinate structures metastable at a mesoscopic level that must be considered to influence the structure of the electron levels possible inside the same systems. This influence has already been observed in the study of luminescence of solid: from the end of the 19th century, it was noted, for certain materials, that the afterglow intensity diminished following not an exponential time trend, typical of independent excitation levels, but more or less a hyperbolic one. This decay law was called Bequerel's law [E.Bequerel, 1868]. With the evolution of techniques, it has been seen that this behaviour regards numerous materials in the crystalline state and has been described as due to a decay of collective excited states called 'excitons' (M.D.Galanin 1966).

The principal objection that can be made against this type of approach is that, in general, these models have been proposed regarding the band theory of solids and are therefore valid in the case of condensed systems that present a well organised structure in domains of rather large dimensions.

Within living systems, despite there are not being stable ordinate structures of relatively large dimensions, many biological molecules form metastable super molecular organisations that are very similar in structure to liquid crystals.

Biological molecules, in fact, do not form real solutions, in which the solutes are distributed randomly within the solvent, but rather form a sort of colloid, in which large molecules are connected by transverse bonds so as to lose the characteristics of liquid and assume almost those of solids, characterised by long distance orientational orders, like lyotropic liquid crystals. In these conditions the short distance between nearest neighbours atoms and the regularity of the structures present contribute to the determination of electron configuration, and therefore of optical characteristics, unusual for molecular systems [M.W. Ho, 1998]

This possible approach to the determination of the mechanism which lies at the base of luminescence seems to be well based both from the experimental and theoretical points of view and will be discussed later in this chapter comparing the behaviour of the DL in biological systems and those in solid state before analysing the possible applicative aspects.

2. MAIN CHARACTERISTICS OF DL IN BIOLOGICAL SYSTEMS

The DL of biological systems depends notably on the nature and the state of the system and not least on the features of the light source.

However certain characteristics are present in all the systems analysed. In particular, experience shows that:

a) The DL kinetics corresponds to the phenomenological Becquerel law, given by the hyperbolic curve

$$I(t) = (a+bt)^{-\alpha} \qquad (1)$$

which can be written more conveniently as:

$$I(t) = \frac{I_0}{\left(1 - \frac{t}{t_0}\right)^\alpha} \tag{2}$$

This fact is shown in fig. 1 for different biological systems. It can be seen that, in effect, all systems decay following the hyperbolic trend described in equation (2). An idea, of how good this approximation is, can be had comparing the trend of the experimental data with those of the relative nonlinear fits obtained according equation (2).

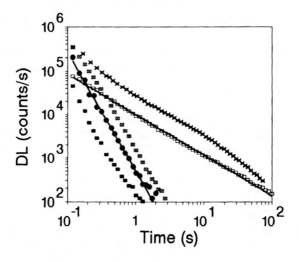

Figure 1: DL of different biological systems (■ yeast culture, □ *Acetabularia acetabulum*, ⊠ soya seeds, ● pepper seeds, ⚹ tomatoes), illuminated with light of the same intensity characterised by a continuous spectrum with a wavelength between 400 e 450 nm. The continuous lines that overlie at the symbols represent the trend foreseen by eq. (2) corresponding to acetabularia and pepper seeds.

In both cases shown in figure 1 the agreement is extremely high (r^2 equal to 0.999 for the acetabularia with I_0=10000, t_0=0 and α=0.94 and equal to 0.996 for the pepper seeds with I_0=520, t_0=0 and α=2.72).
b) The emission spectrum of the systems analysed, as far as can be seen using current apparatuses that work using pass band filters, is a continuous spectrum quite wide (width at half height of the order of some hundred nanometres). This spectrum differs from system to system, as does the

excitation spectrum. That which, instead, is generally found is that the various components of the excitation spectrum all have the same time trend.

This important property of the emission spectrum, which strongly contrasts with the first hypothesis on the phenomenon, is shown in figure 2 that reports, in the case of the Acetabularia sample, the DL of the various components of the spectrum versus the total DL. It is evident that all the experimental data lie on parallel lines and then that the time trend of all spectral components is identical. The same behaviour has been observed for all the biological systems examined

That shown in figure 2 makes it difficult to explain the DL through a mechanistic approach that describes it as a sum of microscopic almost independent sub-phenomena describable in terms of a series of biochemical processes unrelated to living systems.

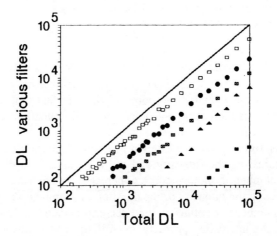

Figure 2: DL of the various components of the emission spectrum of a living acetabularia cell versus the total DL (— no filter, ■ 580-620 nm, ☒ 630-670 nm, ☐ 680-720 nm, ● 750-800 nm, ▲ 800-850 nm).

c) Experiments show that there is (Fig. 3) a strong dependence of the decay slope on the intensity of illumination.

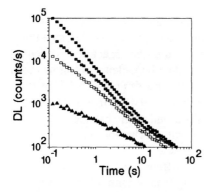
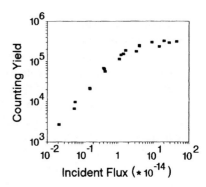

Figure 3 a: DL trend of Acetabularia for different intensities of exciting light at the same light wavelength (425 ± 25 nm): (■) 5 10^{14} photons/ s / cm^2, (⊠) 2.5 10^{14} photons/ s / cm^2, (□) 1.2 10^{14} photons/ s / cm^2, (▲) 6 10^{13} photons/ s / cm^2

Figure 3 b: Total number of photons reemitted by an *Acetabularia acetabulum* vs. the intensity of the incident light beam (illumination time 5s, wavelength 660 ± 20 nm)

The behaviour, shown in figure 3a, of the intensity (connected to I_0) and of the slope (connected to α) on varying the intensity of illumination is fairly general from a qualitative point of view: in fact both the parameter values increase on increasing the intensity of illumination, at least for intensity values far from the saturation ones.

For some systems, as shown in figure 3b in the case of the Acetabularia, on increasing the intensity and the duration of the illumination source the total number of photons emitted increases up to a saturation value; on further increasing the intensity or the duration of illumination, it can happen that the total number of photons emitted decreases again [A. Scordino et al, 1993].

d) An interesting property, even if, until now, not found in all biological systems, probably due to the low intensity of the excitation source used, is the presence of oscillations in the time trend of the DL [F.Musumeci et al., 1992].

These oscillations, shown in figure 4a for an *Acetabularia acetabulum*, appear in proximity to the conditions of saturation of the curve in figure 3b.

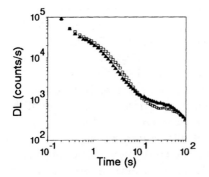
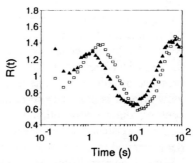

Figure 4 a: DL trend of an *Acetabularia acetabulum* for high intensity and/or long duration of exciting light.

Figure 4 b: Relationship R(t) between the experimental value of the DL and its best fit obtained according to eq.2

In these conditions if the non-linear best fit of the DL is calculated following equation 2, and one divides instant by instant the experimental value of the DL for the best fit of DL values vs. time, one obtains the quantity:

$$R(t) = \frac{I_{sper}(t)}{I(t)} = \frac{I_{sper}(t)\left(1 - \frac{t}{t_0}\right)^\alpha}{I_0} \qquad (3)$$

Figure 4b shows that, in general, *R(t)* results as being an almost periodic function of ln(*t*). It is interesting to observe that this behaviour can be described [F.Musumeci et al., 1992] still with a hyperbolic decay which follows (2) in which, however, the coefficient α is a complex number of which the real part describes the temporal decay of the DL while the imaginary part describes the oscillations.

3. LUMINESCENCE OF CONDENSED SYSTEMS

The optical properties of solid state systems have been widely studied in the last decades, also in relation to the technological importance assumed by different types of semiconductors and the expansion of optoelectronics. In this way the luminescence of condensed systems has also been studied, both for systems in which it seems to be intrinsic, as molecular crystals or some types of semi-conductors, and for systems in which it is due to the presence of impurity levels, as crystallophosphors. In this second case, the presence of the luminescence is connected to the presence of impurities, even if the matrix, which can be crystalline or amorphous, can greatly influence the energy, the intensity and other characteristics of original electronic levels of the impurities themselves. The first case, instead, often regards systems in which the luminescence depends on the topological characteristics of the matrix, which presents a more or less organised structure, and arises from the presence of collective excitations inside the structure. To make this point of view clear it is necessary to remember that, from the beginning of the 20th century, optical absorption spectra were observed, for pure solids, with bands inexistent in the spectra of the atoms of origin. This fact has demonstrated that optical absorption and emission processes in solids cannot be described simplifying the problem and treating the solid as if it were a very dense gas.

The difference between the two systems can be better understood considering the problem, analogous to this, of the thermal response of solids. It is well known that the thermal behaviour of a solid can not be described fully through the description of the movement of the single constituent atoms, calculated as though they were not inter correlated. The translation symmetry present in the lattice imposes the existence of those normal modes of vibration that we call phonons, each of which belongs to a vector of the reciprocal lattice. Analogously it is not possible to characterise fully an excited electron state of a perfect crystal through localised excited electron states, instead one must describe it as a derivation of the superimposition of normal modes each of these connected to a vector of reciprocal lattice.

These normal modes of excitation have been called "excitons" [J. Frenkel, 1931] and are often described as a bound electron-hole pair created by an intraband transition due to the absorption of an impinging photon.

Numerous theoretical and experimental studies have been made as regards the influence of the excitons, not only on the optical absorption spectra of numerous systems, but also on the luminescence.

However, up till now, there has not been much data relative to DL of such systems and, as a consequence, there are no deeper analyses.

The few data reported are probably those that our group has obtained though a special series of measurements [A. Scordino et al., 2000].

These measurements were made on various types of crystal, in particular molecular crystals, in which there should be the presence of localised excitons or Frenkel excitons, and crystals of semiconductors, in which there has been hypothesised the existence of delocalised excitons or Wannier-Mott excitons.

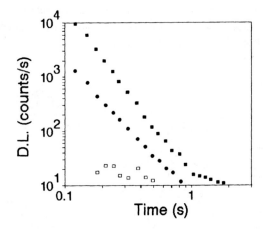

Figure 5: Effect of phase structure on the DL of salol at 42°C. (■) crystals of salol at 42 °C, (●) glass cuvette, (□) liquid salol at 42 °C. The signal relative to the liquid phase, much less than that of the container, is probably due only to noise.

One of the more interesting aspects of DL in solids that distinguish it from analogous phenomena such as fluorescence in liquids and gases is that it is determined essentially by the structure of the system and it exists also in absence of specific fluorophores.

Figure 5 shows this aspect for Phenyl Salicylate (salol). This substance, like other molecular crystals, can maintain itself for quite long times in the liquid state even below melting temperature, which is equal to 43 °C; and therefore it is possible to measure the DL of the two phases, solid and liquid, at the same temperature. As can be seen the destruction of the organised structure brings a substantial reduction of the DL.

Another important fact is that interesting analogies have been found with the results obtained for the DL of biological systems. Figure 6a shows, in fact, the DL from some different crystals in which the physical mechanisms that are at the basis of the phenomenon are different [M.D. Galanin, 1966]:

a) Ruby in which luminescence is due to impurity centres, i.e. atoms in small concentration, present in solids is often transparent in the region of the electronic transition of the impurity. In this case the characteristics of the luminescence are determined by the spectroscopic characteristics of the atoms of impurity and feel only indirectly the structure of the host material. For this fact the intensity of the luminescence, differently to other cases, is almost independent of temperature (data not reported).

b) Silicon in which the main mechanisms that give origin to DL is the recombination of a couple electron-hole after an interband transition due to the absorption of an impinging photon.

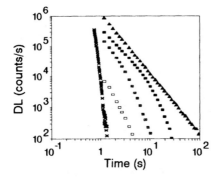

Figure 6 a: Trend of DL decay for some different systems. (▲ polycrystalline CdS, ■ polycrystalline anthracene, ⊠ polycrystalline salol, □ silicon monocrystal, ✦ ruby monocrystal), illuminated with the same source as Fig. 1. It can be seen that, for some systems, the time trend can still be described by eq(2)

Figure 6 b: The DL of the various component of the emission spectrum of Cadmium Sulphide versus the total DL (-no filter, ⊠ 630-670 nm, □ 680-720 nm, ● 750-800 nm, ▲ 800-850 nm, × 850-900 nm)..

c) Cadmium sulphide whose luminescence is described by the recombination of Wannier-Mott excitons.
d) Salol and antracene whose luminescence is described by the recombination of Frankel excitons

It is possible to see that for most systems the trend is still hyperbolic, even if, in the case of ruby, the time trend of the DL is better described by an exponential decay rather than by a hyperbolic trend.

Another important analogy is shown in figure 6b in which it can be seen that, in some cases, as that for Cadmium Sulphide, the various components of the emission spectrum all demonstrate the same time trend.

This property is not general as in the case of biological systems and seems to be possessed above all by systems whose luminescence is described by the recombination of Wannier-Mott excitons.

Figure 5 has shown that, in crystalline condensed systems, the DL is determined essentially by the structure of the system and that disappears if the structure is destroyed. Beginning from this data, the influence that the order parameters of the structure have on the DL have been studied measuring it on samples constituted by micro crystals of CdS of high purity. The different samples were of the same origin and were different only in their average dimensions of micro crystals. The result of this analysis is shown in fig. 7, which shows the total number of re-emitted photons as a function of the average dimension of CdS grains.

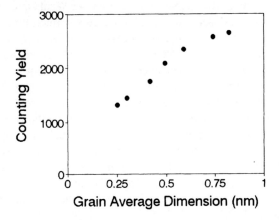

Figure 7: Total number of emitted photons as a function of the average dimension of grains for CdS samples.

As it was possible to foresee, the intensity decreases rapidly (in first approximation is linear, with $r^2 = 0.97$) with the reduction of the dimension of the grains. However the signal is even well measurable for grains having dimension of the order of hundreds of nm, comparable to the dimension of the ordered structures existing inside the cells.

This result makes vain one of the main objections that is made to the possibility of studying biological systems using the formalisms of solid state physics, objections based on the lack, in the former, of a long range topological order.

In fact, at least regarding the DL, it requires metastable ordinate structure of dimensions of the order of tens of nanometers and is a consolidated fact [S.L. Wolfe, 1994] that biological systems also have ordered metastable structures of these dimensions and this fact could influence the behaviour of the electronic orbital relative to the single molecules.

It is not easy to analyse the dependence of the DL on the dimensions of the ordinate structure present in biological systems. To do this a particularly suitable biological system was used: *Acetabularia acetabulum*. In this case, these ordered structures constitute the cytoskeleton and typically have diameters that vary from 7 to 30 nm and lengths up to several microns. To provoke artificially in *Acetabularia acetabulum* a "fragmentation" of the ordered structure certain general anaesthetic can be used. General anaesthetics seem to work, indeed, above all at the level of the cytoskeleton organisation. In fact, because these substances seem to have no chemical relationship with one another, it has been suggested [R. Penrose, 1995] that they may act through the agency of their Van der Waals interactions with the conformational dynamics of the tubulin proteins in microtubules.

This fact has been confirmed in an experiment [R. Van Wijk et al., 1999] in which the effect of Chloroform on *Acetabularia Acetabulum* was studied finding that the use of this substance influenced strongly both luminescence and protoplasm streaming and showing therefore again that DL is connected to the cytoplasm structure. An analogous effect was measured with Thiopental sodium, which is stable in aqueous solutions and permits the measurement of DL changes on varying anaesthetic concentration. In presence of low concentrations of such solutions, it can be assumed that the average volume V_0 of unaltered structures would be proportional to the inverse of the concentration, expressed in number of molecules per unit of volume, and the average dimension D_0 of these structures should be proportional to $(V_0)^{1/3}$

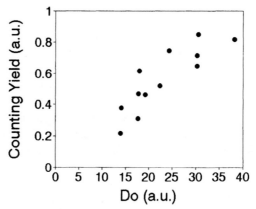

Figure 8: Total number of emitted photons as a function of D_0 parameter for some *Acetabularia acetabulum* samples.

The parameter D_0 is comparable with the average dimension of grains of the aforesaid solid state system. Figure 8 shows, in arbitrary units, the total number of photons emitted from *Acetabularia acetabulum* immersed in some solutions containing Thiopental sodium as a function of D_0. It is possible to see that the trend is linear as the one in figure 7 ($r^2 = 0.82$).

The data presented up to now show that DL could be generated failing specific fluorophores and for structures having typical dimensions of the order of those present in biological systems. In addition it appears that it is a phenomenon connected to the order parameters of the structure both for living and solid state systems. This fact suggests that the base mechanisms could be the same. This means that, in order to understand the origin of DL in biological systems, it should be necessary to study the possibility of the existence of collective electronic states and their properties depending on the organised structure of the system. This possibility will be analysed in the course of the next paragraph.

4. EXCITONS AND SOLITONS IN PERIODIC ONE-DIMENSIONAL MOLECULAR STRUCTURES

We will use a rather simple model, but able to describe what, approximately, happens inside the polymeric chains, which construct the cytoskeleton. From the point of view of electronic structure, proteins are semiconductor-like quasi-one-dimensional systems with the filled valence band and empty conduction band, separated by a gap of finite width [I.I. Ukrainskii and S.L. Mironov, 1979]. These characteristic properties of biological macromolecules favour the existence of coherent collective electron and exciton states, in general, and solitons in particular.

4.1 Excitons

We assume then that the molecules lie along the z axis at a constant distance r and that the internal excited state of a single molecule is characterised by energy ε, and by an electric dipole moment d, which forms with the axis z an angle θ (see fig. 9).

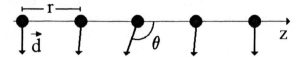

Figure 9: Representation of the one-dimensional model used to explain excitons in a molecular chain.

We are also supposing that the chain is rigid and that, therefore, the distances between molecules do not change when the chain assumes an excited state.

One can hypothesise that the fundamental state of the chain, in which no molecule is excited, has zero energy and we consider only the interactions between the nearest neighbours. Let us suppose, moreover, that the number N of atoms, which form the chain, is big enough to allow us to neglect the boundary effects.

If the molecule n is excited it will undergo two new interactions besides the combined interactions which determine its equilibrium position in the fundamental state. The first will be the resonance interaction between molecule n and its nearest neighbour molecules, and presents energy equal to:

$$J = \frac{\vec{d}^2}{r^3}(3\cos^2\theta - 1)$$

The second, called the deformation interaction, will be due to the modification of previous interactions that determine the form of the chain in the fundamental state and, because a change of this energy generally conducts towards an attractive process, it will be negative and indicated by the quantity $-D$.

The Hamiltonian operator corresponding to an intra-molecular excitation characterised by energy ε is given by:

$$H_e = \sum_n \left[(\varepsilon - D) B_n^+ B_n - J \left(B_{n+1}^+ B_n + B_{n-1}^+ B_n \right) \right] \qquad (4)$$

Of the two operators, B^+ indicates that the molecule is excited, B that it is not. D and J are calculated considering the chain in its equilibrium configuration. It is convenient to convert the operator (4) in its diagonal form through a unitary transformation to the new operators:

$$B_n = \frac{1}{\sqrt{N}} \sum_k A_k e^{iknr} \quad \text{with} \quad A_k = \frac{1}{\sqrt{N}} \sum_n B_n e^{-iknr}$$

k is defined within the first Brillouin zone and can take one of N equally spaced values. Substituting one has:

$$H_e = \sum_k E(k) A_k^* A_k \quad \text{with} \quad E(k) = \varepsilon - D - 2J\cos kr = \hbar\omega(k)$$

and
$$\psi_k(z,t) = \frac{1}{\sqrt{N}} \sum_n e^{i[knr - \omega(k)t]} \phi(z - nr)$$

$$= \left[\frac{1}{\sqrt{N}} \sum_n e^{ik[nr-z]} \phi(z-nr) \right] e^{i[kz - \omega(k)t]} = A_k e^{i[kz - \omega(k)t]}$$

where $E(k)$ represents the energy of the collective excitation of the chain (exciton), $\psi_k(z,t)$ its wave function and $\phi(z-nr)$ is the wave function of the excited state of the n-th molecule. The wave function $\psi_k(z,t)$ has the form of a plane wave with $\lambda = 2\pi/k$ and, if $\lambda >> r$

$$E(k) \cong \varepsilon - D - 2J + \frac{\hbar^2 k^2}{2m^*} = E_0 + \frac{\hbar^2 k^2}{2m^*} \quad \text{with} \quad m^* = \frac{\hbar^2}{2Jr^2}$$

Exciton states described previously, characterised by values defined of k, $E(k)$ and described by the plane wave $\psi_k(z,t)$ represent stationary waves for which the excitation is uniformly distributed along the whole chain and does not travel. To describe the real processes of excitation and the phenomena of energy transmission along the chain it is necessary to consider non-stationary states. If at the instant t the excitation is distributed on a length l_0 we will then have:

$$\psi(z,t) = \int_{k_0 - \Delta k}^{k_0 + \Delta k} C(k) e^{i[kz - \omega(k)t]} dk \quad \text{with} \quad \Delta k = \frac{\pi}{2 l_0}$$

This wave packet propagates at a group velocity:

$$v_g = \left(\frac{\partial \omega}{\partial k} \right)_{k=k_0} = \frac{\hbar k_0}{m^*}$$

And, in propagating, its form changes. If, at instant t, the packet covers a region wide l_t, at instant $t + \Delta t$ we have:

$$l_{t+\Delta t} = \sqrt{l_t^2 + \left(\frac{\hbar \Delta t}{m^* l_t} \right)^2}$$

and therefore, passing the time, the packet dilates and this dilation is greater if the effective mass is smaller. Another fact should be considered: the packet in motion exchanges energy with the lattice through the absorption or emission of phonons and this interaction is describable through a further operator of the form:

$$H_{int} = \chi \sum_n (u_{n+1} - u_{n-1}) B_n^+ B_n \quad \text{with} \quad \chi = \left(\frac{\partial D}{\partial u_n} \right)_{u_n = 0}$$

These processes are possible if the velocity of the excitons exceeds the velocity V_{so} of the longitudinal sound waves along the chain, that is if:

$$\frac{\hbar k_0}{m^*} > V_{so} = r \sqrt{\frac{\gamma}{M}}$$

where γ represents the longitudinal elasticity coefficient of the chain and M the mass of the single molecule. The occurring of this condition is necessary for the applicability of the present model. In fact only if the velocity of propagation of the excitation is greater than the speed of sound the molecules of the chain will not have the time to move, and the hypothesis mentioned at the beginning, that they stay in their initial equilibrium position will be valid.

An exciton can be generated by a photon and decay into a photon. The laws of conservation of energy and impulse, applied to the optical part of the spectrum show that only excitons characterised by a wave vector about zero can interact with the photons. This means that the wavelength of the photons that can interact with excitons is given by:

$$\lambda = \frac{2\pi \hbar c}{E_0}$$

Agranovich demonstrated that there exists a link between the decay times τ_{ex} of an electron and that of the corresponding excitation in an isolated molecule τ_0:

$$\tau_{ex} = \frac{4r}{3\lambda} \tau_0$$

Because the probability that a photon excites an exciton is proportional to $1/\tau$ this means that the probability for a photon to excite a linear chain is several orders of magnitude more than the probability of exciting a single molecule of the chain.

4.2 Solitons

The hypothesis previously made, that the chain behaves like a rigid set in which the single molecules maintain the respective equilibrium positions, may not be justified in biological systems, the polymeric chains of which are in general characterised by low values of elasticity coefficient.

In these conditions the excitation could propagate along the chain with a velocity less than the speed of sound, generating a local deformation of the system in the region of the excited molecules. It is therefore necessary, in the calculation of the collective excitation energy, to take in account the energy necessary to move the molecules from their equilibrium position.

The new Hamiltonian operator then is:

$$H_{so} = H_e + H_{int} + H_{ph} \quad \text{with} \quad H_{ph} = \frac{1}{2}\sum_n\left[\frac{1}{M}p_n^2 + \gamma(u_n - u_{n-1})^2\right]$$

We must calculate how the chain deforms in time and space.

If the position of the molecule placed in z is described by the adimensional parameter $\xi = z/r$ it happens that functions $a(\xi,t)$, whose modulus determines the probability of excitation of the single molecules, are solutions of the non linear Schrödinger equation [A.S. Davydov, 1985]:

$$\left\{i\hbar\frac{\partial}{\partial t} - (\varepsilon - D - 2J + W) + J\frac{\partial^2}{\partial \xi^2} + G|a(\xi,t)|^2\right\}a(\xi,t) = 0$$

$$\text{with} \quad G = \frac{4\chi^2}{\gamma\left(1-\left(\frac{V}{V_{so}}\right)^2\right)} \quad \text{and} \quad W = \frac{1}{2}\sum_n\left[M\left(\frac{\partial\beta_n}{\partial t}\right)^2 + \gamma(\beta_n - \beta_{n-1})^2\right]$$

In these expressions $\beta_n(t)$ represents the average value of the displacement of the single molecules calculated on the possible states of the system. The equation above is satisfied for:

$$a(\xi,t) = \frac{\sqrt{\mu}\left\{i\frac{\hbar v}{2J}(\xi - \xi_0) - \frac{E_v t}{\hbar}\right\}}{\sqrt{2}ch[\mu(\xi - \xi_0 - vt)]} \quad \text{with} \quad v = \frac{V}{r}$$

where:

$$\mu = \frac{\chi^2}{\gamma J(1-s^2)} \quad \text{with} \quad s = \frac{V}{V_{so}} \rightarrow \Delta z = \frac{\pi R}{\mu}$$

$$E_V = \varepsilon - D - 2J - \frac{\chi^4}{3\gamma^2 J} + \frac{1}{2}m_{sol}V^2 = E_0 + \frac{1}{2}m_{sol}V^2$$

$$m_{sol} = \frac{\hbar^2}{2JR^2} + \frac{4\chi^4\left(1+\frac{3}{2}s^2 - \frac{1}{2}s^4\right)}{3\gamma^2 JV_{su}^2(1-s^2)^3}$$

Through the value of $a(\xi,t)$ it is possible to get the value of $\beta(\xi,t)$, representing the displacement of the single molecules, but it is more convenient to use the parameter $\rho(\xi,t) = -\partial\beta(\xi,t)/\partial\xi$ which can be expressed as:

$$\rho(\xi,t) = \frac{2\chi|a(\xi,t)|^2}{\gamma(1-s^2)} = \frac{\chi\mu}{\gamma(1-s^2)ch^2[\mu(\xi-\xi_0-vt)]}$$

Almost always $s \ll 1$. In biological systems, moreover, often γ is small while χ is large and, for this, m_{sol} turns out to be large. It follows that, also at low velocity, the kinetic energy of the solitons can be large.

There is a notable difference in the interaction between photons and excitons and between photons and solitons. In the first case, because both the photons and the excitons are describable through a combination of plane waves, their interaction happens according to the principles of energy and moment conservation. In the second case however, because the solitons are local deformations, the moment may not be conserved.

In addition at the moment of decay of a soliton into a photon, the deformation energy of the chain is dissipated in the form of thermal energy.

This means that, in general, the energy of a photon emitted will be lower than the energy of a soliton of the quantity:

$$\Delta E = \frac{\chi^4}{3\gamma^2 J}$$

ΔE will be quite large in the biological chains, characterised by low γ values and high χ values.

This should give, in the case of processes of absorption and emission, a huge Stokes shift.

The emission probability of a soliton, however, is linked to that of the emission of an exciton by the relation:

$$\frac{1}{\tau_{sol}} = \frac{2\pi^2 r}{3\lambda\mu \ ch^2\left(\frac{2\pi\hbar v}{2\mu J}\right)} \frac{1}{\tau_{ex}}$$

and admits therefore a maximum when its propagation velocity is zero. In these conditions we can write, remembering the definition of Δz previously given:

$$\frac{1}{\tau_{sol}} = \frac{2\pi\Delta z}{3\lambda} \frac{1}{\tau_{ex}}$$

Therefore the probability of decay of a soliton is extremely low, which justifies the very weak intensity of the DL.

A recent paper [F.Musumeci et al., 2001] has shown that the application of this model to the DL of *Acetabularia acetabulum* leads to an excellent accord between the theoretical forecast and the experimental data, especially if one considers the presence of coherent soliton states. This result constitutes a solid base for a full understanding of the mechanisms of the DL, a full understanding that is at the base of possible applicative developments.

5. DL AND THE STRUCTURE OF THE CELLS

The unicellular alga *Acetabularia acetabulum* constitutes a model system for the study of DL. In fact it is possible to measure the DL even from a single individual and so implement tests.

For instance the effect of cations in the culture medium on the parameters of delayed luminescence in connection with morphological changes of *Acetabularia acetabulum* were examined [M.W. Ho et al., 1998]. Some synthetic culture media has been made, characterised by the presence of some of the four more important chlorides commonly contained in the seawater. In this way solutions have been obtained containing, besides sodium, K^+, Mg^{2+} and Ca^{2+}, ions alone or in the different possible combinations, always maintaining constant the content of Na^+ and CL^- present in standard marine water. There were two kinds of response, confirmed from morphological observations: a first one where the algae appear to maintain homeostasis and they do not change appreciably with time of immersion in the test solution, and a second one where drastic changes with time take places and the alga is moving further and further away from homeostasis.

This last behaviour has been attributed to the presence of Ca^{2+} that, entering the cell via Ca^{2+} channels, is responsible for the immediate contraction of the cytoplasmic contents in separate dense zones (Fig.10a).

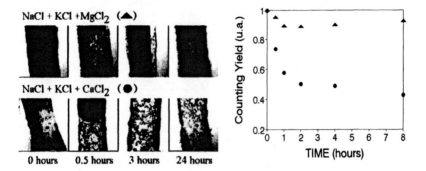

Figure 10 a: Photomicrographs of acetabularia in two different solutions without and with calcium. Note that acetabularia remain relatively unchanged in 24 hours, as consistent with the maintenance of homeostasis and the stability of DL parameters in the first solution, while, in presence of Ca^{++} ions a progressive deterioration of the sample take place.

Figure 10 b: Changes in DL of the same samples: (▲) NaCl + KCl + $MgCl_2$, (●) (NaCl + KCl + $CaCl_2$

We have previously observed, in solid state systems, that the intensity of DL is a function of the size of the grains. The reduction of the cytoplasm to a series of irregular, separate regions is thus expected to reduce DL due to the reduction in the number of excited levels that can be maintained. Fig 10b shows that DL parameters are able to discriminate between these two behaviours.

In order to check the validity of this hypothesis, and to verify therefore if the influence of extra cellular Calcium on DL is connected to the alteration of the cytoplasm structure or to a modification of the emission from chloroplasts, samples of cell-free chloroplasts extracted from intact Acetabularia were prepared. The DL of these extracts, in the absence and presence of Ca^{2+} in a basis solution containing (NaCl + KCl), was measured and was found to remain unchanged. This confirms the hypothesis that the reduction of the DL of the whole cell can be attributed to a reduction of the hierarchical order in the cytoplasm, which is responsible for the coherence of cellular activity.

DL is also connected to the protoplasm streaming that constitutes one of the more evident macroscopic manifestations of the structure of the cytoskeleton. Protoplasm streaming is detected by applying microscopic techniques and it is possible to correlate these observations with the characteristics of luminescence decay. There are chemical and/or temperature treatments able to block rapidly chloroplast motility. Both freezing in liquid nitrogen as well as exposure to chloroform within seconds is able to obtain this effect.

To test whether the effect of freezing in liquid nitrogen on the interaction between chloroplasts affects DL, a test has been performed starting with a preliminary measurement of DL in standard condition (artificial seawater as culture medium, 20°C); and carrying out a new measurement of DL after freezing, thawing and re-incubation at 20°C.

After the freezing-thawing procedure the cell remains intact and keeps its normal green colour. Even at the microscopic level a similar distribution of organelles was observed. The main drastic change is the complete loss of motility.

On the other side it appears that, just after re-incubation, the absolute value of luminescence is drastically lowered. Only 3% of the delayed luminescence remains and this fraction shows a time trend like the unfrozen sample. This trend does not change in the first hours.

This marked effect is not present in prompt fluorescence measured on cells that suffered the same freezing-thawing procedure of those used for DL measurements. It appeared that the total intensity of the prompt fluorescence measured after this treatment remained at 70% of the previous value.

These results show that DL is not simply a long time cue of fluorescence and led to conclude that while fluorescence is essentially related only to the primary events of the photochemical energy conversion in the photosynthetic apparatus, a correlation exists between DL and the organised streaming motility of the chloroplasts.

An analogous conclusion was reached studying the effects of chloroform. These effects were reversible in *Acetabularia acetabulum* cells allowing the evaluation of the consequences of such treatments on the kinetic aspects of luminescence in parallel with monitoring the movements of organelles.

Inhibition of streaming was fully reversible in cells that have been treated with chloroform no longer than 10 minutes, when returned to fresh medium. The intracellular morphology and the streaming behaviour became normal again within some minutes in cells recovered from the chloroform treatment.

Measurements of DL were performed when the alga was in standard condition, at beginning, immediately after the addition of chloroform, some time after the addition of chloroform and immediately after the restoring of normal conditions, obtained by washing and putting the cell in fresh seawater.

There was a strong decrease of DL intensity after the addition of chloroform. In this condition, the alga response was not stable and the total intensity of DL exhibit a decreasing trend and, when standard conditions were restored the DL returned to the initial value (differences less than 20%). This strong influence of the Chloroform on the DL demonstrates again that DL is able to give information on the cytoplasm structure.

A further way to gain indications about the link between the DL and the structure of the cytoplasm is to observe its dependence from the temperature, shown in figure 11.

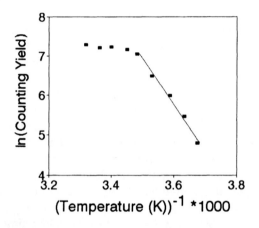

Figure 11: Arrhenius plot of DL total number of counts for a living *Acetabularia acetabulum*. The symbols refer to experimental data; the continuous line represents the trend expressed by eq.5.

In fact there is, increasing the temperature, a phase transition due to the processes of polymerisation of the monomers present into the cytoplasm.

These monomers synthesize in polymeric chains through a hydrolytic polyamidation reaction that occurs increasing the temperature and the system reaches a situation of dynamic equilibrium expressed by the equation:

$$\frac{[\text{products}] + H_2O}{[\text{reagents}]} = B * e^{\frac{-\Delta H_a}{RT}} \qquad (5)$$

The microscopic observation shows actually that, changing the temperature inside the range in which the alga survive, remarkable changes of the protoplasm streaming take place, revealing the existence of modifications in the cytoplasm structure. DL too shows sharp variations as it is shown in figure 11.

It is interesting to notice that DL trend of *Acetabularia acetabulum* could be described through the equation (5) in the range where it changes quickly, as it is indicated in figure 11. This evidence strengthens again the hypothesis that DL is connected to the cytoplasmic structure of the cell.

6. ULTRAWEAK LUMINESCENCE AS A TOOL TO ASSESS THE STATE OF BIOLOGICAL SYSTEMS

The intimate link that exists between ultra weak luminescence and the structure of biological systems makes this phenomenon extremely interesting from the point of view of its applicative potential.

These potentials regard on one hand the global diagnostics of biological systems, for which it is often possible to give a judgement for some properties through a measurement of the DL and on the other the possibility to truly quantitatively analyse the presence of certain substances.

6.1 DL as a Technique of Global Analysis

It has been seen that DL is linked to the structure of biological systems that determine, in a way, which is not simple, the behaviour of the same systems. This fact allows, in principle, to use the measurement of the DL to obtain information on some properties of determined biological systems. There are different examples of this use of the DL as an analysis technique of global quality.

Figure 12 shows, for example, the different luminescence of vegetables of the same cultivars, grown in the same conditions and undergoing the same treatment except for a different technique of initial insecticide treatment of the culture containers, subjected in a case to a thermal treatment called "solarization" and, in another case, to a disinfestation with methyl bromide.

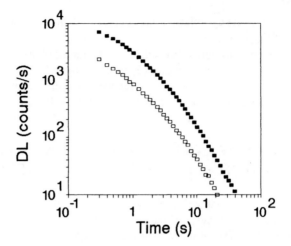

Figure 12: DL of different samples of potatoes submitted to different preparation of the soil. (■ solarized, □ treated with methyl bromide)

Figure 13 shows the existing link between the DL and the germination potential of soya seeds and pepper seeds, both artificially aged. These results show that the DL technique can be considered a good one to evaluate the vitality and viability of seeds, with the advantage that the test, performed on dry seeds, is fast and non-destructive. Moreover this technique is suitable for the preliminary selection of seeds to use in highly prized cultivations.

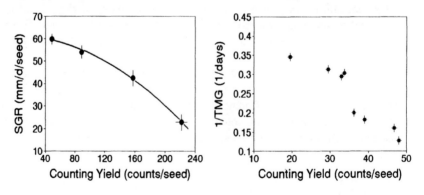

Figure 13 a: Soya seeds: relationship between the seeds growth rate SGR and the corresponding DL counting yield. The symbols refer to experimental data with the relative statistical errors; the continuous line represents a quadratic fit.

Figure 13 b: Pepper seeds: relationship between the inverse of mean time to germination TMG and the corresponding DL counting yield. Standard errors are indicated.

Figure 14 shows, at last, the values of colour modulus, DL total emission and soluble solid concentration for different tomato fruits.

Significant qualitative differences between samples were revealed by the trend of the soluble solids concentration while direct visual observation did not reveal significant differences in colour modulus data in the various samples.

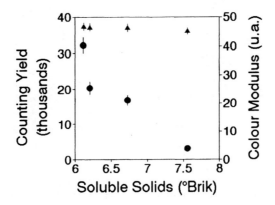

Figure 14: Colour modulus (▲) and delayed luminescence total emission (●) as a function of soluble solids concentration ten days after harvesting for several samples characterized by different ripening stage at harvesting.

However significant differences correlated with the harvest maturity of the samples were observed for DL total emission. These differences suggest that DL measurements may be a valid, rapid and non-destructive method for assessing changes in quality of vegetables.

6.2 DL as an Analytic Technique

In this second case some biological system must be found that functions as a sensor, particularly sensitive to the substance to be detected.

A typical example is that of the possibility of detecting the presence of atrazine in water. Unicellular green algae have been found to be extremely sensitive to the presence of this herbicide in their culture medium.

More precisely, it appears to be a notable variation both in the intensity of the DL in the first part of the decay and in the global slope of the curve, and these variations are shown to be more accentuated for higher concentrations of poison.

The phenomenon can be globally described using the ratio S between the average slope of the decay curve in the presence of poison and the corresponding value measured previously for the same alga placed in its normal culture medium.

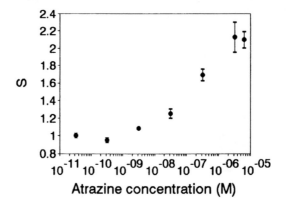

Figure 15: Relative average slope S of the decay curves vs. atrazine concentration. Each symbol represents the mean value of five independent experiments with the relative statistical errors.

Fig. 15 shows the value of the parameter S as a function of different concentrations of atrazine. The parameter presents a monotone trend with the increase of the poison concentration, within a moderate error.

At low concentrations, the differences between S-values are greater than their standard deviation: it could thus be possible to reveal concentrations of poison of the order of 10^{-9} M.

7. REFERENCES

Becquerel, E. (1868) *La Lumiere.* Paris.

Colli, L. and Facchini, U. (1954) Light emission by germinating plants. *Il Nuovo Cimento,* **12**, 150-153.

Davydov, A.S. (1985) *Solitons in Molecular Systems.* Reidel, Dordrecht.

Deveraj, B., Usa, M., Inaba, H. (ed.) 1997) Biophotons: ultraweak light emission from living systems. *Current Opinions in Solid State & Material Science,* **2**, 188-193.

Ezzahir, A., Kwiecinska, T., Godlewski, M., Sitko, D., Rajfur, Z., Slawinski, J., Szczesniak-Fabianczyk, B., Laszczka, A. (1992) The influence of white light on photo-induced luminescence from bull spermatozoa. *ABC*, **2/3**, 133 –137.

Frenkel, J. (1931) On the transformation of light into heat in solids. *Phys. Rev.*, **37**, 17, 1276.
Galanin, M.D. (1966) *Luminescence of Molecules and Crystals*. Cambridge International Science Publishing, Cambridge.

Ho, M.W.,Musumeci, F., Scordino, A., Triglia, A.(1998) Influence of Cations in Extracellular Liquid on Delayed Luminescence of Acetabularia acetabulum. *Journal of Photochemistry and Photobiology B*, **45**, 60-66.

Ho, M.W. (1998) *The Rainbow and the Worm - The Physics of Organisms*. World Scientific.

Lavorel, J., Breton, J., Lutz, M. (1986) Methodological principles of measurement of light emitted by photosintetic systems. In *Light Emission by Plant and Bacteria*, Govindjee et al. eds , Academic Press, New York, 57.

Musumeci, F., Godlewski, M., Popp, F.A. and Ho, M.W. (1992) Time Behaviour of Delayed Luminescence in Acetabularia Acetabulum. In *Recent Advances in Biophoton Research*, Popp, F.A., Li, K.H. and Gu, Q. (eds.), World Scientific, Singapore, 327-344.

Musumeci , F., Triglia, A., Grasso, F., Scordino, A., Sitko, D. (1994) Relation between delayed luminescence and functional state in soya seeds. *Il Nuovo Cimento D*, **16**, 65-73.

Musumeci, F., Scordino, A., Triglia, A., Brizhik, L. (2001) Delayed luminescence of biological systems arising from coherent self-trapped electron states. *Phys. Rev. E*, **64**, 31902

Niggli, H.J. (1993) Artificial sunlight irradiation induces ultraweak photon emission in human skin fibroblasts. *Journal of Photochemistry and Photobiology B: Biology*, **18**, 281-285.

Penrose, R. (1995) *Shadows of the mind*. Vintage, London, 367-371.

Popp, F.A. (1986) On the Coherence of Ultraweak Photon Emission from Living Tissues. In *Disequilibrium and Self-Organization*, Kilmister, C.W. (ed.), D. Reidel Publishing Company, Dordrecht, 207-230.

Popp, F.A. (1989) Coherent Photon Storage of Biological Systems. In *Electromagnetic Bio-Information*, Popp, F.A. et al. (eds.), Urban and Schwarzenberg München, 144-167.

Scordino, A., Grasso,F., Musumeci, F., Triglia, A. (1993) Physical Aspects of Delayed Luminescence in Acetabularia acetabulum. *Experientia*, **49**, 702-705.

Scordino, A., Triglia, A., Musumeci, F., Grasso, F., Rajfur, Z. (1996) Influence of the presence of atrazine in water on in-vivo delayed luminescence of Acetabularia Acetabulum. *J. of Photochemistry and Photobiology B*, **32**, 11-17.

Scordino, A., Triglia, A. and Musumeci, F. (2000) Analogous features of delayed luminescence from Acetabularia acetabulum and some solid state systems. *J. Photochem. Photobiol. B*, **56**, 181-186.

Slawinska, D. and Slawinski, J. (1983) Biological Chemiluminescence (Yearly review). *Photochemistry and Photobiology*, **37**, 709-715.

Triglia, A., La Malfa, G., Musumeci, F., Leonardi, C., Scordino , A. (1998) Delayed luminescence as an indicator of tomato fruit quality. *Journal of Food Science*, **63**, 512-515.

Ukrainskii, I.I., Mironov, S.L. (1979) *Teor. Eksper. Chem.*, 15.

Van Wijk, R., Van Aken, H., Mei, W.P., Popp, F.A. (1993) Light-induced photon emission by mamalian cells. *J. Photochem. Photobiol. B*, **18**, 75-79.

Van Wijk, R., Scordino, A., Triglia, A., Musumeci, F. (1999) Simultaneous measurements of delayed luminescence and chloroplast organization from Acetabularia acetabulum. *J. Photochem. Photobiol. B*, **49**, 142-149.

Wolfe, S.L. (1994) Molecular and cellular biology. *EdiSES, Napoli, chaps.*, 11-12 (in italian).

Chapter 6

BIOLOGICAL EFFECTS OF ELECTRO-MAGNETIC FIELDS ON LIVING CELLS

Jiin-Ju Chang (Jinzhu Zhang)[1,2]
[1] Institute of Biophysics, Chinese Academy of Sciences
Beijing 100101, China
Fax: (0086) 10 6488 8428, E-Mail: changjj@sun5.ibp.ac.cn
[2] International Institute of Biophysics (IIB), Neuss, Germany

Abstract: The significance of studies of the biological effects of electromagnetic fields (EMFs) on living cells is explained using a number of different examples. These clearly show how exposure of living cells to EMFs cause physical stresses and biological effects. Recent studies concerning effects of EMFs on cell signaling and gene expression are reviewed, but it is stressed that the question of the actual mechanisms of the interaction between these EMFs and living systems remains largely unknown, necessitating further investigation. The role of biophoton emission in studying the endogenous EM (electromagnetic) fields of living systems and the role such fields play in understanding life are discussed.

1. INTRODUCTION

With the recent advent of more and more electrical equipment and devices established for civilian and military purposes, the question as to whether they might adversely affect both the natural environment and human health is an extremely important one. On the other hand many experimental results have shown that EMFs can be used positively for medical diagnosis and treatments, and also to increase the yields of certain products in agriculture and in industry. In order to know how to use the beneficial effects of EMFs and to overcome their harmful effects, more and more biologists and physicists have become interested in studying the influences of EMFs on biological systems, especially living cells. For not only is the cell the basic unit of life, but it also is the simplest model of life.

During the past 20 years, however, there have been many seemingly contradictory reports, but often the discrepant results can be traced to subtle differences in experimental protocols, even while epidemiological studies showing an increased risk of leukemia in children living near electric power lines are still controversial. On the other hand, medical studies showing that low frequency electromagnetic fields can accelerate healing of bone fractures are undisputed (see review of M. Blank).[1] An established negative effect of EM exposure is the inhibition of the growth of roots of plants. When exposed to 60Hz electric fields between 350V/m and 450V/m, root growth was inhibited by up to between 70% and 80 %.[2] On the other hand, Simpson et al. [3] reported that ethanol production during growth of yeast cells increased by 25% to 30% when exposed to a pulsed electric field. Electromagnetic fields have also been shown to have potential in medical diagnostics and treatments. By using electric fields, Zeira et al.[4] succeeded in inserting the full length of CD4 protein into the membrane of human red blood cells, thereby offering a new strategy for the treatment of AIDS. In addition, it has proved possible to bring about apoptosis of cancer cells by exposing a mouse with cancer to weak EMFs, illustrating the potential of EMFs in cancer treatment.[5]

This review will focus on some important points that have been the subject of the author's research in recent years. In the interest of promoting a better understanding of the biological effects of EMFs on cells, some fundamental facts will first be briefly summarized.

2. REGARDING ELECTROMAGNETIC FIELDS AND THEIR EFFECTS

According to Blank,[6] electromagnetic fields can be broadly categorized either as ionizing—such as cosmic rays and X-rays, characterized by frequencies between 10^{18}–10^{22} Hz—or as ultraviolet radiation of 10^{16} Hz, both of which damage cells and are non-ionizing. To the latter belong microwaves (10^9-10^{11} Hz), which can also damage cells; these are, of course, used in microwave ovens and in mobile telephony (where there is the possibility that they might be involved in brain cancer). Other examples of non-ionizing radiation include the EM fields used in radio and TV (10^4-10^8 Hz), the effects of which are largely unknown; finally, there are the electric and magnetic fields associated with the transmission of electrical power at 50-60Hz, which is still the subject of much discussion and debate.

Up to now, many kinds of EMFs have been used in the study of living biological effects. According to the WHO, the biological effects of EMFs are classified into the following broad classes: those caused by electric and magnetic fields of extremely low frequency (ELF 0-300Hz and those caused by low intensity radiofrequency/microwave electromagnetic fields (RF from 10MHz to 300MHz, microwaves from 300MHz to 300GHz). Since these low level EMFs do not cause any temperature rise in the exposed organisms, the kind of biological effects they provoke are called nonthermal effects.[7-8] Recently de Pomerai et al[9] reported that prolonged exposure to microwaves induce heat-shock responses in *C.elegans*, despite the fact that the radiation was of sub-thermal intensity: the effect, which appears to be a non-thermal one, indicates that the organism becomes stressed by low-level EM exposure.

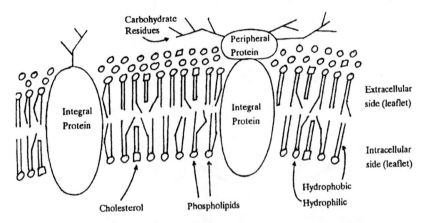

Figure 1. General model of biomembrane (from *Biochemistry,* edited by T. Briggs and A.M.Chandler, 1995, Springer-Verlag, New York-Berlin-Heidelberg).

3. CELL MEMBRANE AND ELECTRICAL PROPERTY OF LIVING CELLS

3.1 The Structure and the Capacity of the Cell Membrane

The membrane, which forms the boundary between a cell and its environment, is called the cell membrane or plasma membrane. Like all other biological membrane (biomembrane), the cell membrane consists of mainly phospholipids and proteins. The phospholipids are arranged as bilayered sheets in which proteins are embedded. The cell membrane is highly selective about what is allowed to pass through it. The membrane phospholipids are amphipathic molecules having both hydrophobic and hydrophilic characteristics. They have a polar "head group" containing a charge or other polar group and a nonpolar "tail" of a long fatty acid chain or other hydrocarbon entity. The bilayered sheets arrange themselves in the membrane with non-polar portions facing each other in the interior, and polar portions facing either extracellular or intracellular sides.

Biomembranes contain 25–75% proteins. Based on the nature of their interactions with membrane lipid and their location, the membrane proteins can be classified into two broad categories: integral (intrinsic) proteins and peripheral (extrinsic) proteins. Intrinsic proteins are stably embedded in the phospholipid bilayer. Peripheral proteins are weakly attached to the membrane, or to other proteins. Usually they are bound to the membrane by interaction with integral membrane proteins indirectly or by interactions with lipid polar head groups directly.

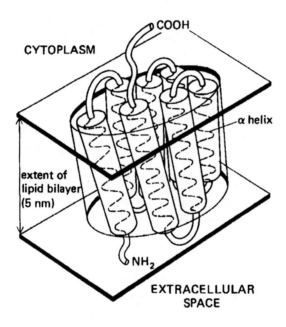

Figure 2. Structure of bacteriorhodopsin reconstructed from electron diffraction analyses of two-dimensional crystals of the protein in the bacterial membrane. The seven membrane-spanning α helices are labeled (R. Henderson et al., 1990, *J. Mol.Biol.* 213:899)

The integral proteins, which span the phospholipid bilayer, contain one or more membrane-spanning domains with the domains extending into the aqueous medium outside of the bilayer. Such protein is called transmembrane protein. Most transmembrane proteins contain multiple membrane-spanning α helices. Some integral proteins are anchored in only one sheet of the bilayer; these proteins contain covalent bound fatty acids that anchor them in the bilayer.

The cell membrane can be regarded as an insulator separating two conducting aqueous phases, i.e. as a dielectric in a parallel plate capacitor. In an equivalent electric circuit, the cell membrane can be represented as a capacitor. The specific capacity of the membrane assumes a value of 1 $\mu F \cdot cm^{-2}$ (Zimmermann).[10]

3.2 Gradients of Ions and Electric Potential Across the Cell Membrane

The cell membrane is a selectively permeable barrier between the cell and the extracellular environment. This permeability ensures a constant internal environment for cells. There is a different transport manner in cell membranes, either by passive diffusion or by active transport. The passive diffusion does not depend on whether the cell is alive or dead; it is driven by the gradient of the concentration of the ions, except that the ion channels are related to the situation of the cells. When a cell is in a rest state, the passive fluxes of ions have to be balanced by the active fluxes (driven by ionic pumps). So the net flux of these ions is zero. For all cells (including animal, plant, and microbial cells), in their resting state (the cytosolic pH is kept near 7.0), the ionic composition of the cytosol (intracellular fluid) usually differs greatly from that of the surrounding fluid (Figure 3).

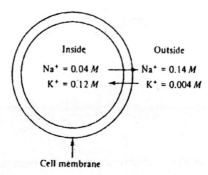

Figure 3. Ion gradients across the cell membrane. The arrows show the active transport direction of these ions. Electric potential is negative inside of the cell with respect to the outside.

The extracellular fluid has a high concentration of Na$^+$ and Cl$^-$, and the cytosol, a high concentration of K$^+$ and of larger organic anions (A$^-$), and the cytosolic concentration of potassium ion (K$^+$) is much higher than that of sodium ion (Na$^+$). The concentration of free calcium ion (Ca^{2+}) in the cytosol is generally about 0.2 micromolar (10^{-6} M), nearly ten thousand times lower than in the extracellular fluid.

From the electrochemical potential of ions, we have

$$\mu = \mu_0 + RT \ln C + zFE \quad (1)$$

where E is the electrical potential given in millivolts, R is the universal gas constant, T is the absolute temperature, z is the charge number of the ion ($z = +2$ for Ca^{2+}, $z = -1$ for Cl$^-$, etc). The net flow is from high to low. For X ion, we have:

$$\mu_{(out)} = \mu_0 + RT \ln C_{(out)} + zFE_{(out)} \quad (2)$$

$$\mu_{(in)} = \mu_0 + RT \ln C_{(in)} + zFE_{(in)} \quad (3)$$

$$\Delta\mu = (\mu_{(out)} - \mu_{(in)}) = RT \ln \frac{C_{(out)}}{C_{(in)}} + zF(E_{(out)} - E_{(in)}) \quad (4)$$

The first term is related to the gradient of concentration of the ions, the second term related to the difference in electrical potential. When the electrochemical equilibrium is $\Delta\mu=0$, we then have:

$$RT \ln \frac{C_{(out)}}{C_{(in)}} = -zF(E_{(out)} - E_{(in)}) \quad (5)$$

therefore

$$E_{(out)} - E_{(in)} = -\frac{RT}{zF} \ln \frac{C_{(out)}}{C_{(in)}} \quad \text{(the Nernst equation)} \quad (6)$$

For an ion, if the equilibrium potential $\neq$ membrane potential, ionic flux occurs,

$$I_k = g_k(E_m - E_k) \quad (7)$$

$$I_{Na} = g_{Na}(E_m - E_{Na}) \quad (8)$$

$$I_{ci}=g_{ci}(E_m-E_{ct}) \tag{9}$$

When the cell is in a steady state in which the transmembrane electrical potential difference is constant, $\Sigma I_i = 0$, then

$$E_m = \frac{g_k}{g_k+g_{Na}}E_{k+} \frac{g_{Na}}{g_k+g_{Na}}E_{Na} \tag{10}$$

This is called the Chord conductive equation. The plasma membrane contains channel proteins which allow ions (Na^+, K^+, Ca^{2+}, and Cl^-) to move through it at different rates. The gradients of concentration of ions and selective movements of the ions through channels create a difference in electric potential between the inside and the outside of the cell. The magnitude of this potential is 20mV to 70 mV with the inside of the cell negative with respect to the outside. What we should notice is that the plasma membrane is only about 7 nm thick. Thus the voltage gradient across the cell membrane is about 0.07 V / 7nm (10^{-7} cm), or100 kV per centimeter.

To generate and maintain these ionic gradients, in addition to the diffusion or passive movement of ions across membranes, active transport mechanisms are needed. These active transports against a concentration gradient require energy. Adenosine triphosphate (ATP) is generally used as the source of energy. For example, to maintain a low intracellular concentration of Na^+, the cell must extrude sodium against the gradient (because of the higher Na^+ concentration outside). The Na^+-K^+ ATPase in the cell membrane pumps K^+ into the cell and Na^+ outwards, thus establishing the high cytosolic concentration of K^+ that is essential for generation of the cytosol-negative potential across the plasma membrane. The Ca^{2+}ATPases pump Ca^{2+} out of the cytosol into the extracellular medium or into intracellular organelles, thereby maintaining the concentration of Ca^{2+} in the cytosol much lower than that in the extracellular medium.

3.3 Electroporation and Electroinjection

Electroporation of cells has been extensively investigated. It is not only a new approach in biotechnology, but also a new topic in the study of cell membranes. Reversible electrical breakdown of cell membrane when the cells are exposed to high voltage with short duration is the primary process of electroporation. In spite of differing explanations as to how exogenous material transfers into the cells, it is commonly understood that when an electric field is applied to a cell suspension, an electrical potential V_g is induced across the cell membrane. When membrane resistance is much higher than the resistance of the medium and of the cytoplasm, according to Laplace formula

$$V_g = 1.5 \, ER \cos\alpha \qquad (11)$$

where E is the field intensity, R is the cell radius, and α is the angle between the field vector and the membrane surface element. The induced transmembrane potential exists only during the pulse application (i.e. when $E \neq 0$), but it can induce longer-lasting effects, such as ion transport, reversible or irreversible poration, and electrofusion.

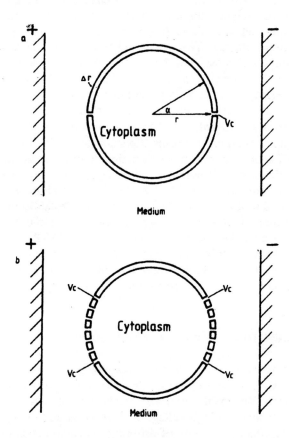

Figure 4. Electrobreakdown of cell membrane when the cell was placed between two parallel electrodes. R is the radius, α is the angle with the electric field direction. $E = V_g - V_m$ (V_m = resting membrane potential, V_g = generated membrane potential), E_c = critical voltage for membrane breakdown. The upper $E = E_c$, the bottom $E \gg E_c$ (Zimmermann, 1988, *J. Electrostatics*, 21:309).

Increasing the intensity of the external field increases the membrane voltage and reaches the critical threshold value at about 1 V. Thus the cell membrane first undergoes electrical breakdown at the sites oriented in the field direction (where $\cos\alpha = 0$) and results in the formation of an aqueous pore in each side of the cell closely to the electrodes. As the field intensity is further increased, more and more membrane sites are subjected to breakdown, and, in consequence, more and more pores are created. The description of electroporation is based on the physical properties of cell membranes. But living cells are different from physical capacitors. The electrobreakdown of cell membrane is reversible. Chang et al.[11] showed that electroporation is a dynamic change in permeability of cell membranes, which is not only related to the applied field but also related to cell properties.

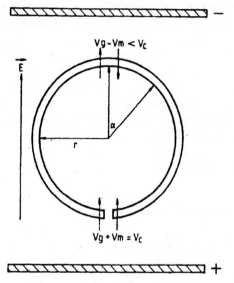

Figure 5. Asymmetric breakdown of a cell placed between two parallel electrodes. R is the radius of the cell, α is the angle with the electric field direction. V_m = resting membrane potential, V_g = generated membrane potential, V_c = critical voltage for membrane breakdown (Zimmermann, 1988, J. Electrostatics, 21:309).

Electroporation has been developed as a new approach in biotechnology to transfer exogenous material into living cells. Electroporation is also called electroinjection, electrotransformation, and electropermeabilization. If the cells were electroporated with either asymmetric breakdown or only one hemisphere having a pore, then there is considerable advantage in electroinjection practice. It can minimize the current flow through the cell. In addition, the resealing properties of the membrane (thus the viability of the treated cells) seem to be greatly enhanced when only one hemisphere of the cell is subjected to electropermeabilization. Asymmetric breakdown is favorable for electroinjection, but not for electrofusion, which needs to break down the cell membrane on both hemispheres of the cells.

3.4 Dielectrophoresis

The movement of particles in a nonuniform AC field has been termed dielectrophoresis (Pohl, 1978, Dielectrophoresis, Cambridge University Press). This phenomenon can be obtained for any polarizable object, no matter if it is a live or dead cell. Dielectrophoresis is used to bring the cells to form the cell chains for cell electrofusion. In an electric field a whole cell can be treated as a neutral particle. Cells suspended in a non-conductive medium are polarized in such way that the positive pole is close to the cathode and the negative to the anode. In a uniform field the field intensities are equal on both sides of the cell, thus there is no net force to the cells, therefore no drifting movement can occur. However, in a nonhomogeneous alternating electric field, the field strengths on both sides of the cells are unequal, and result in a drifting movement of the cells toward the higher intensity of the field along the field lines, finally bringing cells into contact, forming a chain-like aggregate, the so called pearl-chain. In this case living cells appear as physical particles (see Figure 6).

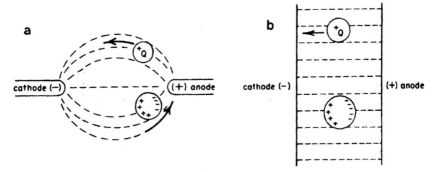

Figure 6. Movement of an electrically neutral (but polarizable) body and a charged body in (a) nonuniform and (b) uniform electric field. No movement of the neutral body is observed in the uniform field (adapted from Pethig, 1979, Dielectric and Electric Properties of Biological Materials, Wiley, New York)

3.5 Electric Charges of the Cell Surface

Because of the structure of membrane protein and its properties, the cell surface possesses both electrical positive and negative charges. Most of the cell surface has a net negative charge, therefore positively charged ions are absorbed on the nearby cell surface forming ionic layers. According to the distance these absorbed ions can be divided into the absorption layer which is close to the cell surface and the diffusion layer (Figure7). These layers of ions are only formed when the cells are in a conductive medium, especially a low concentration of conductive medium.

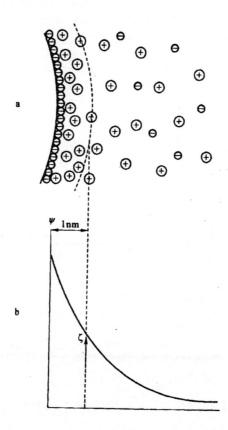

Figure 7. Charges bound on cell surface (from *Biophysics* ed. by Zhao, N. and Zhou, H, 2000. China Higher Education Press, Beijing and Springer-Verlag, Heidelberg).

3.6 Permitivity and Conductivity of Cells

When a cell suspension is applied to an electrical voltage via two parallel electrodes, the characteristics of the cells can be represented by the conductivity σ' and the relative permitivity ε', which are related respectively to conductance G and capacitance C.

The permitivity ε' and the conductivity σ' of the biological system change as a function of the frequency. Typically as the frequency is increased, the permitivity decreases while conductivity increases (Fig. 8). Typically, in a "classical" dielectric spectrum of a biological cell suspension or tissue, there are three major and two minor dispersions, which are usually attributed to: (α) the tangential flow of ions along cell surfaces, (β) the build-up of charge at the cell membrane via a Maxwell-Wagner effect, (γ) rotation of small molecular weight dipoles, especially water, (δ) rotation of macromolecular side-chains and "bound" water, and (β_1) the rotation of intracellular proteins[12].

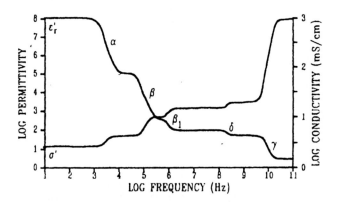

Figure 8. Changes of permitivity ε' and conductivity σ' of biological systems as a function of frequency (from Christopher et al., 1995).

4. EFFECTS OF EMFS ON LIVING CELLS

It has become generally accepted that EMFs cause various biological effects.[7,8] Application of electric pulses to a suspension of cells may cause alteration of cell membrane potential and altering of ions transport, cause electrofusion and electroporation, membrane enzyme activation, membrane lipid phase transition, membrane protein insertion and membrane protein reorganization, membrane skeleton disturbance, alteration in cell signaling and gene expression, protein synthesis; and cause cell apoptosis etc. Among these effects in responding to the external physical factors, living cells appear differently. In some cases, cells appear the same as non-living cells--for instance, electric breakdown of cell membrane where living cells appear like physical capacitors, with the exception of reversibility. Under these conditions, the effects of external electric fields on cells follow the energy relationship and mostly appear in linear relationship. However, in the case of electrical activation of membrane enzymes, where the cell appears as a biological living system, T.Y.Tsong[13] has found that membrane ATPase can be activated by alternating electric fields. The optimum parameters of the field is 20 volts and 10KH for NaK-ATPase. He proposed a coupling model for explaining the electric activation of ATPase. By using NMR, Chang et al.[14] studied changes in concentrations of Na^+ inside human erythrocytes, and found electric activation of NaK-ATPase of erythrocyte membrane under weak pulsed electric fields. The electric activation of membrane protein may induce even more reactions within the cells, and the cells may produce a series of biological reactions in response to external fields. Therefore, the activation of membrane protein by an electric field can be seen as an activation of the membrane receptor binding some sort of chemical signal ligands that switch on the pathway of the signal transmission within the cells. In such cases the interactions between living cells and external fields do not simply follow an energy relationship and cannot be described linearly, because some kind of informational interactions have been involved.

Another example is electro-luminescence from living cells. Comparisons of electro-luminescence of aqueous solution with and without bacterial contamination of at least 10 to 100 /ml of bacteria can be distinguished (Fig. 9). This method is sensitive, fast, and can provide real-time monitoring in an industrial production line (Chang and Popp Chinese patent ZL 96 1 05260.0).

There are two possibilities for the cause of the emission of light from electro-luminescence:

(1) Because in the electrolysis of organic material, there appear cations (A^+) on the anodes whereas anions (A^-) appear on the cathodes. After diffusing into the solution, they move to the opposite electrodes. When they meet each other, discharge takes place and light is emitted, as described by:

$$A^+ + A^- \rightarrow A^* + A$$
$$A^* \rightarrow A + h_v \qquad (12)$$

(2) Reactions took place on the electrodes. For instance, in an NaCl solution, the following reactions happen on the cathode:

$$Na^+ + e^- \rightarrow Na$$
$$2H_2O + 2\ e^- \rightarrow H_2 + 2OH^- \qquad (13)$$

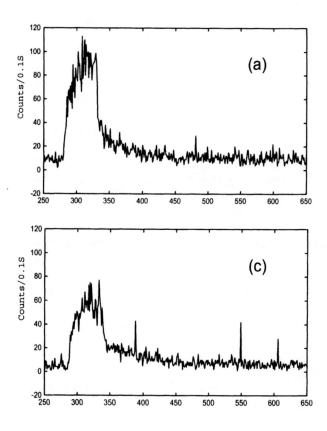

Figure 9. Electro-luminescence from culture medium (a), and the same culture medium with 10^3/ml bacteria of *Rhizobium japonicium*(c). All were stimulated by 30V for 5 seconds (Chang and Popp, 1998, *Biophotons*, Kluwer Publishers).

on the anode

$$2Cl^- \rightarrow Cl_2 + 2\,e^- \tag{14}$$
$$2H_2O \rightarrow O_2 + 4H^+ + 4\,e^-$$

In these reactions, they may reach excited states, and then when they go back to ground states, they may emit light.

Considering very dilute electrolyte solutions in infinite dilution, the ions are independent of one another. When two electrodes are immersed in such a solution, application of an electric potential across the electrodes leads to negative ions (including negative charges from electrolysis of the organic material) moving to the anode and the positive ions to the cathode where electrode reactions take place. As described in section 3.5, living cells appear with negative charges on the surface and are surrounded by an atmosphere of positive charges as well as negative charges. When an electric potential is applied to the solution containing living cells, the free moving charged particles are reduced, resulting in decreasing the above two reactions.

In this case, cells only absorb ions, just like all charged particles. But in other cases, cells behave differently. Chang et al. studied electro-photon emission from living cells induced by electric stimulation. Cultured *P. elegans* naturally emit photons as bioluminescence. Applying 5 to 10 voltages/cm to a cell suspension of cultured *elegans* results in photon emission intensities dramatically increasing at first and then quickly relaxing to delay emission. After stimulation, the cells of *elegans* respond with several oscillation waves (Fig. 10). The behavior of electro-photonemission from *elegans* is related to the parameters of the electric field, but even more to the response of *elegans* to electric field interactions between two biological entities.

Photon emission from the brain cells of a chicken embryo stimulated by a pulsed electric field is even more interesting. At first the intensity of electroluminescence immediately increases but some time later the photon emission intensity increases again, appearing similar to super delayed light emission from Drosophila embryos observed by Ho et al.[15]

It seems that all of the environmental factors including electric fields influence cells in the following two ways: as a stress factor causing some effects and as an informational signal transmitting to cells, in response to a series of reactions happening in the cells.

4.1 Effects of EMFS on Cell Signalling

Cells have their signal transmission systems. There are several pathways of signal transmission pathways in one cell, and different cells have different pathways. One of the important pathways is that Inositol 1,4,5-Trisphosphate causes the release of Ca^{2+} ions from Endoplasmatic Reticulum (ER). Binding of many hormones to cell-surface receptors induces an elevation in cytosolic Ca^{2+} even when Ca^{2+} ions are absent in the surrounding medium. Ca^{2+} is released into the cytosol from the ER and other intracellular Ca^{2+} stores. The signal pathway from a hormone-receptor binding to release a Ca^{2+} store became clear in the early 1980s. We know the following: hydrolysis of PIP2 by the plasma-membrane enzyme phospholipase C (PLC) yields two important products:1,2-diacylglycerol (DAG), which remains in the membrane, and inositol 1,4,5-trisphosphate (IP3). Once IP3 diffuses to the ER surface, where it binds to a specific IP3 receptor opening of the channel of the Ca^{2+} store, it allows Ca^{2+} ions to exit into the cytosol. The overall sequence goes from the binding of hormones to a G protein-linked receptor to Ca^{2+} release.

In addition to the IP3-sensitive Ca^{2+} store, muscle cells and neurons possess other Ca^{2+} stores called ryanodine receptors (RYRs) which are dependent on the influx of extracellular Ca^{2+}. In these cells a small influx of extracellular Ca^{2+} induces binding of these Ca^{2+} to RYRs that results in Ca^{2+} releasing the Ca^{2+} stores.

Experiments have shown that EMFs can alter signal transmission in cells. Chang et al. studied the effects of a pulsed electric field on the brain cells of chicken embryos and found that intracellular second messenger cAMP and Ca^{2+} were influenced by the field and both were shown to be correlated (cross talking). Cytosolic free calcium ions ($[Ca^{2+}]_i$) concentration in brain cells of embryonic chicks showed that $[Ca^{2+}]_i$ increased immediately as the cells were exposed to a single electric pulse. The amplitude and rate of the increases of $[Ca^{2+}]_i$ depend nonlinearly on the intensity of the external electric field. In the presence of Ca^{2+} chelant EGTA or Ca^{2+} channels blocker La^{3+} in the pulsation solutions, the increase of $[Ca^{2+}]_i$ was still observable, suggesting that the electric field activated not only Ca^{2+} channels in the membrane opening but also released intracellular calcium stores.[16] There are other reports on Ca^{2+} signaling affected by EMFs.[17-18] Dibridik et al.[19] and Kristupaitis et al.[20] reported that low energy EMFs cause IP3 increase. But other reports showed that EMFs cannot cause IP3 increases.[21] Ca^{2+} spikes have been observed, but spontaneous calcium spiking was strongly argued in a review paper by Mullins and Sisken.[22] They found this spiking to be completely dependent on extracellular calcium ions and to be independent of the inositol 1,4,5-trisphosphate-sensitive intracellular calcium store. The spontaneous calcium spiking is unrelated to normal events of the cell cycle and most likely result from the damaging effects of excessive loading. Cho et al. [17] using epifluorescence microscopy examined the $[Ca^{2+}]_i$ response to AC electric fields. Application of (Hep3B) cells to AC fields induces a fourfold increase in $[Ca^{2+}]_i$ (from 50 nM to 200 nM) within 30 min. Incubation of cells with the phospholipase C inhibitor U73122 does not inhibit AC electric field-induced increases in $[Ca^{2+}]_i$, suggesting that receptor-regulated release of intracellular Ca^{2+} is not the response in this effect. Therefore the increase in Ca^{2+} may be a direct effect of EMFs.

Concerning the effects of EMFs on cell signaling, the following models are most accepted: According to M. Blank,[6] through resonance, EMFs change biosynthesis of proteins that stimulate G protein on cell membrane; Berg[23] thought that beside resonance, mobile ions surrounding the cell membrane, discharging of cell membrane, and reorganization of water molecular polarity have to be considered. T.Y. Tsong[24] put forward his electroconformational coupling model. In his model, because the changes in membrane electric potential caused by external fields conformational changes take place in some membrane protein, that is the first step toward stimulating cellular signaling. Eichwald and Kaiser's[25] model is based on the process of IP3-induced calcium release and involves a positive feedback loop between intracellular calcium and the activation of phospholipase C(PLC).

4.2 Effects of EMFS on Gene Expression

In recent years, considerable effort has been devoted to finding the mechanism by which electromagnetic fields affect biological systems. As a consequence, there have been growing reports on electromagnetic field-induced alterations in gene expression. Up to now the results are contradictory. Goodman et al.[26] reported that exposure of cells to low-frequency electromagnetic fields resulted in quantitative changes in gene transcripts. A number of reports showed that exposure of T-lymphoblastoid cells to a 60 Hz sinusoidal magnetic field altered the transcription of genes encoding c-fos, c-jun, c-myc, and protein kinase C. (reviewed by Phillips [27]). Lin et al.[28] found there is a specific region of the c-myc promoter responding to electric and magnetic fields. However, Bloberg and Balcer-Kubiczek could not find a differential gene expression in the HNE cell, HBL100 cells, and HL60 cells exposed to a 60Hz magnetic field and to sham exposed cells. These genes are related to cancer. Another is heat shock protein. Lin et al.[29]

showed that following exposure to magnetic fields, HSP70 expression can be altered. But Morehouse et al.[30] found that exposure to low-frequency electromagnetic fields does not alter HSP70 expression.

5. THE MECHANISM OF BIOLOGICAL EFFECTS OF EMFS

Within the past few years studies on biological effects of extremely weak EMFs have increased, but the mechanism still remains unclear. Up to now several approaches have been tried and different theoretical models have been proposed. In principle, these models suggested so far can be divided into two groups: models of influence on special signaling pathways of the cell, especially the pathway connected with calcium, and models in which direct interaction of the external fields is considered.[31]

A lot of reports showed the biological effects of extremely low-level electric and magnetic fields. But calculations made from physics show that low frequency fields do not radiate energy. For example, a 100-meter power transmission line carrying 500MW of 50/60Hz electric power radiates less than 10 μw of energy, which is approximately 2×10^{-8} W/m^2 at one meter from the power line. On the earth, at noon time the sun light radiates 1400W/m^2, and moonlight at the full moon time radiates 2×10^{-3} W/m^2, which shows that the radiation of the power line is less than 10^5 times the moonlight radiation.[32]

Adair[33] thought that protein molecules are 3-dimensional structures, because conformational changes do not happen with very weak fields. He calculated the effects of moving energy and potential energy of typical radio frequency fields with 10mW/cm^2 (200V/m) to a cell in diameter of 20 μm. The results show that small fields do not cause obvious changes in cells. He also calculated resonance energy, and the results showed that the maximum of resonant absorption energy was much less than several orders of heat vibrations. From the point of reference of frequency selectivity, it seems that resonance cannot take place.

Different approaches have been considered, including the cyclotron resonance model and Stochastic resonance.[34] However, none of them give a satisfying full description of the mechanism. The general problem with these models is that they cannot satisfactorily explain how the very weak external fields interact with a biological system effectively despite the presence of noise. A big mystery is why such low energy can cause biological effects.

The interference model[35] can solve this problem. In this model, the effects of external fields on the living were treated as the interactions between EMFs and living things in the manner of interference. Only when living systems exist in certain conditions where the interference with external fields can take place does the external EMF influence the living by forming destructive interference outside of the cells, while at the same time constructive interference takes place inside of the cells. The interference (interaction) between the living and external EMFs depends upon both rather than on the situation of the living only. The living system is not passively receiving energy from external EMFs but actively guiding the moving energy. Based on this model, the direct action of EMFs on DNA can be understood, and certain EMFs can be expected to damage cancer cells selectively.

6. INTRINSIC ELECTROMAGNETIC FIELDS IN LIVING SYSTEMS AND BIOPHOTON EMISSION

Because biological macro-molecules possessing electrical charges or dipoles and forming different highly ordered structures with a variety of functions, the most fundamental phenomenon is that cells have their intrinsic electrical field including localized electric fields and non-localized wholly electromagnetic fields. Cell membrane potential is an example of the localized field. Calcium waves and oscillations observed in cells may be considered as an example of the non-localized field.

Fröhlich[36] elucidated that because of the properties of cell membranes, because of certain bonds (such as H bonds) recurring in bio-macromolecules, and probably because of the existence of non-localized electrons, biological systems were expected to have a long-range phase correlation, i.e., coherence. Popp[37] explained that within cells a typical transition time is some nanoseconds, since within this period the phase relations in traveling wave packets will not break down. The electromagnetic field in living systems is coherent, which maintains the systems in a steady state far from thermal equilibrium, in a state where there are no waves or particles. What have you then? In this state, time and distance no longer have meaning.

Actually, the intrinsic field of the living appears finally as a very weak one. Therefore it is very difficult to measure the intrinsic field. Using an external field to influence the living by observing dynamic reactions of the living under the external field, one can learn about the intrinsic field. At present biophoton fields can be studied by measuring photon emission. So far it may be the only window for observing the intrinsic fields in living systems.

It is also one of the most important methods for studying microprocesses in biological systems by observing the visual events.

Biophoton emission from biological systems is a universal phenomenon. It concerns weak light emission of a few up to some hundred photons per cm^2 of surface per second. The frequencies cover up to 800 nm. According to F.A. Popp[37] biophoton emission originates from non-localized coherent electromagnetic fields within the living, in particular, from their optical modes. The coherence of biophoton fields has been proved in the following two ways. Statistical distribution of biophoton emission counts follow a Poissonian distribution but not a geometrical one; delayed luminescence from the living system follows a hyperbolic relaxation function.[38] Experiments show that living things also emit non-classic (squeezed) light. These photons play an important role in cell communication and biological regulation.

Theoretically, long-distance, cell-to-cell communication may be accomplished by an electromagnetic field. Electrical signaling takes place between different leaves in the same tomato plant.[39] Albrecht-Buhler[40] reported that electromagnetic radiation can work as the carrier of signals. In his experiments he cultivated BHK cells on face A of a glass slide. After several day's growth of the cells, he inoculated BHK cells onto the opposite face B of this slide and continued to cultivate in darkness. He found that the new inoculated BHK cells on face B were not randomly distributed. Instead, the distribution and the orientations of these cells were influenced by the cells in face A, indicating that there was cell-to-cell communication probably by electromagnetic radiation on faces A and B. Shen et al.[41] have found non-substantial communication between blood cells. Chang et al.[42] made experiments using *Dinoflagellates*. They placed two samples of *Dinoflagellates* in different chambers A and chamber B of a double-chamber system which is constructed either to connect the two samples or to disconnect them by opening or closing the shutter between them. Two photomultipliers, each connected to one of the chambers, register the photon emission from the two samples separately. Experiments showed that when sample A was stimulated by applying an electric current, sample A and sample B synchronously responded to the stimulation when the shutter was opened (Fig. 10).

The coincidence counting system (CCS) is a new method[43] that has been used in Popp's laboratory since 1995. The idea of the measurement is based on the registration of the counts in both sample channels plus the coincident counts c which are compared with the random coincidence rate r. If the measured c is equal to r, then we know that the photon emission is random. The CCS system can be used for investigating whether the samples emit coherent or chaotic light according to the difference in the numbers of random coincidences between a coherent field and a chaotic one (where the coherence time is not much smaller than the preset measuring time intervals).

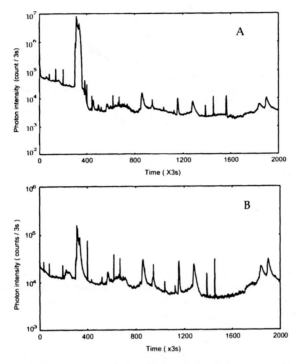

Figure 10. Photon emission from samples A (upper curve) and B (lower curve), where the shutter is opened. Note almost perfect synchronization of photon emission from A and B. Sample A was exposed to electric stimulation of 5V/cm at 900-120 s (Chang et al. 1995)[42].

By using this method, *Dinoflagellates*, embryonic chick brain cells, tomato seedlings, and bacteria have been measured. Some of the experimental results will be presented here as examples to show the characteristics of biophoton fields and their functions.

Fig.11 to Fig. 13 show the relationship between measured coincidences and theoretical random coincidences. The solid line shows a theoretical Poisson distribution that should correspond to a fully coherent photon field. The broken line shows the geometrical distribution of a completely chaotic photon source. The circles show the results measured when the shutter is open, and the dark spots show the results when the shutter is closed.

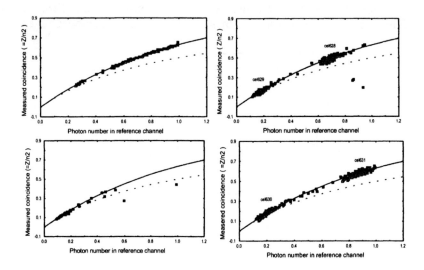

Figure 11(left) and Figure 12 (right). Relationship between measured coincidences and random coincidences showing that the measured coincidences are fixed with theoretical Poisson distribution rather than geometrical distribution and that there are some differences between the shutter open and closed (see the text). Measuring parameters: $\Delta t=10^{-4}$s, $\tau =3 \times 10^{-5}$ s, for Fig. 11; $\Delta t=10^{-5}$ s, τ = minimum for CEL 28 and CEL 29, $\tau =10^{-5}$ s for CEL 30 and CEL 31 (Chang and Popp, 1997 Conference of International Institute of Biophysics in Neuss, Germany, unpublished).

From Figure 11 and Figure 12, we can see that the measured coincidences are well fixed within the theoretical Poisson distribution. That means that in these conditions the photon emission is random and coherent rather than chaotic. The measured coincidences deviate from theoretical Poisson distributions and were shown to be lower than the random coincidences, specifically when the photon intensities of the reference channel were higher. As shown in Fig. 12 during the 4 continuous measurements (from CEL 628 to CEL 631), of which each measurement took 5 min., as soon as the shutter was closed the photon intensities dropped down, although the measured coincidences were still fixed within the Poisson distribution.

Some measurements show that the measured coincidences are not completely random, that they deviated from theoretical random distributions (Fig. 13). In order to describe the deviations quantitatively, a factor F was introduced,

$$F = (-r)/(1-r) \qquad (15)$$

When the shutter is open, F often shows a positive value, but it shows a minus value when the shutter is closed.

By changing the interval time Δt from 10^{-3} to 10^{-5} seconds, the measured coincidences were shown to deviate from the theoretical Poisson distribution when Δt gets smaller. Sometimes as the delay time τ increases there is more deviation in the measured coincidences compared with the random rates.

Fig. 13 shows an example from which distinct deviations of the measured coincidences from random ones can be identified and also the differences between open and closed shutter.

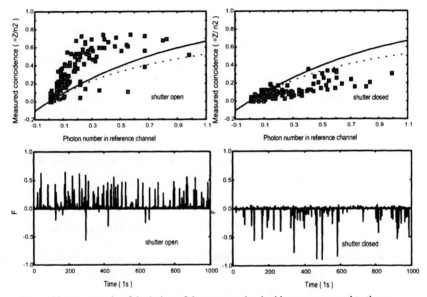

Figure 13. An example of deviation of the measured coincidences compared to the random coincidences showing the differences between the shutter open and closed. Measuring parameters: $\Delta t = 10^{-5}$ s, τ is minimum (Chang and Popp, 1997 Conference of International Institute of Biophysics, Germany).

In a cell scale even chaotic light may display a coherence time. Actually, an allowed optical transition time is 10^{-9} s. During this period emitted light can transfer a distance of 10 cm which is 10^4 larger compared to the size of a cell. Therefore, in this period the phase relations of light cannot get lost. In other words, over the distance of a cell even if the light field is not coherent, the phase relations of light cannot get lost. The dimension of a cell is smaller than the coherence length even of chaotic light. Hence, it is impossible that a cell "sees" a photon. In the coherent state, including the non-classical squeezed state, perfect localization and perfect delocalization may take place at the same time, and the spatio-temporal structure of a cell may be partially or even completely ordered or disordered. The coherent photon field may fluctuate around phase transitions between "chaos" and "order" with dependence on the compensation of either chaotic or ordered influences. The basic property of all biological tissues was represented as optically "dense" matter, which means that the distances of sub-structures in cells are small compared with the wavelength of light. According to Dicke, spontaneous reemission of absorbed light is impossible. Under these conditions we have to understand the interaction of light and matter in terms of interference. As soon as the structural distance is significantly smaller than the wavelength, the photons split into two new "regimes" of superradiance and subradiance. Superradiance corresponds to constructive interference of light waves cumulating up to coherent light flashes that are then emitted in relatively short time intervals. Subradiance is defined as the destructive interference of light waves (shown by Fig. 13). When the two samples have established destructive interference, any stimulation or disturbance to one of the two samples will be responded to by the whole field but only by sample A (Fig. 10). There is no difficulty in understanding the coincident flashing of Dinoflagellates. As soon as they are connected, an electromagnetic field of a complicated interference structure may build up. These are essential points for understanding what information is carried by biophoton emission.

7. APPLICATIONS AND BIOLOGICAL STUDIES

Recently efforts have been made in our laboratory to show the possibility of apoptosis of cancer cells caused by EMFs. Mice inoculated with Ehrlich ascites cells were exposed to a weak electromagnetic field for a long time and sham exposed cancer mice as well as healthy mice were used as the controls. During the treatment, growth curves of mice were measured and compared between exposed and sham exposed mice. The results show that the growth curves of healthy mice are fixed well with the ideal curve of logistic growth, but the growth curves of cancer mice deviate from the ideal curve. There is no difference in growth curves between exposed and sham exposed healthy mice and, they both fit well with the ideal curve. However, a notable difference in growth curves between exposed and sham exposed cancer mice was obtained. Moreover, the growth curves of sham exposed mice deviate even more than those of the exposed mice, in other words, the growth curves of Ehrlich ascites mice deviate from the ideal curve of healthy mice but tend toward it with the EMF treatments.

After treatment with EMFs, apoptosis of Ehrlich ascites cells from inoculated mice was analyzed by several methods, including flow cytometry, fluorescence microscopy, and DNA gel electrophoresis. Statistical analysis from flow cytometry shows that the apoptotic ratio of cells from exposed Ehrlich ascites mice was significantly higher than those from sham exposed mice. Our experimental results showed that under certain treatments by electromagnetic fields, apoptosis of cancer cells from mice were significantly increased. The experimental results provide a potential of EMFs for cancer treatment.

It is interesting to compare these results with delayed luminescence emission from cancer cells and normal cells with increasing cell density (Schamhart and van Wijk[44]). The delayed luminescence from tumor cells increases non-linearly, but decreases in healthy cells. Although the two experiments described different features of biological systems, they were both based on the same principle and showed loss of coherence among cancer cells.

These two different experiments showed that electromagnetic fields in living systems including biophoton fields play a very important function. They exist in living biological systems although we cannot see them. They are some kind of structure with specific patterns, but they are not real matter. They are only fields that bring the living system to coherence, even to squeezed states. In such a state within coherent space there is no difference between particles and waves, therefore distance has no meaning. This state provides ideal conditions for the communication that is the basis for biological regulation.

The 21st century is a new century for biology. In the coming thousand years there will be some big changes in the understanding of objects: real space and dummy space, substantial and non-substantial existence, and informational networks should be considered simultaneously to generate new discoveries in biology. About 50 years ago, Watson and Crick established the DNA double helical structure model, and at the same time Perutz and John Kendrew solved 3-D structures of protein molecules. These two achievements opened a new era in biology. Although biology has made much progress in molecular biology and genetics and their applications, the essential question of what is life and how it is regulated still remains as a mystery. Based on the coherence of the intrinsic field of living things, a number of biological problems may be explained properly in terms of modern physics, for instance the functions of DNA, biological organization and biological regulation, oncogenesis, even biological evolution.

ACKNOWLEDGMENTS

The author would like to thank Prof. Dr. Shizhang Bei for his many years' guidance. Thanks are also due to all my colleagues for their many various contributions, in particular to Prof. Dr. Fritz-Albert Popp for his cooperation in biophoton research and valuable discussion of this paper.

REFERENCES

1. Blank, M. (1995) Biological effects of environmental electromagnetic fields: Molecular mechanisms. *Biosystems,* **35,** 2-3.

2. Brayman, A.A., Miller, M.W., and Cox, C. (1987) Effects of 60-Hz electric fields on cellular elongation and radial expansion growth in cucumber roots. *Bioelectromagnetism* **8,** 57-72.

3. Simpson, J., Brady, D., Rollan, A., Barron, N., Mchale L., Mchale, A.P. (1995) Increased ethanol production during growth of electric-field stimulated. Kluyveromyces marxianus IMB 3 during growth on lactose-containing media at 45 degrees C. *Biotechnology Letters* **17,** 757-760.

4. Zeira, M., Tosi, P.F., Mouneimne, Y., Lazaite, J., Sneel, L.,Volsky, D. and Nicolau, C. (1991) Full length CD4 electroinserted in the erythrocyte membrane as a long-lived inhibitor of infection by human immunodeficiency virus. *Proc. Nat. Sci. Am.,* **88,** 4409-4413.

5. Jiao, H. L., Wang, Y., Chang, J.J. (2001) Apoptosis of cancer cells by long term exposure of Ehrlish Ascites mice to weak electromagnetic fields. *Electro and Magneto Biology,* **20** (3), 299-311.

6. Blank, M. (1995) *Electromagnetic field: Biological interactions and mechanism.* Washington, DC: American Chemical Society Press, 1-[?].

7. Repacholi, M. H. (1998) Low-level exposure to radiofrequency electromagnetic fields: Health effects and research needs. *Bioelectromagnetics,* **19,** 1-19.

8. Repacholi, M.H. and Greenebaum, B. (1999) Interaction of static and extremely low frequency electric and magnetic fields with living systems: health effects and research needs. *Bioelectromagnetics,* **20,** 133-160.

9. Pomerai, D., Daniells, C., David, H. et al. (2000) Non-thermal heat-shock response to microwaves. *Nature,* **405,** 417-418.

10. Zimmermann, U. (1982) Electric field-mediated fusion and related electrical phenomena. *Biochim. Biophys. Acta,* **694,** 227-277.

11. Chang, J.J., Liu, Y., and Kong, K.J. (1994) Dynamic cell-membrane events following the application of single-pulse electric fields. In *Bioelectrodynamacs and Biocommunication,*

Ho, M.W., Popp, F.A, Warnke, U. (eds.), Singapore-London: World Scientific Publishing, 251-268.

12. Christopher, L. D., and Douglas, B. K. (1995) *The low-frequency dielectric properties of biological cells in Bioelectrochemistry of cells and tissues.* Waltz, D. Berg, H. and Milazzo, G. (eds.), Basel, Switzerland, Birkhäuser Verlag, 159-207.

13. Tsong, T.Y. (1990) Electrical modulation of membrane proteins. *Annu. Rev. Biophys. Chem.,* **19,** 83-106.

14. Chang, J.J., Sun, T., Chen, Y., Zhou, S., Chen, Y., and Pang, S. (1997) Studies on Changes of Intracellular Na+ Concentration in Erythrocytes Effected by Pulsed Electrical Field. *Science in China (Series C),* **40(5),** 488-495.

15. Ho,M.W., Xu, X., Roos, S., and Saunders, P.T. (1994) Light emission and rescattering in synchronously developing populations of early Drosophila embryos. In *Recent Advances in Biophoton Research and its Applications* Popp, F.A., Li, K.H., and Gu, Q. (eds.). Singapore-London, World Scientific Publishing, 1992, 287-36.

16. Chen, Y., Wang, Y., Sun, T., Chang, J.J et al. (2000) Dynamic changes of [Ca2+]i in cultured cerebellar granule cells exposed to pulsed electric field. *Science in China* (C), **43,** 77-81.

17. Cho, M.R., Thatte, H.S., Silvia, M.T., Golan, D.E. (1999) Transmembrane calcium influx induced by AC electric fields. *FASEB J.,* **13(6),** 677-83.

18. Galvanovskis, J., Sandblom, J. (1997) Amplification of electromagnetic signals by ion channels. *Biophys J.,* **73(6),** 3056-65.

19. Dibirdik, I., Kristupaitis, D., Kurosaki, T., et al. (1998) Stimulation of Src family protein-tyrosine kinases as a proximal and mandantory step for SYK kinase-dependent phospholipase Cγ2 activation in lymphoma B cells exposed to low energy electromagnetic fields. *J. Biol. Chem.,* **273(7),** 4035-4039.

20. Kristupaitis, D., Dibirdik, I., Vassilev, A., et al. (1998) Electromagnetic field-induced stimulation of Bruton tyrosine kinase. *J. Biol. Chem.,* **273(20),** 12397-12401.

21. Miller, S. C., Furniss, M.J. (1998) Bruton's tyrosine kinase activity and inositol 1,4,5-trisphosphate production are not altered in DT40 lymphoma B cells exposed to power line frequency magnetic fields. *J. Biol. Chem,* **273(49),** 32618-32626.

22. Mullins, R.D., Sisken, J.E. (2000) Mechanisms underlying spontaneous calcium spiking in aequorin-loaded ROS 17/2.8 cells. *Bioelectromagnetics,* **21(5),** 329-37.

23. Berg, H. (1995) Low-frequency electromagnetic field effects on cell metabolism, in *Bioelectrochemistry of cells and tissues*, Walz, D. & Berger H. (eds.), Birkhäuser Verlag, Basel, Boston, Berlin, 285-301.

24. Tsong, T.Y. (1994) Electroconformational coupling: Enforced conformational oscillation of a membrane enzyme for energy transduction. In *Biomembrane electrochemistry*, Blank, M. and Vodynoy (eds.), I. Washington, DC: American Chemical Society, 561-578.

25. Eichwald, C., Kaiser, F. (1995) Model for external influences on cellular signal transduction pathway including cytosolic calcium oscillations. *Bioelectromagnetics,* **16,**75-85.

26. Goodman, R., Wei, L.X., Xu, J.C., Henderson, A. (1989) Exposure of human cells to low-frequency electromagnetic fields results in quantitative changes in transcripts. *Biochim Biophys Acta,* **1009(3)**, 216-20.

27. Phillips, J.L. (1993) Effects of electromagnetic field exposure on gene transcription. *J Cell Biochem,* **51(4)**, 381-6.

28. Lin, H., Goodman, R., Shirley-Henderson, A. (1994) Specific region of the c-myc promoter is responsive to electric and magnetic fields. *J. Cell Biochem,* **54(3)**, 281-8.

29. Lin, H., Head, M., Blank, M., Han, L., Jin, M., Goodman, R. (1998) Myc-mediated transactivation of HSP70 expression following exposure to magnetic fields. *J. Cell Biochem.,* **69(2)**, 181-8.

30. Morehouse, C.A., Owen, R.D. (2000) Exposure to low-frequency electromagnetic fields does not alter HSP70 expression or HSF-HSE binding in HL60 cells. *Radiat Res.,* **153(5 Pt 2)**, 658-62.

31. Blank, M. and Goodman, R. (1997) Do electromagnetic fields interact directly with DNA? *Bioelectromagnetics,* **18,** 111-115.

32. Valberg, P.A., Kavet, R. and Rafferty, C.N. (1997) Can low-level 50/60 Hz electric and magnetic fields cause biological effects? *Radiat. Res,* **148,** 2-21.

33. Adair, R.K. (1994) Effects of weak high-frequency electromagnetic fields on biological systems. In *Radio frequency radiation standards: Biological effects.* Klauenberg, B.J., Grandolfo, M., Erwin, D.N. (eds.), New York: Plenum Press, 207-221(2-96 R811-532 R12).

34. Bezrukov, S.M. and Vodyanoy, I. (1997) Stochastic resonance in non-dynamic systems without response thresholds. *Nature,* **385,** 291.

35. Popp, F.A., Chang, J.J. (2000) Mechanism of interactions between electromagnetic fields and living organisms, *Science in China* (Series C) **43(5),** 507-518.

36. Fröhlich, H. (1968) Long Range Coherence and Energy Storage in Biological Systems. *Int.J.Quant.Chem,* **2,** 641-649.

37. Popp, F.A. (1992) Some essential questions of biophoton research and probable answers. In *Recent Advances in Biophoton Research and Its applications,* Popp, F.A., Li, K.H., Gu, Q. (eds.), Singapore-London: World Scientific Publishing Co. Pte. Ltd., 1-46.

38. Popp, F.A. and Li, K.H. (1993) Hyperbolic Relaxation as a Sufficient Condition of a Fully Coherent Ergodic Field. *International Journal of Theoretical Physics,* **32,** 1573-1583.

39. Wildon, D.C., Thain, F., Minchia, E.H. (1992) Electrical signaling and systemic proteinase inhibitor induction in the wounded plant. *Nature*, **360**, 60-62.

40. Albrecht-Buehler, G. (1992) Rudimentary from of cellular "vision." *Pro. Natl. Acad. Sci.* (USA), **89**, 8288-?.

41. Shen, X., Mei, P., Xu, X. (1994) Activation of neutrophils by a chemically separated but optically coupled neutrophil population undergoing respiration burst. *Experientia*, **50**, 963-968.

42. Chang, J.J., Popp, F.A. and Yu, W.D. (1995) Research on cell communication of P. elegans by means of photon emission. *Chinese science bulletin,* **40**, 76-79.

43. Popp, F.A., Shen, X. (1998) The photon count statistic study on the photon emission from biological systems using a new coincidence counting system. In *Biophotons*, Chang, J.J., Fish, J. and Popp, F.A. (eds.), Dordrecht-Boston-London, Kluwer Academic Publishers, 87-92.

44. Schamhart, D.H.J. and van Wijk, R. (1987) Photon Emission and the degree of differentiation. In *Photon Emission from Biological Systems*, Jezowska-Trzebioatowska, B., Kochel, B., Slawinski, J., and Strek, W. (eds.), Singapore, World Scientific, 137-152.

Chapter 7

DETECTORS FOR THE QUANTIZED ELECTROMAGNETIC FIELD

John Swain
Department of Physics
Northeastern University, Boston MA 02115
john.swain@cern.ch

Abstract   After a quick review of classical electromagnetism, the concept of the photon as the quantum of the electromagnetic field is introduced, followed by a discussion of the interaction of photons with matter. Specializing then to the case of visible and near-visible radiation, detectors which are capable of registering single quanta are described, showing the wide range of possibilities which are open to the interested experimenter. Advantages and disadvantages of each as well as practical suggestions are provided. Finally we take a look at relatively novel devices with bright futures.

Keywords: Photodetectors, Single photon detection

1. CLASSICAL ELECTRIC AND MAGNETIC FIELDS

The classical electromagnetic field is, as far as we know, completely described by the classical Maxwell equations (for a very readable text, see [1]) which are differential equations governing how electric and magnetic fields change in space and time. Together with the Lorentz force law, which describes the effects of these fields on point charges, this is all one really needs to know about electricity and magnetism at a classical level, and all the advanced textbooks on the subject are largely concerned with working out the consequences of these equations. For completeness, let's have a quick look at Maxwell's equations:

First we have the electric and magnetic fields $\vec{E}$ and $\vec{B}$. These are the fundamental degrees of freedom of the electromagnetic field, and are *vector fields*. That is to say that at each point of space and at each time, they each define a *vector* - a quantity with geometric meaning independent of any coordinate system, although we often find it convenient to express vectors in terms of their components along some specified axes.

Now these quantities can be measured by looking at what forces they exert on a point charge. (We take point charges for simplicity. More complicated distributions of charge can be thought of as the sums of many little pointlike charges.)

If a particle of charge q at rest feels a force (as evidenced by its suffering an acceleration - the definition of a force is something that produces an acceleration $\vec{a}$ of a mass m via Newton's equation $\vec{F} = m\vec{a}$) then we define the electric field where the charge is via

$$\vec{F} = q\vec{E} \qquad (1)$$

If the particle is in motion and it experiences an additional force, which experiment shows always to be perpendicular to its velocity $\vec{v}$, then we write

$$\vec{F} = q(\vec{E} + \vec{v} \times \vec{B}) \qquad (2)$$

where "×" represents the usual vector product. This is the Lorentz force equation and serves to *define* what the electric and magnetic fields are.

Of course you may think that something is wrong with this picture in terms of relativity, for we have said nothing about the frame in which motion (or lack thereof) is supposed to be determined. In fact, the picture is completely consistent with relativity as changing reference frames changes not only the velocity of the particle, but also changes the magnetic and electric fields, so that what is a purely electric field in one frame of reference is a mixture of both electric and magnetic fields in another. For example, if you run next to a charge at rest you will see something that looks like an electrical current, and this will be accompanied by a magnetic field. Since relative motion mixes up electric and magnetic fields, one usually speaks not so much of electric and magnetic fields, but of a union of the two which we call the *electromagnetic field*. Again, while the values of $\vec{E}$ and $\vec{B}$ are well-defined for any one observer, they will not in general agree with those measured by another in relative motion.

Now just as Newton's equations $\vec{F} = m\vec{a}$ describe how a mass m will move under the influence of a given force once the initial position and velocity are given, we have Maxwell's equations which govern how the electric and magnetic fields vary once one starts them off. There are four Maxwell equations. Two are rather similar to Newton's equations in that they tell you how the electric and magnetic fields vary in time after you initially set them up. The other two are of a rather different character, as they are *constraints*. They tell you that you can't just have whatever initial electric and magnetic fields you'd like, but rather that they must satisfy certain equations which constrain how the vary

Detectors for The Quantized Electromagnetic Field

in space. In this sense they represent constraints on the sort of free will you can exercise as an experimenter - some initial conditions are just impossible to get.

The two constraint equations are

$$\text{div } \vec{E} = \rho/\epsilon_0 \tag{3}$$

and

$$\text{div } \vec{B} = 0 \tag{4}$$

The expression "div" represents a combination of rates of change of the components of the fields and physically corresponds to a notion of "spreading out". Equation 3 is equivalent to Coulomb's law which says that the electric field around a charge is directed radially and falls off as the square of the distance from it, and here ρ represents the density of charge per unit volume. (*i.e.* the electric field "comes out of" charges). The ϵ_0 is a constant which we'll return to later. Equation 4 states that there are no charges for magnetism (*i.e.* you only get magnetic fields from currents and there is no notion of them "spreading out" from point "magnetic charges").

The other two Maxwell equations control the dynamics of the fields and are

$$\text{curl } \vec{E} = -\frac{\partial \vec{B}}{\partial t} \tag{5}$$

$$\text{curl } \vec{B} = \mu_0 \vec{J} + \mu_0 \epsilon_0 \frac{\partial \vec{E}}{\partial t} \tag{6}$$

where $\vec{J}$ is the current density (charge per unit area per unit time) and μ_0 and ϵ_0 are constants called the permeability and permittivity of free space, respectively. The operator "curl" is a combination of space derivatives that represents the degree to which integrals around closed curves of the quantity it operates on fail to vanish. For example, curl$\vec{E} \neq 0$ says that there are closed paths around which one can move an electric charge so that the electric field does work on it – something which the corresponding Maxwell equation tells us cannot happen if there is no time-varying magnetic field present.

1.1 A Few Words About Units

The system of units used in the above section is the usual MKS system used by engineers and most people engaged in practical work. Units are a cause of enormous confusion to many, and you can find Maxwell's equations in all sorts of different and confusing forms in various books, so a brief discussion of units is probably in order.

There are basically two things you have to decide in order to get a set of units to work with electromagnetism. The first one is that for the magnitude of the force F between two charges q_1 and q_2, separated by a distance r you have the inverse square law (Coulomb's law):

$$F = k\frac{q_1 q_2}{r^2} \tag{7}$$

where k is some constant you have to pick depending on how you define charge. If you choose $k = 1$, then you need no new units for electric charge. We choose $k = \frac{1}{4\pi\epsilon_0}$ where $\epsilon_0 = 8.854 \times 10^{-12}$ with units so that force is in Newtons, distance is measured in metres, and charges are in Coulombs where one electron carries 1.6×10^{-19} Coulombs of charge. This may sound like a perverse choice for k, but the thing is you have to pick something, and there is a sort of conservation law of messiness that pretty much forces some ugly constants to turn up somewhere, and the MKS system makes them turn up here. Other systems choose $k = 1$ or $k = \frac{1}{4\pi}$.

The second thing you need to do is to figure out how you're going to measure speeds. The speed of light c turns out to be $\frac{1}{\sqrt{\mu_0 \epsilon_0}}$ so the constants c, μ_0, and ϵ_0 are not all independent. The MKS system we're using here has the speed of light measured in metres per second, and then μ_0 turns out to be $4\pi \times 10^{-7}$ in our units.

With forces in Newtons(N), distances in metres(m), masses in kilograms (kg), time in seconds (s), speed in metres per second (m/s), charge in Coluombs (C), and μ_0 and ϵ_0 as given above, we measure electric fields in Newtons/Coloumb(N/C) which are equivalent to Volts/metre (V/m), and magnetic fields are measured in Tesla (T). A flashlight battery which produces 1.5V, for example, will produce an electric field of 15V/m if connected across two metal plates .1 m apart. A kitchen magnet will typically have a magnetic field of a few milliTesla (a few tens of Gauss; 1 Gauss is 10^{-4}T).

1.2 Electromagnetic Waves

One of the most remarkable features of Maxwell's equations is that they admit solutions which represent waves in empty space which are made of electric and magnetic fields which change in space and time in such a way as to continuously reproduce each other and travel arbitrarily far from whatever source produced them. They are only produced if source charges are accelerated. Moving at a constant velocity will change part of an electric field into a magnetic field, but it won't tear part of the field away as a wave that radiates off to infinity. Maxwell's equations *predict* the speed of these waves to be $\frac{1}{\sqrt{\mu_0 \epsilon_0}}$, and the remarkable fact that this number, which can be calculated in terms of quantities μ_0 and ϵ_0 which can be measured in the laboratory turns out to be the same as the speed of light was the first indication that light might be a form of electromagnetic wave. The fact that Maxwell's equations predicted a speed which seems to be independent of who measures it was also the first indication that there might be something wrong with Newtonian views of time and space. Indeed, as Einstein showed, just as relative motion mixes up electric and magnetic fields, it also mixes up space and time in such a way that the speed of light (which is how much distance it covers in a given time) remains the same no matter who does the measuring.

An electromagnetic wave of any frequency (or wavelength) is a possible solution of Maxwell's equations, and we now understand that it is simply a matter of frequency that distinguishes between radio waves, microwaves, visible light, ultraviolet light, X-rays, and gamma rays! Like all waves, the frequency f, wavelength λ and speed c are connected by $c = f\lambda$ - a pretty trivial formula which just says that the speed of a wave is its length divided by how long it takes to go past! It has the important implication that high frequencies correspond to short wavelengths and vice-versa.

The electric field in electromagnetic waves is always perpendicular to the direction of propagation, and corresponds to the *polarization* of the electromagnetic wave. There are just two degrees of freedom, which correspond to two perpendicular directions in the plane perpendicular to the direction in which the wave goes.

2. QUANTIZATION OF THE ELECTROMAGNETIC FIELD

Now the classical electromagnetic field goes a long way towards explaining an enormous range of phenomena, but it runs into trouble when it comes to the photoelectric effect.

The simple experimental fact was the following: if you shone a light on a piece of metal in a vacuum you could get electrons knocked loose. This is the photoelectric effect. Measuring the energies of the electrons that came off and the wavelength of the light that hit the metal, the following rather puzzling behaviour was observed:

- There exists a frequency below which no electrons get kicked off.

- The higher the frequency (shorter the wavelength) of light the *more energetic* the electrons that come off are.

- A brighter light makes *more* electrons but doesn't affect the energy with which the electrons are ejected.

For example, depending on the material, you might find that blue light just manages to release electrons, while violet light has them come off with more energy. Red light, on the other hand, no matter how bright, might release none.

This is very hard to explain classically, since you'd expect that the brightness of the light would correspond to the strength of the electric field carried by the wave, and it's the electric field that ought to pull electrons out of the metal, so colour should have little to do with the matter, and a sufficiently bright light of any wavelength ought to do the job!

In 1905 Albert Einstein made the bold suggestion that Planck's earlier idea that the electromagnetic field could come in *quanta*, or little chunks, might allow for a simple explanation. He proposed the equation

$$E_e = hf - W \qquad (8)$$

where E_e is the energy of the ejected electron, f the frequency of the incident light, h Planck's constant 6.626×10^{-34} J · s, and W the "work function", a characteristic of the material being used. The physical interpretation of this is the following:

- A beam of light of frequency f is actually made up of a stream of photons, each carrying energy $E = hf$.

- Higher frequency light is made of more energetic photons, while lower frequency light is made of less energetic ones. The brightness of the light is just a measure of how many there are, but *not* of the energy they carry.

- The photoelectric effect is easy to understand now: the work function represents how much energy you need to get an electron out of the metal. If the frequency of the incident light is too low, the chunks that comprise it can't knock any electrons loose. If the photons carry energy greater than W then the electrons emerge with the energy they get from a photon less what it cost to get out of the metal.

For this work, Albert Einstein received the Nobel Prize in 1921 (and not, as many people imagine, for his work on relativity!).

Incidentally, for the experts reading this, you can actually get away with explaining a good part of the photoelectric effect by taking the matter to be quantized but not the radiation - just have certain quantized *amounts* of radiation absorbed from the field, but it's more than a little clumsy, and there are now plenty of more solid reasons for believing in photons! Pedagogy sometimes requires simplifying the story a little from time to time! A very interesting account of the concept of the photon is in reference [2].

What about the polarization of the electromagnetic wave from the quantum perspective? Photons, as it turns out, can be thought of as spinning about their direction of flight, either parallel or antiparallel to it. These two degrees of freedom correspond to the two classical degrees of freedom of polarization.

3. QUANTIZING THE ELECTROMAGNETIC FIELD

A full description of the quantized electromagnetic field is far beyond the scope of these lecture, but for those of you with a smattering of quantum mechanics, the following may help alleviate any trepidation you have in attacking the literature. If you find any good introductory book on quantum mechanics (for a very readable introduction, see [3]) and get as far as the simple harmonic oscillator, you can read this section. If you're more inclined to just deal with practical matters, you can skip this section without any loss of continuity.

You can write the Hamiltonian (that is, the energy) for the electromagnetic field as $\frac{1}{2}\epsilon_0|\vec{E}|^2 + \frac{1}{\mu_0}|\vec{B}|^2$ added up at every point of space (i.e. integrated over volume).

Now what you do is realize that this looks an awful lot like $p^2 + \Omega^2 x^2$ for a suitable choice of Ω and definition of what "momentum" p and "position" x are. This is especially easy to see since each electric fields look like time derivatives of magnetic fields (and vice versa) so you can think of one them as being like a position. But this is just the Hamiltonian for a simple harmonic oscillator (mass on a spring) so we just treat it as such., promoting p and x to operators and imposing $[x,x] = 0$, $[p,p] = 0$, $[x,p] = i\frac{h}{2\pi}$, where $i = \sqrt{-1}$ and turn the crank.

Even better, we could replace the sum over all points in space that we started off with by a sum over all frequencies via a Fourier transform - in other words, expand the fields in terms of plane waves: $\sum_{\vec{k}} a_{\vec{k}} \exp(i(\vec{k} \cdot \vec{x} - wt)) + a_{\vec{k}}^* \exp(i(\vec{k} \cdot \vec{x} + wt))$ Now the total energy is the sum of energies of a lot of harmonic oscillators, each at a different frequency. Again, you can follow the textbook procedure for the simple harmonic oscillator with the a_k's promoted to annihilation operators and the a_k^*'s to creation operators. so that $[a_i, a_j^\dagger] = \delta_{i,j}$, which is just a rewrite of $[x,p] = i$. The standard textbook treatment then leads to states labelled by integers: $|n\rangle = \frac{(a^\dagger)^n}{\sqrt{n!}}|0\rangle$ where $|0\rangle$ is the vacuum state with no photons present.

If you now identify each excitation of an oscillator of wavenumber $\vec{k}$ as a *photon* of wavenumber $\vec{k}$ (where $k = \frac{2\pi}{\lambda}$). These excitations are *photons*. Now the classical electromagnetic field which could represent any linear combination of waves of any frequencies or wavelengths, not interacting with each other, is expressed quantum mechanically as a superposition of states with well-defined numbers of photons, all not interacting with each other.

We glossed over the fact that there are polarization degrees of freedom, but there is no major change in the procedure - just repeat everything including a sum over two possible polarizations.

To add interactions, add a Hamiltonian for matter to the original free Hamiltonian and deal with its effects as a perturbation of almost free fields.

4. THE INTERACTION OF PHOTONS WITH MATTER

As described earlier, the interaction of the classical electromagnetic field with matter is described by the Lorentz force law. The quantum description is actually much simpler. Photons (wiggly lines) actually hit electrons (solid lines) as shown in the "Feynman diagram" of figure 1, where the incoming and outgoing momenta of the electron are indicated as p_1 and p_2. Much more than just doodles, these diagrams are short-

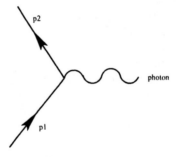

Figure 1.   Feynman diagram of interaction of a photon with a charged particle.

hand for mathematical expressions which determine the probability that various interactions between charges and photons take place. In fact, the photons need not always be photons in the sense that we have thought of them until now. By allowing them to travel slower than the speed of light, such diagrams actually contain the full Maxwell equations including the laws of electrostatics and magnetostatics, where these forces are mediated by such "virtual" photons.

Depending on the energies carried by photons, their effects on matter vary wildly. While we will concentrate in the following section on detectors for visible photons which carry energies of a few electron volts, a summary of these effects is given in the following list:

- Energies less than ~ 1 eV
  - very hard to detect as single quanta
  - few places where that sort of small gap appears naturally
  - noise is bad ($k_B T_{room}$, the energy characteristic of thermal fluctuations at room temperature, is $\sim \frac{1}{40} eV$)
- A few eV (visible light from 400 to 800 nm or so)
  - photoelectric effect
  - cross section varies rapidly with nuclear charge Z and photon energy E ($\sim Z^5$ and E^{-3}) making this the dominant effect out to $\sim$ keV
  - in solid state devices can cause electron-hole pair creation
- keV
  - the Compton effect dominates, whereby electrons are simply kicked with an energy so great that the atomic binding of them is negligible
- MeV and up
  - photons carry enough energy to make electron-positron (e^+e^-) pairs provided there is an atomic nucleus against which to recoil. While this process dominates, photonuclear reactions can also take place.

5. PHOTOMULTIPLIERS

Photomultipliers, or "PMT's" as they are often called, are the traditional workhorses of single photon detection, and are based on the clever use of the photoelectric effect, together with an analogous effect where an electron impinges on a metal surface and knocks loose several more.

Myriad geometries are possible, but the design is always pretty much based on the following basic principles:

- You get an evacuated glass tube with a transparent window allowing light to hit a photocathode.
- The photocathode is a material chosen to have a work function low enough so that light of the interesting wavelengths can release electrons from it, and it is biased with a negative voltage relative to the rest of the tube ensure that electrons which are knocked out of the photocathode are repelled and get away from the surface.
- Nearby, another electrode of a similar material, called a "dynode" is biased at a somewhat more positive voltage than the photocathode. The idea is that electrons liberated from the photocathode

will be drawn towards it by the electric field between the two and gain enough energy to be able to knock loose several "secondary" electrons.

- Repeated dynode stages biased at consecutively more positive potentials produce multiplications so that after, say, 8 stages with multiplications of about 10, one find that one single electron liberated from the photocathode will produce 10^8 electrons at the end. This signal is now large enough to be detected with standard electronics and amplified into a measurable pulse.

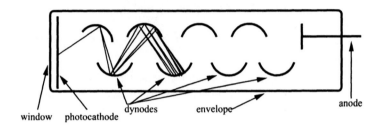

Figure 2. Schematic diagram of a PMT showing electron multiplication at the first few dynodes. The photocathode is negatively biased and each dynode must be biased to a higher positive potential than the first one, and the output signal appears at the anode.

Note the key idea is that a single incident photon can give rise to a large pulse of current at the end. Now that we've sketched how photomultipliers work with broad brushstrokes, let's look in a little more detail at the various bits and pieces that make them up.

In many ways, photomultipliers are like living things. There is a wide "zoology" of them, and they exhibit many almost biological features, including the ability to be wounded, to heal, to age, and to die!

5.1 Spectral Sensitivity and Quantum Efficiency

In the absence of noise (a subject to which we will return later) there are two key things you'd want to know about a PMT before almost anything else:

- Given an incident photon of a given wavelength, what are the odds that it gets detected – that is, that it produces an output signal?
- How big is the output signal for an incident photon that gets detected?

The spectral response of a PMT determined primarily by

- photocathode material at large λ
- window material at small λ

The idea is that at long wavelengths, photons carry so little energy that it's hard to find a photocathode material whose electrons are loose enough (whose work function is low enough) that it will work. At shorter wavelengths you tend to run into trouble finding window materials that are transparent enough.

We measure spectral response in terms of *quantum efficiency* (QE), which is quoted as a function of wavelength.

$$QE = \frac{\text{Number of Photoelectrons Produced}}{\text{Number of Photons In}} \times 100\% \qquad (9)$$

It is not often particularly high, with a couple of tens of percent being pretty good for PMT's, and we will return to the issue of how to get high quantum efficiencies later when we discuss avalanche photodiodes. It's not too hard to see why very high quantum efficiencies are difficult to obtain in almost any device based on a photocathode. If you make a photocathode too thick, you may find that the photons get absorbed deep in the material and photoelectrons don't make it it out. If you make it too thin, photons may just pass through unabsorbed. Finally, even if you do knock an electron loose, it's not by any means obvious that you'll managed to guide it to the first stage of amplification without any losses.

An alternative, and frequently quoted measure of quantum efficiency, is the radiant sensitivity S:

$$S = \frac{\text{Photoelectric Current(A)}}{\text{IncidentPower(W)}} \qquad (10)$$

5.2 The Tube Itself

The tube itself must hold a very good vacuum, on the order of 10^{-5} Pa. (Standard atmospheric pressure at sea level is over 10^5 Pa.) A bad vacuum in a PMT is bad for a lot of reasons. Air molecules can stop electrons from reaching dynodes, reducing both sensitivity and gain. It can also give rise to additional noise, about which we'll see more in section 5.9.

5.3 Photocathode and Dynode Materials and Geometries

There are many possible choices for photocathode and dynode materials, and some of the popular ones and their key characteristics are listed below.

Photocathode materials include:

1. Cs-I : solar-blind, good for detecting vacuum ultraviolet (VUV) at $\lambda < 200$nm
2. Cs-Te : solar-blind, good for detecting VUV at $\lambda < 300$nm
3. Sb-Cs : UV-visible, used often for reflection type photocathodes
4. Bialkali (Sb-Rb-Cs,Sb-K-Cs) :
    - fits emission spectra of many scintillators (NaI(Tl))
    - higher sensitivity and lower dark current thatn Sb-Cs
5. High Temperature, Low Noise Bialkali (Sb-Na-K) :
    - good to 175 C (usual photocathodes only good to 50 C)
    - good for low noise applications (low dark current at room temperature)
6. Extended Green Bialkali : specially sensitive in the green
7. Multialkali (Sb-Ba-K-Cs) : good from UV to IR
8. Ag-O-Cs : good from 300 nm to 1000 nm but with extra IR sensitivity
9. GaAs(Cs) :
    - good from 300 nm to 850 nm but with sharp IR cutoff (~ 900 nm)
    - very susceptible to damage from excess light
10. InGaAs(Cs) : similar to GaAs(Cs) but with better IR reach $9 \sim 1000$ nm)

There is some choice in the photocathode geometry as well, perhaps most noticeably whether the photoelectrons are ejected from the side which faces the light or the side opposite to it.

Dynode materials are often made of alkali antimonide, BeO, MgO, GaAs, GaAsP and typically release 5-10 electrons per incident electron, although this is clearly voltage-dependent. The basic principle of PMT operation allows for an enormous range of possible (and usually rather self-descriptive) dynode geometries, some of which are listed below:

1. Linear focused: an electric field guides the electrons in nearly straight lines giving fast response and good linearity
2. Box and grid type: good collection efficiency and uniformity, but slow
3. Box and linear focused type: good collection efficiency, good uniformity

    4 Circular cage type: fast, compact
    5 Venetian blind type: good uniformity, but slow
    6 Mesh type: good uniformity and pulse linearity; position detection

    7 Multichannel plates: a very novel design where a plate with a large number of holes, or channels, are coated on their inside surfaces with dynode material to make what is effectively a huge array of tiny PMT's providing position sensitivity
    8 Channel PMT's: which are like one channel of a multichannel plate, but often with an improved snake-like geometry so that what is really one continuous dynode looks more like the usual discrete set facing each other. This design offers very easy biasing since the role of a voltage divider network to supply ever more positive voltages to later and later dynodes is played by the one continuous dynode itself.

5.4   Window Materials

There are numerous choices of window materials, including the following:

1 MgF_2
   - good down to 115 nm
   - suffers from deliquescence

2 Sapphire (Al_2O_3)
   - good down to 150 nm

3 Synthetic silica
   - good down to 160 nm
   - less UV absorption than fused silica
   - very difference thermal coefficient from glass; can't be used with kovar

4 UV-transmitting glass
   - good down to 185 nm

5 Borosilicate glass
   - good down to 300 nm
   - available in special "K-free" form to avoid noise from ^{40}K β decays due to potassium normally present in glass

You should keep in mind that windows produce losses in quantum efficiency both due to their reflection of incident light (which can be polarization dependent, and depending on the angle of incidence of the incoming light, two factors which are often forgotten!) as well as due their absorption of light on its way to the photocathode.

5.5 Biasing

Biasing simply means providing each successive dynode with a somewhat more positive voltage than the previous one in order to make sure that electrons move from dynode to dynode and amplification takes place. It is usually done by a passive chain of resistors which acts as a voltage divider, with each dynode connected to a resistor in a long chain.

Two important things to keep in mind are:

- The PMT is not a passive element – make sure the current in the divider network is at least 20 times the maximum current you expect from the tube. One implication is that you can find in bright light that the gain is lower than you would have expected due to the PMT itself loading the voltage divider network so much that the voltage across it starts to drop with the concomittant loss of gain. Even if the voltage divider network is up to the job, you should also be aware that space charge effects can kick in at high tube currents and also drop the effective gain.

- Capacitors across the resistors can help offset the effects of sharp pulses of current, caused, for example, by momentary exposure to brighter-than-average light. For linearity better than x%, choose $C > \frac{100}{x} Q_{max}$ where $Q_{max} = It/V$, with I the current in the pulse, t its width, and V the voltage across the capacitor.

Gain is quite a sensitive function of voltage and typically grows as a power law. This is easy to see, since each stage of multiplication will provide a gain which is roughly proportional to the voltage V, so that the overall gain for n stages will go like V^n. Note that this means that depending on your application, you may well need to have quite a well-regulated power supply in order to avoid unacceptable gain variations. Most PMT's will exhibit a plateau of relatively stable signal to noise (about which more later) ratio where there voltage sensitivity is reduced, and operating a tube in such a region is usually a good idea.

There are a couple of special cases for biasing that are worth keeping in mind. One is the channel PMT, which requires no special voltage division network, that task being provided by the one long dynode itself. The other is the use of what is called "Cathode Potential". The idea is to

Detectors for The Quantized Electromagnetic Field

ground the anode and have successively more negative voltages applied to the dynodes leading up to the photocathode. This alternative way to set up bias sounds sort of trivial at first blush, but has some important differences:

- The grounded anode makes potential difference between readout point and other things (meters, oscilloscopes, *etc.*) small
- You really have to be careful of HV between the cathode and grounded housings, shielding *etc.* – stray currents even through glass (!) can cause electrolysis of the photocathode, leading to flashing and noise!

A useful technique for grounded anode biasing is to paint a PMT with black conductive paint (a so-called "HA coating") and connect it to the cathode supply. This helps repel stray electrons from the glass so they don't strike it and cause noise flashes – a useful and non-obvious result of what might otherwise seem like a small modification in biasing.

One final issue to keep in mind is that the timing response of a PMT is affected by the applied voltage. The speed of light is about 1 foot per nanosecond, so you can imagine that even at the speed of light, there is a delay between a photon coming in and a signal coming out which is not necessarily negligible, depending on the application. Electrons in a PMT never get near the speed of light (they're relativistic at a few hundred thousand electron volts where their kinetic energies are comparable to their rest energy of .511 MeV, but in PMT's, voltages are orders of magnitude lower between dynodes). In general, higher voltages mean faster responses.

5.6 Readout Electronics

Space considerations preclude a detailed description of the readout electronics which has to be associated to a PMT (or, for that matter, any other photodetector), but it's something that's worth being aware of and thinking about carefully. For single photon counting you'll usually want to have a very high gain final output stage that "decides" if a pulse from the PMT was large enough to be considered a photon. There will always be some noise, which we discuss later in section 5.9, and any amplifier will itself contribute a certain amount of noise. Other factors that you will want to consider when thinking about readout electronics is what sort of timing response the amplification has, and how it affects the shape of output pulses. In many applications the readout electronics is gated so that one only looks for pulses given some other condition, which reduces the noise, but is only rarely applicable to bioluminescence studies.

5.7 Life, Aging, and Death of PMT's

Photomultipliers don't last forever, and there are many reasons for them to fail. They can gradually lose their vacuum, but by far the main reason for tubes to fail is that they are exposed to bright light while biased. A photocathode is usually rated for life in terms of the total current that it delivers. Little by little over time, a photocathode will get poorer and poorer at delivering photoelectrons, so the less photoelectrons you ask for, the longer your tube will last and the more slowly its performance will deteriorate with time.

Even brief exposure to bright light will make a photomultiplier behave poorly for a period of several hours, which is something you should keep in mind when doing single photon counting. A wise idea is to keep a shutter in front of the photocathode ensuring that it stays in the dark as much as possible and isn't exposed to bright light unnecessarily. Exposure to bright light at high gains can cause so much current to flow that the tube itself (not just the photocathode) will be permanently damaged, so, again, keep these things in the dark!

5.8 General Warnings

Just as a couple of odds and ends that don't always get discussed, you should be aware that photomultipliers can be affected by both temperature and magnetic fields.

Temperature can either raise or lower the signal you get, while magnetic fields can deflect the paths of electrons in the tube and give rise to huge changes. You can get around both of these problems with temperature control (and we'll have a few more words to say about temperature later when we discuss noise) and magnetic fields can be shielded with soft (in the magnetic sense) ferromagnetic materials like mumetal. Sensitivity to magnetic fields can be largely controlled by careful tube design and may be one of the parameters that you want to look into when you're thinking of purchasing a tube for applications where you think magnetic fields might be an issue.

5.9 Noise in PMT's

The main idea of this section is: *"Just because it went click doesn't necessarily mean there was a photon!"*

The following is a list of some of the main reasons that a signal can appear even though there was no incident photon:

1 Thermionic emission: Room temperature corresponds to an average thermal energy scale of about $\frac{1}{40}eV$, but this is just an average.

There will be plenty of times that fluctuations well in excess of this amount occur, and in fact if they're about the same as the work function, they can eject photoelectrons just as an incident photon would. This "Richardson" current, $I_{\text{Richardson}}$ is proportional to $T^2 \exp(-W/k_B T)$ where W is the work function in eV, T is the absolute temperature measured in Kelvin (*i.e.* degrees Celsius from absolute zero, or about -273 C) and k_B is Boltzmann's constant, 8.63×10^{-5} eV $\cdot$ K This thermionic emission can take place at the cathode or any of the dynodes, and

- is obviously worse the more gain it gets (*i.e.* the closer to the photocathode that it happens), and
- can be alleviated by cooling the tube if possible. This is not as trivial as it sounds. First of all rapid or non-uniform cooling can crack a PMT. Even uniform and gradual cooling of materials with different coefficients of thermal expansion can lead to cracks, and cooling below the dew point can cause condensation on the PMT face, making a vacuum insulation jacket a necessity.

2 Ionization of residual gas in the PMT by photoelectrons. This:
- can show up as "afterpulsing"
- can be worked around by rejecting pulses closely following a signal pulse. This is obviously a bit dangerous as one might be rejecting a physically interesting pulse that was due to a real incident photon.

3 Glass scintillation:
- which can occur when electrons hit the glass PMT envelope and make it glow,
- gets worse at higher voltages, so reducing voltage (at the cost of gain) may help, and
- can be minimized by operating the PMT with anode at HV and cathode at ground and possibly using the HA coating described above.

4 Ohmic leakage:
- which can contribute to constant dark current,
- can be due to bad bases and sockets, and
- can be worsened by dirt and humidity.

5 Field emission, whereby electrons are pushed out the photocathode in regions where the electric field just happens to be high. This can be reduced by running tube at least 200-300 V below its specified maximum

6. Radiation, which can come from both
   - internal sources (components, glass) and
   - external sources (radon, cosmic rays, ...)
   - and may necessitate a careful choice of components and the use of shielding (which itself should not be radioactive!)

7. Electronic noise:
   - since even passive devices like resistors will generate noise, and
   - you may need to minimize passive and stray impedances and try cooling electronics

8. Mechanical noise, since even just plain shaking and vibration can cause false signals.

6. MORE GENERAL COMMENTS ON NOISE

This section is somewhat more mathematical (following [4]) than others and is meant to deal with noise in photodetectors as shown in figure 3 in a somewhat more general way without necessarily referring to PMT's. Here we have:

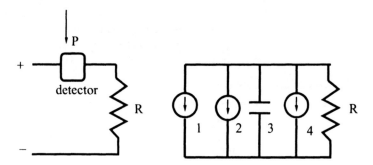

Figure 3. Photodetector and equivalent circuit with noise sources.

1. Current from detector σP which is proportional to the incident power P.

2. Intrinsic (shot) noise due to photons being discrete: $I_i^2 = 2e(I_{ph} + I_d)B$, where I_{ph} and I_d are the signal current (due to photons) and dark current respectively.

3. Capacitance of the detector which leads to a high frequency cutoff of $B = \frac{1}{2\pi RC}$

4 Johnson (thermal) noise current given by $I_R^2 = 4k_BTB/R$

The total RMS (root-mean-square) fluctuation is then:

$$I_n^2 = 2e(I_{ph} + I_d)B + 4k_BTB/R \tag{11}$$

If there is a fluctuating gain M with variance σ_M, we can introduce an *excess noise factor* $F = 1 + \frac{\sigma_M^2}{M^2}$ so that

$$I_n^2 = 2e(I_{ph} + I_d)BM^2F + 4k_BTB/R \tag{12}$$

and we have a signal-to-noise ratio:

$$S/N = \frac{I_{ph}}{\sqrt{2e(I_{ph} + I_d)BF + 4k_BTB/RM^2}} \tag{13}$$

These equations can be used to provide quick estimates of how noise will affect a given photodetection setup.

7. PHOTODIODES

Regular photodiodes are actually not well-suited to single photon detection as they lack gain and this, combined with too much noise, makes them almost hopeless as single photon detectors. Nevertheless, in conjunction with other components (as in a hybrid photodiode) or with some design modification (as in an avalanche photodiode) they are actually quite important, so we we describe them briefly here.

A solid state diode is basically made of two pieces of semiconducting material, one doped to be "n-type" so that it carries currents of electrons, and the other doped to be "p-type" so that it carries currents of holes - vacancies in the crystal lattice where electrons ought to have been. These are placed in contact with each other, and if the "n-type" side is connected to a negative potential and the "p-type" to a postive one, then electrons in the "n-type" material flee the negative potential and are collected on the "p-side" and vice-versa so a current flows. In this situation, the diode is said to be "forward-biased". If the electrical connections are reversed, then ideally no current flows, all the carriers being swept out of a so-called "depletion region". The diode is then said to be "reverse-biased". This difference in the way it behaves for two different biasings leads to the name "diode", for "two ways".

A photodiode is basically just a reverse-biased solid state diode which has been designed so that photons can reach it and liberate electron-hole pairs which then can contribute to a small current. There are, of course, thermally generated electron-hole pairs, so there is always some current in a reverse-biased diode, and it is all but impossible in a simple diode to distinguish such electron-hole pairs from a pair produced by a single photon. That being said, photodiodes do perform admirably for relatively bright light sources, and there are at least two ways to modify or extend them to make them suitable for single photon detection.

8. HYBRID PHOTODIODES

A hybrid photodiode (HPD) is made simply a regular photodiode placed in an evacuated tube with a photocathode in front of it and across which a high voltage (a few thousand volts typically) is applied. An avalanche photodiode can also be used to provide additional gain.

Single photons can liberate electrons from the photocathode and then, rather than multiplying them through interactions with successive dynodes, they simply gain a large kinetic energy by passing through the potential difference between the photocathode and the diode.

These devices suffer the same sort of low quantum efficiency which is associated with devices that have photocathodes, but there is no fluctuation associated with the multiplication at each stage – each photoelectrons passing through 1000V, say, gets 1000 eV of energy and can then liberate a large number of electron-hole pairs in the diode.

9. AVALANCHE PHOTODIODES

Avalanche photodiodes are essentially just reverse-biased photodiodes with a high electric field region in which carriers can gain enough energy to release new electron-hole pairs. If the electric field is high enough, this process repeats leading to an avalanche. A schematic figure of how one particular APD is made is shown in figure 4. This is not the only possible design, and as is the case for PMT's, many are possible. In fact, there are numerous parallels between APD's and PMT's, although APD's offer a number of distinct advantages. First of all, APD's are quite compact devices, which require only a few hundred volts to run. They can be very mechanically robust, having neither bulky glass envelopes, nor a vacuum. They also run quite well in enormous magnetic fields, and can be made to have rather good tolerance to radiation. (In fact, in the figure shown, silicon nitride is used for the window in place of the more common silicon dioxide since it blackens less on exposure to radiation).

Probably the single most spectacular feature of APD's is that they can offer enormous quantum efficiencies, going over 90% . This sort of performance, which is quite impossible for PMT's or in fact any device with a photocathode, is intrinsic to their design: a photoproduced carrier is essentially already where it has to be to get multiplied!

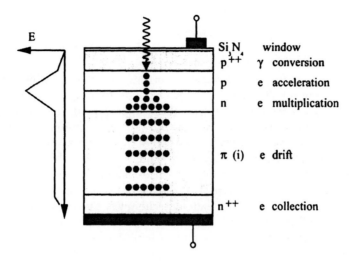

Figure 4. Schematic picture of an APD.

While APD's are just two-terminal devices, like so many other photodetectors they are amazingly subtle and complex. Their parameters vary with voltage and temperature, and include capacitance, noise, and excess noise factor in addition to the more obvious ones such as dark current and the wavelength-dependent quantum efficiency and gain.

The downside of APD's is that they still have rather large dark currents (which have Richardson-like temperature dependence), but recent work by my colleagues and me suggests that this may be reduced to acceptable levels by cooling, while retaining high quantum efficiency.

10. POSITION SENSITIVE DETECTORS

There are many ways to achieve some position sensitivity with single photon detectors, but they all basically boil down to making a lot of little subdetectors. For example, you can:

- Replace a single anode in PMT with multiple anodes
- Replace the single channel in channel PMT with many making a multichannel plate
- Replace a single avalanche photodiode with an array of photodiodes
- In a single HPD replace the photodiode with an array of photodiodes. The electrons from the photocathode go in pretty much straight (or at least, non-intersecting) lines so this works fine.

A fine point to keep in mind with position-sensitive devices, and in fact even with devices with one single detection surface is that response is never completely uniform across the detector face, and data may have to be corrected to take this into account.

One detector which I regret is getting rather scant attention here is the charge-coupled device, or CCD. Aside from space limitations, the main reason for not discussing these important devices a bit more is that in general they don't manage to do single photon detection. This criticism is actually a bit harsh as they can detect several.

CCD's are based on marching charges liberated by photons across a specially made semiconductor devices "shift-register style". They can contain very large numbers of tiny pixels, and offer excellent position resolution with very simple readout techniques, making them a mainstay for a lot of popular applications like tiny TV cameras. Their sensitivity to light can also be increased by placing an image intensifier in front, which is basically a microchannel plate or hybrid photodiode type of device where rather than having the multiplied electrons collected electronically, they are made to strike fluorescent plate and emit light. Aside from their relatively high threshold of around 10 photons per pixel, they have the disadvantage of being rather slow to read out, as in general you have to shift the measurement in one pixel across the whole device to read it out. On the other hand, I suspect that the technology is likely to improve and that these devices, or variants of them, will make it to the single photon counting stage at some point in the future.

10.1 Localization of Photons?

While it's nice to talk about trying to get position information for photons, it is a not very well-known fact that photons are not localizable with arbitrary precision. This is a quantum mechanical effect, and not likely to be a limiting factor in any presently-conceivable practical devices, but it's perhaps interesting to know a little about it.

The key point is that photons are massless and have spin, and all such particles are *NOT* localizable. You can find a derivation of this fact in Schwinger [5] where he shows that position operators for a photon cannot commute, much as the operators for position and momentum in quantum mechanics do not. This is probably the simplest example of a "noncommutative geometry" in position space.

An easier argument against localizability of photons follows from the observation that Maxwell's equations have no length scale (remember that they described *any* wavelength of electromagnetic radiation).

How hard is it to localize photons? It turns out that you can only localize a photon to an area $\sim \lambda^2$ in the plane perpendicular to the direction in which it's going. The idea that it "was" anywhere better defined than that is a violation of the uncertainty principle.

11. VLPC's

"Visible Light Photon Counters", or "VLPC's" were developed by Rockwell International originally with military applications in mind, and have been applied at the DØ experiment at Fermilab to detect the faint optical signals from scintillating fibers that flash as charged particles pass through them. They offer quantum efficiencies of 90% and gains of 40,000, with timing responses of 1 nsec, but must run at 7K, which makes them very difficult to use for many applications, and they do have rather high dark count rates. They are actually an outgrowth of an earlier, similar technology of "Solid State Photomultipliers" or SSPM's which had good performance quite a ways into the infrared, but were not available for general civilian use.

They are rather strange devices, very similar to APD's but based on impurity band conduction. This occurs when a semiconductor is heavily doped with shallow donors or acceptors. This makes an impurity band 0.05 eV below the conduction band and electrical transport is by charges hopping from impurity site to impurity site. The normal 1.12 eV Si valence band is used to absorb photons, while the .05 eV gap used to make an avalanche of electrons and "D^+" holes. The small gap means only a few volts is needed, but also clearly means that thermal fluctuations are going to be problematic except at very low temperatures. Similarly to APD's, they have a gain region with a linear field gradient and a drift region with a constant field.

12. SUPERCONDUCTING TUNNEL JUNCTIONS

Is it possible to imagine a device which gives photon counting/timing AND wavelength measurement AND position information?

As incredible as this sounds, it looks like there is. Developed only recently, with a lot of work continuing today and a strong interest from the astronomy community, superconducting tunnel junction detectors look like almost perfect devices (aside from their requirement of extreme cryogenics)[7, 8].

First it's worth recalling what a superconductor is. Superconductors are materials in which electrons can exert weak attractive forces on other electrons through the intermediary of the lattice. You might imagine a heavy bowling ball on a soft sofa making a dent that would tend to drag another bowling ball towards it. If the lattice is cold enough so that thermal fluctuations don't overwhelm this weak attractive force, electrons can pair up into *Cooper pairs* and form spinless current carriers with a charge twice that of a single electron. Lots of these can condense together and form a macroscopic quantum state which will carry current perpetually with no losses whatsoever, and this is what one calls a superconductor.

A Superconducting Tunnel Junction (STJ) is made of two thin films of a superconducting metal such as niobium, tantalum or hafnium separated by a thin insulating layer. When operated at temperatures well below the superconductor's critical temperature (typically below 1 K), the equilibrium state of the junction is easily perturbed by any visible photon striking it - such photons carry energy well in excess of the energy needed to break up the Cooper pairs that make up the superconducting state. The greater the energy of the absorbed photon, the more of these pairs are broken up, so if one measures the amount of charge released from the superconducting state one can measure the energy of the incident photon. Because the binding energy of a Cooper pair is so tiny, even infrared photons can release thousands of charges, and it is relatively easy to measure the energy carried by an individual photon from infrared (a couple of microns in wavelength) to ultraviolet with a few percent resolution!

Detectors based on arrays of these devices could well offer single photon detection with high quantum efficiency where not only would you know where and when a photon hit, but also what colour it was!

13. OUTLOOK

While PMT's remain the main workhorses of photodetection, several devices are catching up quickly. Keep in mind that all single photon counting devices have their own little idiosyncrasies, and while I've tried to point out a reasonable number of them here, you'll likely find the help and advice of experienced colleagues to be invaluable.

High quantum efficiencies are probably only attainable with solid-state devices, and while noise remains a problem, prospects are improving and there are a lot of very interesting developments on the horizon!

14. FINDING OUT MORE

The references given are by no means exhaustive, but should be enough to get you started. If you want to read more, there are many excellent texts, and I would particularly recommend [9] and [4], as well as keeping careful tabs on new devices. You might also find a look at manufacturer's web sites such as [10] of interest.

ACKNOWLEDGMENTS

I would like to thank all my colleagues at the IIB as well as the students at the school for stimulating discussions, as well as my colleagues in high energy physics - especially Yuri Musienko and Stephen Reucroft with whom I do much of my APD work. I would also like to thank the US National Science Foundation for its continued and generous support.

REFERENCES

[1] David J. Griffiths (1998), *Introduction to Electrodyamics, 3rd Ed.*, Prentice Hall PTR, Upper Saddle River, New Jersey.

[2] M. O. Scully and M. Sargent III (1972), *The Concept of the Photon*, Physics Today, March 1972, 38.

[3] David J. Griffiths (1995), *Introduction to Quantum Mechanics*, Prentice Hall Inc., Englewood Cliffs, NJ.

[4] S. Donati (2000), "Photodetectors", Prentice Hall PTR, Upper Saddle River, New Jersey.

[5] J. Schwinger (1989 reissue), "Particles, Sources, and Fields, Volume I", Addison-Wesley Publishing Company, Redwood City, California, equation (1-3.56) on p20.

[6] J. Kim et al.(1997), "Noise-free avalanche multiplication in Si solid state photomultipliers", Applied Physics Letters **70** 2852-2854;
J. Kim et al.(1999) "Multiphoton detection using visible light photon counter", **72** 902;
M. Atac, "Visible light photon counters (VLPCs) for high rate tracking medical imaging and particle astrophysics", proceedings of 7th ICFA School on Instrumentation in Elementary Particle Physics, Leon, Guanajuato, Mexico, 7-19 July 1997.

[7] S. Kraft et al. (1999), *On the development of superconducting tunnel junctions for use in astronomy*, Nuclear Instruments and Methods in Physics Research Section **A 436** (1-2) 238-242

[8] P. Verhoeve (2000) *UV/optical imaging spectroscopy with cryogenic detectors*, Nuclear Instruments and Methods in Physics Research **A 444** (1-2), 435

[9] H. Kume (Chief Editor) (1994), "Photomultiplier Tube", Hamamatsu Photonics.

[10] http://www.hamamatsu.com/
http://opto.perkinelmer.com/index.asp
http://www.electron-tubes.co.uk/menu.html
http://www.advancedphotonix.com/
http://www.rmdinc.com/index2.html

Chapter 8

DETECTION OF PHOTON EMISSION FROM BIOLOGICAL SYSTEMS

Xun Shen
Institute of Biophysics,
Chinese Academy of Sciences
Beijing 100101, P.R.China

1. INTRODUCTION

Since photon emission from living organisms is extremely weak (the photon flux is the order of less than 10^{-15} W) in comparison with a normal light source, some very sensitive methods and devices have been established to detect biophotons both in integral emission intensity and the two-dimensional pattern of the emission on living organisms. This chapter will discuss the modern technology and systems used by various research groups to perform the detection, spectral analysis and two-dimensional imaging of ultraweak photon emission from living organisms such as cells, tissues and even organs. In addition, photon emission from biological system has some intrinsic features, such as coherence, differing with that from non-living systems. Thus, some particular methods are needed to characterize the physical and biological features of photon emission from biological systems.

2. HIGH SENSITIVITY PHOTON COUNTING SYSTEM

2.1 The Detection Methods

Three kinds of photoelectric methods have been employed in the spectral region accessible to a photomultiplier (PM). The first two are DC and AC methods, based on analog schemes, where the light intensity information is extracted as the DC component and the associated AC component shoot noise into the PM output respectively. Since the sensitivity of either the DC method or the AC method is not high enough to detect the spontaneous photon emission from living organisms, the third method, which is often called single-photoelectron counting or simply single-photon counting (SPC), has been developed [1,2]. In the SPC method, the contribution from photoelectrons is resolved in time so that the signals in the form of electron pulses are detected by means of a pulse-counting electronic system. Table 1 summarizes the signal-to-noise ratio for the three methods [3]. Evidently, the sensitivity of the single-photon counting method is much higher than the other two methods. From the formula listed in Table 1, it can be seen that increase of the detection time, decrease of the dark pulse rate N_n or selection of suitable discriminator level will improve the detection sensitivity of the SPC method.

Table 1: Classification and comparison of signal-to-noise ratio of DC, AC and SPC methods for detection of ultraweak photon emission

Classification	Analog detection	
	DC method	AC method
Effective range	$<N_S> > 2B\mu s$	$<N_S> < 2B\mu s$
Signal to noise ratio	$\dfrac{<N_S>/\sqrt{2b}}{\{[<N_S>+<N_n>(<G_n^2>/<G_S^2>)]\mu s\}^{1/2}}$	$\dfrac{<N_S>/\sqrt{2b}}{\{[<N_S>+<N_n>(<G_n^4>/<G_S^4>)]\mu s'\}^{1/2}}$

Classification	Digital detection
	SPC method
Effective range	$<N_S> < 0.1B^*$
Signal to noise ratio	$\dfrac{<N_S>\sqrt{T}}{\{<N_S>+<N_n>\xi(V)\}^{1/2}}$

It is assumed that the output pulse from a photomultiplier ranges in the single photoelectron event with the probability, which is given by $\exp\{-(<N_S>+<N_n>)/2B\}$, of 95%.

$<N_S>$ and $<N_n>$ are average rate for signal photoelectron and noise pulse respectively. B and b are bandwidth of amplifier and low pass filter. $<G_S^2>$

and $<G_n^2>$ are average gain of photomultiplier for signal and noise. μ_S, and μ_S' are excess noise factor: $\mu_S = <G_S^2>/<G_S>^2$, $\mu_S' = <G_S^4>/<G_S^2>^2$. $\xi(V)$ is signal-to-noise counting efficiency ratio as a function of selected pulse-height discriminator voltage V, which is defined as. $\xi(V) = \int_v^\infty P_n(v)dv / \int_v^\infty P_S(v)dv$, where $P_S(v)$ and $P_n(v)$ represents the pulse-height distribution function of single-photoelectron and dark pulse respectively. T is detection time.

2.2 Single-Photon Counting

In order to achieve the highest sensitivity, the photomultiplier (PM) must work in a single-photoelectron detection model. This means that when one photoelectron is released from the cathode of PM, the formed electron pulse on the anode of the PM can be detected and discriminated from shot noise generated in dynodes. With this working model of a PM tube, even one photon impacted onto the cathode of a PM can result in a registered pulse. Fig.1 shows the noise-pulse amplitude distribution for the photomultiplier *EMI-9659QB* working in a single-photoelectron counting model.

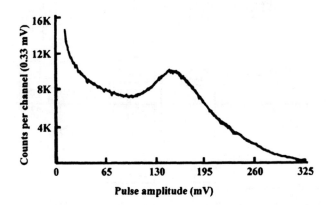

Figure 1: The noise-pulse amplitude distribution of a photomultiplier EMI-9659QB applied with 1400 V. The distribution is measured by a multichannel pulse-height analyzer at room temperature..

The peak of the distribution represents the pulse amplitude corresponding to one photoelectron emitted from the cathode. Thus, in the single-photon counting system the discrimination level for registering the pulse from a PM must be set to allow all the pulse, which is higher than the single photoelectron pulse height, to be registered. The SPC system used by Inaba's group is shown in Fig.2. In the system, the living organisms are put in a sample cell (usually a quartz cuvette), which is placed in a sample compartment controlled at a suitable temperature. The emitted photons are effectively gathered on the photocathode of the PM, which is cooled, by liquid nitrogen or cold ethanol by means of an ellipsoidal reflector. The spectral measurement of ultraweak photon emission from living organisms is accomplished with the successive insertion of colored glass filters arranged on a rotating disk into the optical path between the reflector and the PM. For spectral analysis, interference filters and cut-off filters can also be used. In order to get high-sensitivity, the photomultiplier used in the single-photon counting apparatus must be cooled down. A typical design of the detector of the single photon counting apparatus developed by F.A.Popp's laboratory [4] is shown in Fig.3.

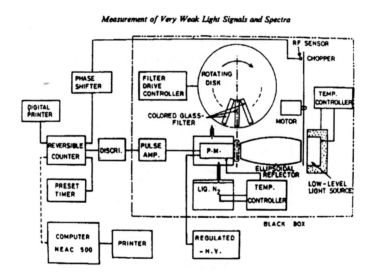

Figure 2: Block diagram of the single-photon counting apparatus. The apparatus can either measure the intensity of photon emission from a sample or analyze the spectrum of the photon emission.(cited from H.Inaba et al.[3])

2.3. Spectral Analysis of Biophoton Emission

The simple way to measure the spectrum of photon emission from living organisms is to use colored glass filters (or called band-width filters)

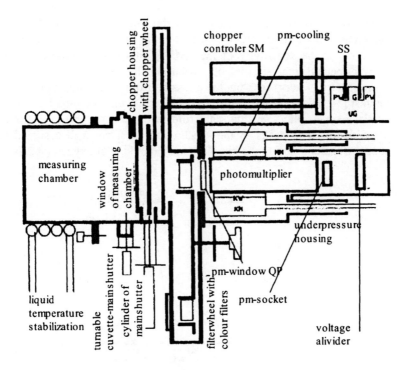

Figure 3: The detector of the single-photon counting apparatus used in the International Institute of Biophysics (Neuss, Germany).

covering the wavelength region between 280 and 650 nm with a single-photon counting system. The transmittance of the colored filters is usually in the range of 40% to 60%. The spectrum is given as the count rate of the photons passed through the individual colored filters divided by the transmittance of the filter and the bandwidth, then corrected against the sensitivity of the PM photocathode in the colored filter-corresponding wavelength band. However, the cut-off filters are recommended for more precise measurement of the photon emission spectrum. The spectral analysis is carried out by calculating the count rate for the individual spectral window defined by the subtraction of two transmission curves with different sharp short-wavelength cut-offs corresponding to the successive colored filter, followed by the calibration of the spectral sensitivity of the photocathode. That is, the average optical intensity P_{ij} between the ith and jth successive colored filters is expressed as

$$P_{ij} = (N_i - N_j) / \int S_k(\lambda)[F_i(\lambda) - F_j(\lambda)]d\lambda$$

where $N_i - N_j$ is the subtraction of the count rate, $F_i(\lambda)-F_j(\lambda)$ is the subtraction of the transmission curve for the ith and jth successive filters, and $S_k(\lambda)$ is the sensitivity of the PM's photocathode respectively. Using such a spectral analysis method, Inaba and his co-workers have obtained some spectra of very weak photon emission from some samples of biological interest [5]. Those obtained spectra demonstrate that the bimolar reaction of singlet oxygen plays an important role in the photon emission from lipid peroxidation. The peaks in the spectra can be fairly well assigned to the energy transition during the bimolar reaction of singlet oxygen (see Table 2)

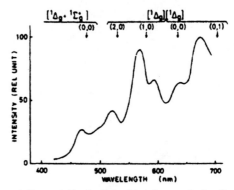

Figure 4: Spectral distribution of the ultra-weak chemiluminescence measured in the reaction of linoleic acid hydroperoxide with ceric ions.

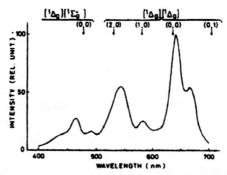

Figure 5: Measured spectrum of the ultra-weak chemiluminescence generated in the reaction of xanthine oxidase-acetaldehyde system

Table 2: Spectral data of the ultra-weak chemiluminescence observed during the oxidation of lipid hydroperoxide

Electronic upper state and vibrational transition of $^1O_2$ to ground state $[^3\Sigma_g^-][^3\Sigma_g^-]$	λ_{max}(nm) (relative intensity)			λ_{max} (nm) Ref.
	System 1	System 2	System 3	
$[^1\Sigma_g^+][^1\Sigma_g^+]$ (0,0)	None	None	None	400*
$[^1\Sigma_g^+][^1\Delta_g]$ (0,0)	480 (0.45)	470 (0.26)	None	478† 480*
$[^1\Delta_g][^1\Delta_g]$ (2,0)	520–530 (1.00)	520–530 (0.43)	512–520 (0.30)	520*
$[^1\Delta_g][^1\Delta_g]$ (1,0)	570 (0.62)	570 (0.90)	595 (0.85)	578; 580*
$[^1\Delta_g][^1\Delta_g]$ (0,0)	620–640 (0.66)	640 (0.64)	630 (1.00)	633‡,§
Unspecified transition	670 (0.67)	670 (1.00)	680 (0.80)	670‡
$[^1\Delta_g][^1\Delta_g]$ (0,1)				703‡

*Furukawa, K., E. W. Gray and E. A. Ogryzlo (1970) *Ann. N.Y. Acad. Sci.* **171**, 175.
†Khan, A. U. and M. Kasha (1966) *J. Am. Chem. Soc.* **88**, 1574.
‡Seliger, H. H. (1964) *J. Chem. Phys.* **40**, 3133; Brown, R. J. and E. A. Ogryzlo (1964) *Proc. Chem. Soc. London*

3. THE EXPERIMENTAL APPROACH TO STUDY PHOTON COUNTS STATISTICS

3.1. Photon Counts Statistics

Theoretical work [6,7] and some experimental evidence [8,9] suggest that either spontaneous or photon-induced photon emission from a biological system is at least partially coherent. This was verified by statistical distribution of photocounts measured in a short time interval such as 50 to several hundred milliseconds. It has been firmly established by quantum optics that the photons from a fully coherent field obeys Poissonian distribution, but the photons from a single-mode chaotic field are subjected to geometrical distribution [10,11]. The photon count statistics (PCS) measures the probability, $p(n, \Delta t)$, with which n photocounts are registered by a single-photon sensitive detector in a preset time interval, Δt. A fully coherent field in a stationary state is always subject to a Poissonian distribution:

$$p(n, \Delta t) = \frac{<n>^n}{n!} e^{-<n>} \quad (1)$$

$$\sigma^2 = <n>$$

where $<n>$ is the average count registered within the preset time interval, and σ^2 is the variance of the distribution. For the single-mode chaotic field, the photocounts follow the geometric distribution:

$$p(n, \Delta t) = \frac{\langle n \rangle^n}{(1+\langle n \rangle)^{n+1}} \quad (2)$$

$$\sigma^2 = \langle n \rangle (1 + \langle n \rangle)$$

For a multi-mode chaotic field, i.e. a chaotic field with $\Delta t \gg \tau$ (τ is the coherence time of the field), the photocounts registered are subject to the following distribution:

$$p(n, \Delta t) = \frac{(n+M-1)!}{n!(M-1)!} \left(1 + \frac{M}{\langle n \rangle}\right)^{-n} \left(1 + \frac{\langle n \rangle}{M}\right)^{-M} \quad (3)$$

$$\sigma^2 = \langle n \rangle \left(1 + \frac{\langle n \rangle}{M}\right)$$

where the number of modes, $M = \Delta t / \tau$, is a measure of the "degree of freedom" of the chaotic field. Using the above theoretical criteria, the coherence of the ultraweak photon emission from living organisms can be checked experimentally by measuring their PCS.

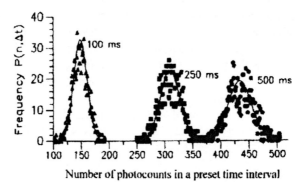

Figure 6: The photocounts distribution of the spontaneous photon emission from the soybean Rhizobium (ATCC 10324) bacteroids in 2 ml of 0.06M phosphate buffer (1.2×10^8 cells/ml). Each distribution consists of 1000 samplings corresponding to one of the three preset time intervals, 100 ms, 250 ms and 500 ms. The solid lines represent the theoretical Poissonian distributions with the same mean values of each measured distribution.

In experimental studies on PCS, each measurement usually consists of a

1000 sampling of registered photocounts for a preset time interval. The frequencies with which k photocounts are registered, $F(k)$, are scored. If the sampling size is big enough, $F(k)$ can fairly well represent the $p(k, \Delta t)$. Fig.6 shows the measured photocount distributions of the spontaneous photon emission from the soybean rhizobium bacteroids in phosphate buffer [9]

Since it is difficult to judge quantitatively how close a particular distribution is to the Poissonian just by looking at the distribution profile as shown in Fig.6, Popp suggested to take the value $\delta = (\sigma^2 - <n>)/<n>$ to measure how close a particular distribution is to a Poissonian distribution. $<n>$ and σ^2 can be calculated from the observed frequency $F(k)$:

$$<n> = \sum k\, F(k)/\sum F(k) \tag{4}$$
$$\sigma^2 = \sum (k-<n>)^2\, F(k)/\sum F(k) \tag{5}$$

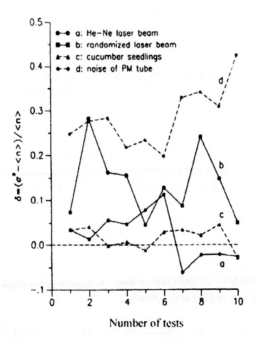

Figure 7: The photocount statistics (PCS) measure for He-Ne laser, the randomized laser beam, cucumber seedlings and the noise of the photodetector at –18 °C respectively. Each test consists of 1000 samplings in the time interval of 100 ms.

Agreement with a Poissonian distribution is expressed as $\delta = 0$, while δ

>0 as a bunching effect indicated a chaotic source, and $\delta < 0$ means that antibunching takes place. Fig.7 shows the δ values of ten independent tests for an He-Ne laser, the randomized laser beam, cucumber seedlings and the noise of the photodetector at -18 °C respectively [9]. It can be seen clearly that the δ values of the photocount distributions for a laser beam and for cucumber seedlings are close to zero, while those of the other two sources are relatively distant from zero. This means that the photocounts distribution for a laser beam and the living organism are close to Poissonian, but the distribution for a randomized laser beam of chaotic nature and the thermal emission from the cathode of the photomultiplier is relatively far away.

3.2. Coincidence Counting System

In the photocounts statistics experiment performed with a conventional counting system, two experimental conditions must be met in order to obtain the real statistical distribution of the photon counts from the source under investigation. First, a stationary or a quasi-stationary photon field is required. However, this requirement is hardly satisfied in most circumstances due to the continuous metabolic changes in living systems and slowly decaying fluorescence. Second, the noise of the counting equipment should be much lower than the net photon counts from the living organism in order to prevent the measured photon counts statistics from distortion caused by the noise. This requirement is even more difficult to be fulfilled in an experimental study on living systems, since the spontaneous photon emission from living organisms, such as cell suspension and tissues, is rather weak and is often not one order of magnitude higher than noise. In order to overcome the above difficulties, a new coincidence counting method has been developed[12]. The new system and the coincidence counting method can be used to determine the photocounts statistics of a non-stationary photon field with an ever-wanted high accuracy.

3.2.1. Principle of the Coincidence Counting Method in Determining the Photon Counts Statistics

Consider two photon sources X_1 and X_2 respectively or one photon source measured by two independent detectors associated with two channels. Channel 1 measures the photons from the source X_1, and channel 2 those from the same source X_2. The coincidence counting system (CCS) is constructed in such a way that a photon, which is counted in channel 1, is registered as a coincident one as soon as at least one other photon has been counted in channel 2 in a time interval Δt after the photon counting happens in channel 1(see Fig.8). The number of random coincidences, Z_j, in the j-th time interval ΔT_j is then:

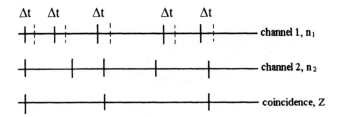

Figure 8: The correlation of coincident counts with the counts in two channels.

$$Z_j = n_{1j}\, P_2(\Delta t, n_{2,\Delta t} \geq 1) \qquad (6)$$

where n_{1j} is the number of counts in channel 1 within the j-th time interval ΔT_j, and $P_2(\Delta t, n_{2,\Delta t} \geq 1)$ is the probability of counting at least one photon in channel 2 in the time interval (or coincidence time) Δt. Since $P_2(\Delta t, n_{2,\Delta t} \geq 1) = 1 - P_2(\Delta t, 0)$, where $P_2(\Delta t, 0)$ is the probability of counting no photon in the time interval Δt in the channel 2. From (6) we get:

$$Z_j = n_{1j}[\,1 - P_2(\Delta t, 0)] \qquad (7)$$

Since the CCS allows measuring as well Z_j as n_{1j}, we can determine $P_2(\Delta t, 0)$ of the photon source. Let us compare a fully coherent and a completely chaotic photon source. The photon count statistics of a fully coherent source follows a Poissonian distribution, while that of a completely chaotic field is subject to a geometrical distribution. This means that for a coherent field the probability, $P(\Delta t, n)$, counting n photons in a time interval Δt is:

$$P(\Delta t, n)_{coherent} = \frac{<n>^n}{n!} e^{-<n>} \qquad (8)$$

and for a chaotic monochromatic field:

$$P(\Delta t, n)_{chaotic} = \frac{<n>^n}{(1+<n>)^{n+1}} \qquad (9)$$

It is obvious that we can fairly well distinguish a coherent from a chaotic field, since two completely different probabilities for $n = 0$ are obtained:

$$P(\Delta t, 0) = e^{-\langle \hat{n} \rangle \Delta t}$$

$$P(\Delta t, 0) = \frac{1}{1 + \langle \hat{n} \rangle \Delta t}$$

where $\langle \hat{n} \rangle$ is the expected value of the photon count rate. This leads obviously to different values of the coincidence, Z_j:

$$Z_{j\,coherent} = n_{1j}\,(1 - e^{-\langle \hat{n}_2 \rangle \Delta t}) \tag{10}$$

$$Z_{j\,chaotic} = \frac{N_{1j}\,\langle \hat{n}_2 \rangle \Delta t}{1 + \langle \hat{n}_2 \rangle \Delta t} \tag{11}$$

where $\langle \hat{n}_2 \rangle$ is the expected value of the photocount rate in channel 2. Equations (10) and (11) can also be rewritten as:

$$\frac{Z_{j\,coherent}}{n_{1j}} = 1 - e^{-\langle \hat{n}_2 \rangle \Delta t} \tag{12}$$

$$\frac{Z_{j\,coherent}}{n_{1j}} = \frac{\langle \hat{n}_2 \rangle \Delta t}{1 + \langle \hat{n}_2 \rangle \Delta t} \tag{13}$$

If N measurements on the photon counts in channel 1, channel 2 and corresponding coincident counts were made with time interval ΔT, and the Z_j/n_{1j} ($j=1, 2, \cdots, N$) was plotted against the $(n_{2j}/\Delta T)\cdot \Delta t$, the data points will be very close to the curve $1 - e^{-\langle \hat{n}_2 \rangle \Delta t}$ for the coherent field, but to the curve, $\langle \hat{n}_2 \rangle \Delta t / (1 + \langle \hat{n}_2 \rangle \Delta t)$ for the chaotic field. The data points from a partial coherent photon field will appear between these two curves in such a plot. From the above-presented formulas, it should be noticed that there is no limitation regarding how stationary the photon field is. The equations (12) and (13) provide an extremely powerful means to determine the photon count statistics of a photon source by registering the photon counts in two channels and their coincident.

3.2.2. Experimental Measurements on the Photon Count Statistics of the Delayed Photon Emission

Mungbean seedlings with a length of about 2 cm (two days after germination) and leaflets from the elder bush were used as the investigated living organisms. 20 mungbean seedlings or one leaflet were placed in each of the two quartz cuvettes ($24 \times 24 \times 40\ mm^3$), then the two cuvettes were placed in two dark chambers corresponding to the two photon counting channels after exposing to daylight for ten seconds. The counting was immediately started as soon as the chambers were closed in order to get a fast decayed photon emission from the seedlings or the leaflets. The kinetics of the photon counts in channel 1, channel 2 and their coincident registered for mungbean seedlings and leaflets are shown in Fig.9 and Fig.10 respectively. The time interval for the measurement of the photon counts in each channel was 100 ms, and the coincidence time was 0.5 ms for mungbean seedlings and 0.2 ms for leaflets. In each case, 1000 registered photon counts in each channel and the corresponding coincident counts were analyzed. The ratios of the coincident counts to the counts in channel 1 for each 100 ms time interval were then plotted versus the count rate in channel 2 *(i.e. $n_2/\Delta T$)* times coincidence time Δt.

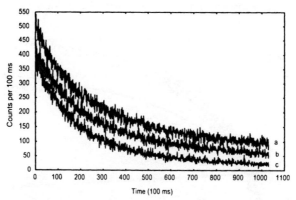

Figure 9: The decay of the photon emission from Mungbean seedlings after exposing to sunlight. a: counts in channel 2; b: counts in channel 1; c: coincident counts.

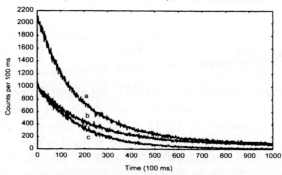

Figure 10: The decay of the photon emission from leaflet of elder bush after exposing to sunlight. a: counts in channel 2; b: counts in channel 1; c: coincident counts.

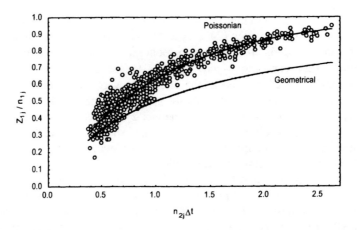

Figure 11: The probabilities of measuring at least one photon from the mungbean seedlings after exposing to sunlight. The time interval for photon counting and the coincidence time are 100 ms and 0,5 ms respectively. 1000 measured counts are analyzed.

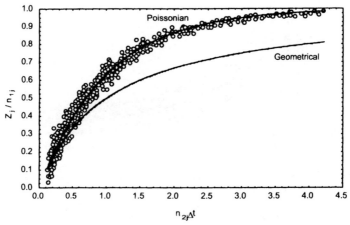

Figure 12: The probabilities of measuring at least one photon from the leaflet of elder bush after exposing to sunlight. The time interval for photon counting and the coincidence time are 100 ms and 0.2 ms respectively.

According to formulas (12) and (13), this plot gives the probabilities of measuring at least one photon in channel 2 in the time interval Δt at various count rates in channel 2. Fig.11 and Fig.12 show the probabilities of measuring at least one photon for the delayed photon emission from mungbean seedlings and leaflet, as well as the theoretical-calculated probabilities for Poissonian and geometrical distribution.

It can be seen clearly that the measured probabilities of measuring at least one photon from the two investigated living systems are in very good agreement with the probabilities calculated based on the Poissonian distribution.

The new coincidence counting method provides a powerful tool to determine the photon count statistics for a photon field, in particular for the photon emission from biological systems. Since there is no requirement for a stationary field, this method can be used to determine the photon count statistics for a delay luminescence or the photon-induced photon emission from living systems. It endows this new method with another great advantage over the conventional photon counting method, in that the distortion of the measured photon count statistics for the photon emission from a biological system by the unavoidable noise of the counting system can be actually ruled out. Besides the determination of photon count statistics, this coincidence counting technique can also be used for the systematic investigation of the correlation between two photon sources, in particular the biocommunication between two biological systems.

4. TWO-DIMENSIONAL IMAGING OF ULTRAWEAK BIOPHOTON EMISSION

In order to analyze spatial-temporal properties of biophoton emission, to calculate intensity kinetics at several regions or space-time correlation between two regions, to know the spatial distribution of biophoton emission on the surface of the organisms, it is necessary to have two-dimensional images of the biophoton emission from biological systems such as cells, seeds, tissues and even animal bodies. Inaba's group obtained the first two-dimensional images of biophoton emission in 1989 [13]. They obtained the images of germinating soybeans using microchannel plates (MCP) and a position sensitive detector. Since then, great technical advances have been made in this direction.

4.1. Photon Counting Imaging System Based on MCP

The first generation of the single photon imaging system is illustrated schematically in Fig.13. Light from a sample is imaged onto the photocathode by the lens. Photoelectrons ejected from the photocathode strike a stack of microchannel plates (MCP), which provides amplification of the order of 10^8. A burst of secondary electrons emerging from the last MCP then impacts a sensitive anode position detector (PSD). From the relative pulse amplitudes measured at the corners of the PSD, the position of the

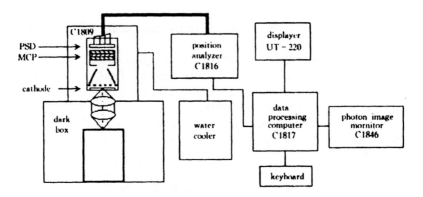

Figure 13: Diagram of photon-counting image acquisition system (PIAS). MCP: microchannel plate; PSD: position sensitive detector.

photoevent is determined. In this way, the two-dimensional image of the photon emission from organisms is measured and the information displayed on a monitor and stored in computer for processing. In order to improve signal to noise ratio and reduce the time for getting sufficient data on the image, the photocathode and MCP are usually cooled down to –20 °C.

Fig.14 shows the imaging of germinating rice seed detected by a Chinese research group in Zhejiang University using such a single photon imaging system [14]. It may be seen that the most intensive emission comes from the root and the area near the root. The photon emission from the embryo is relatively weak.

4.2. Photon Counting Imaging System based on Cooled CCD Camera

With the development of a charge-coupled device (CCD), this new device has been used in the two-dimensional imaging of biophotons. It replaced the position-sensitive detector about 4 years ago. Recently, as the cooled very low noise CCD has developed, a CCD can even replace both an MCP and position sensitive detector. The sensitivity of a cooled CCD can be as good as that of a photomultiplier. The conventional imaging systems used for biophoton imaging uses a two-dimensional photo counting tube (2D PMT) that consists of a photocathode, an MCP and a position-sensitive detector and has excellent sensitivity. However, the quantum efficiency of the

(A) Photon-counting image of a rice seed at drier condition of germination

(B) Reflected image of the rice seed at drier condition of germination

(C) Photon-counting image of a rice seed at wet condition of germination

(D) Reflected image of the rice seed at wet condition of germination

Figure 14: The images of spontaneous photon emission from the rice seed at a drier (A) and wet condition (C) of germination. (B) and (D) are the reflected image of the rice seed under dim illumination.

photocathode is very low, approximately 20% at peak wavelength of 400 nm and less than 5% in the red or near-IR regions, hence is not useful for a wide variety of biophoton emission measurements. The minimum detectable optical power at each pixel of a 2D PMT with a signal-to-noise ratio of unity can be defined as[15]:

$$(P_{min})_{PMT} = (N_{dl}T)^{1/2} \; hc/\eta_1 \lambda TS \qquad (14)$$

where N_{dl} is the number of dark counts per unit time at a single element of 2D PMT, T is the measurement time, h is Planck's constant, c is the velocity of light, η_1 is the quantum efficiency of the PMT photocathode, λ is the wavelength and S is the area of pixel. In the case of a CCD camera, the minimum detectable power is defined as:

$$(P_{min})_{CCD} = (N_{d2}T + N_r^2)^{1/2} hc/\eta_2 \lambda TS \tag{15}$$

where N_{d2} is the number of electrons contributing to the dark current per unit time at a single element of the CCD, η_2 is the quantum efficiency of the CCD and N_r represents the electron mean square root of the readout noise of amplifier circuit per pixel. From the above equations it can be deduced that if the CCD were operated at extremely low dark current and longer exposure time to reduce the contribution of the readout noise, the signal-to-noise ratio of the CCD camera would be superior to the conventional 2D PMT under similar conditions of wavelength and measurement time.

REFERENCES

1. Beall, H.C. and Haug, A (1973) A photon counting device for the measurement of nanosecond and microsecond kinetics of light emission from biological systems. *Anal Biochem.* **53**, 98-107.

2. Inaba, H., Shimizu, Y. and Tsuji, Y. (1975) Measurement of very weak light signal and spectra, *Japan.J.Appl.Phys.*, **14** (Suppl.14), 23-32.

3. Mieg, C., Mei, W.P. and Popp, F.A. (1992) Technical notes to biophoton emission. In *Recent Advances in Biophoton Research*, Popp, F.A., Li, K.H. and Gu, Q. (eds.), World Scientific, Singapore, 197-205.

4. Inaba, H., Shimizu, Y., Tsuji, Y. and Yamagishi, A. (1979) Photon counting spectral analyzing system of ultra-weak chem.- and bioluminescence for biochemical application. *Photochem. Photobiol.*, **30**, 169-175.

5. Popp, F.A. and Li, K.H. (1993) Hyperbolic relaxation as a sufficient condition of a fully coherent ergotic field, *Int.J.Theoretical Phys.*, **32**,1573-83.

6. Popp,F.A., Gu,Q. and Li, K.H.(1994) Biophoton emission: Experimental background and theoretical approaches. *Modern Phys. Lett. B*, **8**, 1269-1296.

7. Popp, F.A., Li, K.H., Mei, W.P., Galle, M. and Neurohr, R. (1988) Physical aspects of biophotons. *Experientia*, **44**, 576-85.

8. Shen, X., Liu, F. and Li, X.Y. (1993) Experimental study on photon count statistics of the ultraweak photon emission from some living organisms. *Experientia*, **49**,291-95.

9. Arecchi, F.T. (1969) Photon count distribution and field statistics In *Quantum Optics*, Glamber R.J. (ed.), New York, Academic Press.

10. Perina, J. (1985) *Coherence of Light*. New York, Reidel Publ. Co, Dortecht.

11. Popp, F.A. and Shen, X. (1998) The photon count statistics study on the photon emission from biological system using a new coincidence counting system. In *Biophotons*, Chang, J.J., Fisch, J. and Popp, F.A. (eds.), Dordrecht, Kluwer Academic Publishers.

12. Scott, R.Q., Usa, M. and Inaba, H. (1989) Ultraweak emission imagery of mitosing soybeans. *Appl.Phys.B*, **48**, 183-85.

13. Bao, C. (1998) A study on ultraweak photon-counting imaging of rice seeds at the stage of germination, *Acta Biophysica Sinica*, **14**, 772-776.

14. Kobayashi, M., Devaraj, B., Usa, M., Tanno, Y., Takeda, M. and Inaba, H. (1997) Two-dimensional imaging of ultraweak photon emission from germinating soybean Contents of the templates seedlings with a high sensitive CCD camera. *Photochem.Photobiol*, **65**, 635-537.

Chapter 9

PHOTON EMISSION FROM PERTURBED AND DYING ORGANISMS – THE CONCEPT OF PHOTON CYCLING IN BIOLOGICAL SYSTEMS

Janusz Slawinski
Department of Radio-& Photochemistry,
Institute of Chemistry and Technical Electrochemistry,
Faculty of Chemical Technology, Poznan University of Technology.

1. THERMODYNAMICS OF THE "LIVING STATE" – HOMEOSTASIS

Biological organisms are open systems as they exchange energy E, information Inf and mass m with their environment (Fig. 1). They achieve the degree of spaciotemporal order (infrastructure) necessary for survival through the flow of metabolic free energy ΔG. To maintain this order and a steady-state (a dynamic equilibrium) requires that the rate of entropy S production inside the open system dS_i / dt is balanced by the rate of external entropy S_e production as shown in Fig. 1:

The most economical steady-state is characterized by the minimum value of the entropy source σ:

$$\sigma = dS_i/dV\, dt = const \longrightarrow min \qquad (1)$$

for $\sigma > 0$. Here V is the volume of the entropy-producing compartment. For controlling the functions of a self-regulating living system in a perturbing environment some, kind of feedback control system involving communication with biosensors and biotransmitters is necessary. After a stimulus exceeding a threshold value, the living system will respond by stress (perturbation of homeostasis), producing a reflex reaction with the

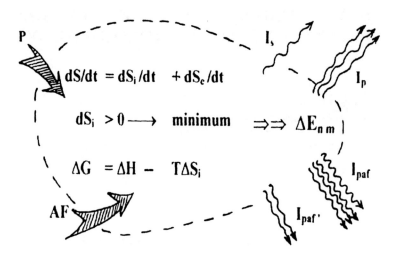

Figure 1: A schematic simplified model of the open system far from thermodynamic equilibrium and its photon emission response under a steady state I_s, (an intact organism), when perturbed by stress (P, I_p) and additionally affected in the perturbed state(AF, I_{paf}). Explanations in the text.

effect of cancelling any perturbation produced by the stimulus. This process consumes free energy ΔG, the sources of which are catabolic reactions. Therefore the rate of these reactions $J = \xi$ and fluxes J_i of reactants (i) increases:

$$\sigma = \sum_i J_i X_i > 0 \qquad (2)$$

where $X = A/T$ is a thermodynamical force or gradient (grad, chemical affinity $A = \Sigma \, v_\gamma \, \mu_\gamma$ and $\mu_\gamma = (\Delta G/dn_\gamma)_{n',p,T}$, n being a molecular fraction of a reactant γ). Any perturbation of homeostasis, in turn, increases energy dissipation σ. Part of the total $\Delta G < 0$ can be used for the chemiexcitation and generation of an electronic excited state (*) of a reactant and following radiative relaxation, i. e. chemiluminescence (photon emission):

$$A \xrightarrow{|\Delta G| \geq h\nu} \to \to A^* \to A + h\nu \qquad (3)$$

This is symbolically shown in Fig 1 as the term $\Delta E_{n\,m}$ and will be discussed further.

2. PERTURBATIONS OF HOMEOSTASIS AND ULTRAWEAK PHOTON EMISSION

All disturbances of homeostasis lead to disturbance of the steady state. This, in turn, is associated with changes in the rate and energetics of those exergonic reactions which produce excited molecules A*, due to the relationship (3):

$$-k_B T \ln K^{\#} = \partial G^{\#}/\partial \xi = \partial H^{\#}/\partial \xi - T \partial S/\partial \xi \tag{4}$$

ξ is the degree of advancement of a chemical reaction. The derivative $(\partial G^{\#}/\partial \xi)_{p,\,T}$ is the difference in free energy ΔG between product(s) and reactant(s) at a given point during the course of reaction and as such is an instantaneous rate of ΔG with respect to ξ. $K^{\#}$ is the equilibrium constant associated with the equilibrium between reactants in the transition state ($^{\#}$). $\Delta H^{\#}/\partial \xi$ and $\partial S^{\#}/\partial \xi$ represent enthalpy and entropy change associated with motion into the transition state. The value of ΔG contributes to the excitation energy (*) and its distribution among the products of an elementary light-producing reaction according to the obvious relationship:

$$|\Delta G| = h\,c/\lambda_c \tag{5}$$

where λ_c is a short wavelength limit of the emission spectrum of luminescence. Values of the total quantum yield Φ of a chemiluminescent reaction with the intensity $I = \Phi J$ and $I(\lambda)$ depend on ΔG and ΔH-distribution among the reaction products as well as on their spectroscopic properties (energy level configurations, radiative lifetime rate constant of quenching processes etc).

It is very important to realize that thermodynamics is only a formal macroscopic description of biosystems. Intrinsic, real reactions at the molecular and submolecular levels are accomplished by electromagnetic forces. As a matter of fact every living system is endoved with an endogenous electromagnetic field, the parameters of which depend upon physiological and environmental factors. In a perturbed state, a biological system may become more sensitive to external and internal stimuli than in an unperturbed (intact) state, and this behavior is reflected in parameters of photon emission. This synergetic behavior is expressed in parameters of photon emission, e.g. its intensity that is shown in Fig. 1 by the number of waves emitted under the influence of additional factor (AF) acting simultaneously at the perturbed (P) state.

3. EXPERIMENTAL DATA ON THE PERTURBATION-INDUCED PHOTON EMISSION

There is a large number of publications on this subject [for review see e.g. 1-7]. In this paper, however, we will focus on the irreversible perturbation of homeostasis leading to the death of an organism. Such a phenomenon was observed for lower organisms and published in a few works [8-14]. Almost all organisms tested so far emit enhanced PE with the signal-to-noise ratio *(S/N) >1* when affected by stress factors. The value S/N increases when a perturbing/lethal factor is applied under the oxygen atmosphere and at *pH > 8*. Evidently, these conditions facilitate a "trivial chemiluminescence" associated with highly exergonic radical reactions involving superoxide ion-radical $O_2^-\bullet$, $OH\bullet$, H_2O_2. Thus, the "trivial" chemiluminescence "contaminates" degradative or necrotic radiation and contributes to the total luminescent signal from the dying organism. In the light of these findings, a new methodological requirement of experiments on necrotic radiation has been formulated: the set of experimental conditions should maximally eleminate those stress factors and conditions which induce direct exergonic oxidative reactions. The above requirement was not taken into consideration in the majority of previous experiments. Therefore in our recent investigation, "soft" chemical stress factors which do not stimulate direct oxidative degradation of biological tissue were used. Highly dehydrating and lipid-dissolving substances such as acetone, ethanol, protein denaturating compounds like formaldehyde and trichloracetic acid (TCA) and ionic non-oxidative *HCl*, which bring about weak molecular and ionic interactions, were applied.

In a series of model reactions it was checked whether these chemicals give a measurable PE when reacting with certain aminoacids like trp, gly, cys, amines L-dopa, proteins (bovine serum albumine, calf thymus DNA, egg yolk, phospholipids, linoleic acids, sugars and Na-dodecylsulfate in aerated aqueous solutions at pH = 4-7 at 293 K. In all cases no statistically significant or only spurious PE with *S/N <1* was recorded. However, the increase of pH above 8.0 elicited a measurable PE with *S/N>1*. These experiments are described elsewhere [4,6,15-18]. As an example, only necrotic PE resulting from the interaction of yeast cells with TCA [19] and of a fly + acetone are shown (Fig.2 and 3). An electrostatic fast interaction between the electric charge of the cell membrane and anions of TCA *(pH ≅*

4.0) leads to the death of the cells and precipitation of the dead cells with concomitant enhancement of PE. This interpretation is supported by the analysis of the excitation and emission fluorescence spectra of trp residues in the yeast proteins. Ultraweak PE occurs in the broad spectral range 440-850 nm with the main maximum centered around 600 nm.

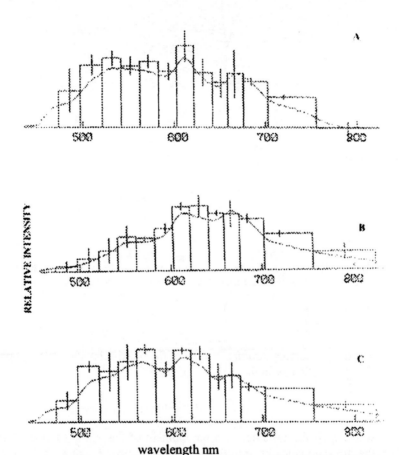

Figure 2: Ultraweak photon emission spectra of dry (A), water (B) and 5% trichloracetic acid (C) -treated bakery yeast cells *Saccharomyces cerevisiae Harvest Gold, GB* (10^7 cells/ml in PBS).

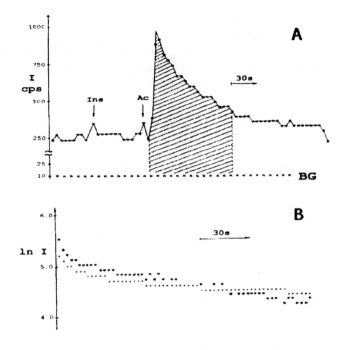

Figure 3: Necrotic emission from a single fly *Musca domestica* treated with 0.1 ml acetone (AC, A) and the fit of the experimental data (dashed part) to the theoretical hyperbolic curve (B .. . . .). BG - background PE of the apparatus, Ins - a short red illumination for the inspection of fly.

The decay kinetics $<dn/dt> = I f(t)$ of the quasi-relaxation (decay) phase fits the hyperbolic law in the majority of cases:

$$I = A (t - t_0)^{-\beta} \tag{6}$$

where t_0 is the time between the moment of the action of stress factor and the first statistically significant change in the I value, A is a constant (amplitude) and β is a decay constant. Unambigous assignment of the decay kinetics requires, however, the change of I and t-values over at least two decades of the I and t-scale which is hardly attainable because of a low S/N ratio.

Comparison of the PE response from alive (I_{max}) and dead (I'_{max}) organisms previously treated with the same or other lethal factor shows that the dead organism responds to the second treatment with a much lower value of $(I_{max}) / (I'_{max})$ than that of the alive one (the first treatment). Moreover, the kinetic pattern $<dn/dt> = I f(t)$ is significantly different. Therefore, one can conclude that the irreversible destruction of intact structures and metabolism of the living state results in enhanced PE. Since "soft" stress factors and experimental conditions significantly eliminate direct exergonic chemiluminescent reactions, the measured PE signals can originate mainly from weak molecular and supramolecular interactions. These may involve: 1/ the reorganization of hydrogen bonds and Van der Waals complexes within vital structures and 2/ a radiative relaxation of metastable energy-rich structures (states) in membranes, the cytosol, cytoskeleton and in superhelical structures of DNA in the nuclear chromatine. However, the energy liberated in the single elementary step of these interactions (ΔH, ΔG) is too low to excite electronic energy levels corresponding to the observed spectral range of PE (400-800 nm), i.e. $|\Delta H, \Delta G| < hc/\lambda_0$ according to (5). Therefore it seems to be necessary to assume that the molecular interactions collectively accumulate small portions of $|\Delta H, \Delta G| < hc/\lambda_0$ until the threshold value of $E > hc/\lambda_0$ is reached. Thermodynamical aspects of this problem are similar to e.g. crystaloluminescence, when weak molecular interactions form the structure of an organic crystal and the requirement (5) is not fulfilled. For biological systems this problem is still more complex and requires a separate theoretical treatment.

Particular considerations should be devoted to the action of anaesthetics, e.g. lignocaine on algae [20, 21,23]. Lethal doses of these compounds induce irreversible depression of action potentials conduction and membrane excitability, complete cytoplasmic streaming cessation and concomitant increase in PE with the $S/N \leq 5$. The sudden increase in the PE induced by 15 mM procaine (a lethal dose) occurs after about 120 min exposure and its gradual decay reveals damped oscillations [21,23].

Recent experiments on the action of lethal doses of hydrophobic solvents and oxygen/nitrogen atmosphere on mouse in vivo performed with a photon counting imaging technique [24] should bring further clarification of the necrotic PE's phenomenon.

There is an interesting question whether the enhanced PE is also expressed during apoptosis - a genetically coded, programmed cell death. Apoptosis can be induced by an oxidative stress. A strong disproportion between oxidant production and antioxidative defence is observed: mitochondria of apoptotic cells produce more reactive oxygen species (ROS) and the glutathione metabolism is downregulated [25]. However, signals of PE from apoptotic cells may be much weaker than those observed in the above cited works [1-24] because of a low reaction rate w according to the basic equation:

$$I = \Phi w \tag{7}$$

where Φ is the total quantum yield of PE. Highly sensitive imaging and spatiotemporal analysis of ultraweak PE may be very useful in studies on the apoptosis.

If necrotic PE appears to be a universal phenomenon strictly related with the irreversibly perturbed homeostasis, some kind of biophysical "singularity" announcing the end of biological life, then its measurements could be a new criterion of biological death. Perhaps the flat response of an EEG, i.e. the lack of measurable electromagnetic activity, follows the necrotic peak in PE. In such a case one can envisage a very broad application of the phenomenon in medicine, toxicology, ecology etc. Recent studies on the correlation between PE and brain activity in intact and an oxidative stress-perturbed rats' brain [3,26,27] as well as correlation between action potential conductance and circulation of cytoplasm and "necrotic peak" of PE [20, 21-23] will shed light on this problem.

There is however, a more general philosophical aspect of the necrotic radiation. It regards the very fundamental question: is radiation merely a one bit message announcing the irreversible "phase transition" of a biological entity from life to a dead state, a kind of the last "SOS" or the Omega Point ? Or, does it contain a hidden subtle message about the dying entity coded in the parameters of the emitted electromagnetic field? Speculations on the possible information contents in the necrotic radiation and its association with evolution of the Universe (the Omega Point Theory) have been considered elsewhere [28] and attempts to answer the questions are vitally dependent on further progress in spatiotemporal analysis and extraction of maximum information from ultraweak light signals.

4. MODELS OF ULTRAWEAK PHOTON EMISSION

Two alternative models: biophysical and biochemical have been proposed to account for the source and mechanism of PE and related to the stress-induced perturbation of homeostasis resulting in changes in ultraweak PE. The biophysical model assumes the existence of a delocalized coherent electromagnetic field within the cell which plays a regulatory role as a triggering and information-carrying factor. The DNA double helix of nuclear chromatine is proposed as a biological resonator of a high "quality factor", i.e that is able to store DNA-exciplex energy (a photon storage hypothesis). This model is theoretically well elaborated and published elsewhere [7, 29, 30, 31, 32].

The biochemical model presented here attempts to unite thermodynamics of the perturbation of homeostasis with the bioenergetics of redox processes (metabolism) and the electromagnetic radiation- response from living systems. More correctly, it should be called the biophysicochemical model.

The physical and electrochemical properties and the ability of cooperative conformational phase transition of the lipid (RH) bilayer with embedded proteins are essential factors determining the sensitivity of the living cell to external influences. Moreover, free-radical reactions of lipid peroxidation produce enough energy ΔH, ΔG or $\Delta H^{\#}$, $\Delta G^{\#}$ to generate electronic excited states:

$$RH + O_2 \longrightarrow ROO\bullet + H\bullet$$

$$2\ ROO\bullet \longrightarrow \begin{cases} {}^3(>C=O)^* + ROH + O_2({}^3\Sigma^-_g) & (8a) \\ {}^1(>C=O) + ROH + O_2^*({}^1\Delta_g, {}^3\Sigma^-_g) & (8b) \end{cases}$$

Thus, the intensity of the PE is proportional to the square of [ROO•]:

$$<dn/dt> = I = \Phi\ k_6\ [ROO\bullet]^2 \qquad (9)$$

where k_6 is the rate constant of the dismutation reaction (8a and 8b). Under homeostatic conditions the flux J_R of peroxyradicals $ROO\bullet$ is minimalized due to the ROO• -scavenging reaction:

$$ROO\bullet + InH \xrightarrow{k_7} ROOH + In\bullet \qquad (10)$$

where InH is an inhibitor of the radical chain reaction and it represents a variety of antioxidant defense systems. Thus, the oxidative homeostasis in a cell is maintained at the proper level. The stationarity condition $dS/dt = const \longrightarrow$ minimum requires that the fluxes J_R, J_{InH} and J_{RH} are counterbalanced and the reaction rate:

$$d[ROO\bullet]/dt = d[InH]/dt = d[RH]/dt \longrightarrow minimum \qquad (11)$$

The above part of a biophysicochemical model was analyzed in terms of the phenomenological equations of Prigogine - Nicolis - Onsager' thermodynamics of the open system and published elsewhere [4,5,17]. Due to the quadratic term(s) $[ROO\bullet]^2$ in these equations, the solution gives a parabola or paraboloid as a geometrical representation of the solution (Fig 4 A and B). The plane section of the paraboloid in the 3-D coordinates σ_I, x_I,

x_2 gives a 2-D parabola σ_2, x_1, x_2, i.e. Fig. 4A. The most stable system is represented as being at the bottom O of a paraboloid valley. Small perturbations caused by weak stimuli (stresses) cannot rise or lower it much. On the side of a steep hill, a small stress can initiate a large perturbation (Fig. 4A). When a bioregulatory system is stressed, it is pushed up the hillside, and will give larger responses to stimulus (stress). This situation corresponds to the PE flux I_P and I_{PAF} or I_{PAF}' in Fig. 1.

In the stationary state the condition:

$$_i\Sigma \, \delta J_i \, \delta A_i \geq 0 \qquad (12)$$

is fulfilled, where δ is the fluctuation of J, A and reactant, intermediate and product concentration. Far from thermal equilibrium δ and an extra entropy production, i. e. an energy dissipation σ increase:

$$\delta \sigma = 1/T \, _i\Sigma \, \delta J_i \, \delta A_i \qquad (13)$$

When values of δ exceed the threshold values and the evolution of a reaction system O-N approaches the critical value of a generalized reaction coordinate R_{cr}, an instability arises (Fig. 4C). The reaction system jumps (N → P) into a new stationary regime PR.

The above model is based on the heterogenous metabolic chemiluminescence associated with compartmentalized peroxidative reactions and cooperative processes in biomembranes. It predicts an oscillatory behavior of a perturbed biosystem (jumps N → P and P → N depending on the delivery of metabolic energy). It seems to be in agreement with recent experimental data [6,12, 21, 23, 24].

The model may be better fitted to a real feature of PE by analyzing physical aspects of transformation of the metabolic energy ΔG into electronic excitation and consecutive radiative emission. For this purpose we consider the system consisting of three parts:

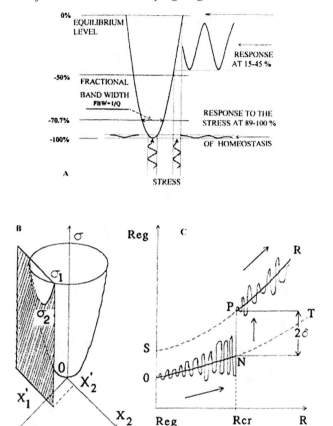

Figure 4: A two (A) and three (B) -dimensional representation of homeostasis in a self-regulating system and a nonlinear nonequilibrium dissipative system σ (C). Explanations in the text.

- coherently oscillating units of macromolecules and their complexes or supramolecular ensembles, e. g. sections of the membrane of cells or parts of H-bond helix of nucleic acids;

- the remainder of the biological system which formed the heat bath, and

- an external energy (E) and information (Inf) source which couples to the oscillating units (Fig.5). The long-range Coulomb interaction between the oscillating units will produce a narrow band of energy levels with eigenfrequency ω_i ($\omega_0 < \omega_i < \omega_{max}$) which corresponds to the normal modes of electromagnetic oscillations. When two or more units are built into a larger system, a new oscillation is created with a degenerated frequency $\omega_{dg} \neq \omega_i$, i.e. a degenerate state or an oscillation made of two or more oscillations ω_i which have come together and oscillate in harmony at common ω_{dg}. The greater the degeneracy coefficient, the higher the internal order (coherence) of the system, and the greater is its control over subsystems (units).

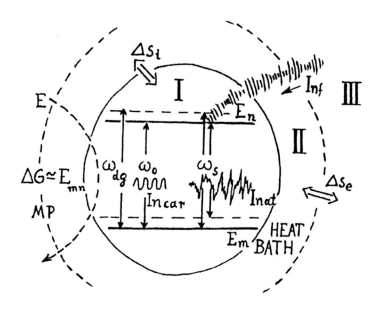

Figure 5: Schematic representation of a functional model of a living - system an attempt to unify thermodynamic and electronic features of metabolism and photon emission. Explanations in the text.

To be an electromagnetic (EM) oscillating circuit (unit), an inductance L must be: connected to the capacitance C:

$$\lambda \quad 2\pi c / \omega_0 \;=\; 2\pi c \sqrt{LC} \qquad (14)$$

λ is the length of the EM wave in a vacuum corresponding to a circular frequency ω_0, $c = 1/\sqrt{\varepsilon_0 \mu_0}$ is the light velocity in a vacuum.

Every oscillating unit - an ensemble of subcellular structures contains 3 elements:

1) The membrane of the live biological cell is analogous to a platelike condenser with the capacitance:
2)
$$C = \chi \varepsilon \varepsilon_0 \pi \rho^2 / d \qquad (15)$$

where χ is the portion of the platelike surface carrying electric charges, d is the distance between surfaces and ρ is platelet radius. The free energy G^0 is related to the vector of the electric field $\boldsymbol{E}$:

$$\Delta G^0 = \varepsilon \varepsilon_0 E^2 / 2 \qquad (16)$$

2) The helical coils of proteins (P) and helices of nucleic acids (NA) with inductance

$$L = \mu \mu_0 d / 8\pi \qquad (17)$$

3) A neuronal network with highly nonlinear neuron interconnections which contributes to the C and resistance R values of the oscillating unit:

$$Z = \sqrt{R^2 + (\omega L - 1/\omega C)^2} \qquad (18)$$

The impedance Z is the smallest when the circuit is in resonance, i.e. $\omega L = 1 / \omega C$, and, under a constant electric energy supply from an external source (~, MP) stationary oscillations with ω_0 and corresponding λ_0 (nonmodulated carrying oscillations) are generated. An ensemble of oscillating units forms a higher order integrated synergetic system with the eigenfrequency

$$\omega, \neq {}_n\Sigma \, \omega_0 \qquad (19)$$

Due to interconnections and biofeedbacks operating in biosystems this ensemble may be considered as a holistic sustainable "representative oscillator" (Fig. 6).

Changes in energy levels $E_{n,m}$ ($n \leftrightarrow m$) of the oscillating unit (the RLC circuit) coupled to the energy flow from the source III through a metabolic pump MP are characterized by the oscillator strength

$$f_{n\,m.} = 2\omega_{n\,m} \, m_e / h \, |r_{n\,m}|^2 \qquad (20)$$

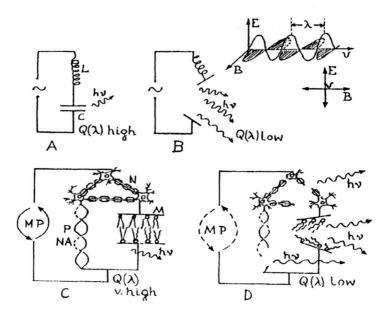

Figure 6: A simplified outline showing how the energy-information storage may be related to the functional and structural intactness of technical and biological oscillatory systems.

$|r_{nm}|^2$ is the matrix element of the probability of the $n \leftrightarrow m$ quantum transition associated with the change of the dipole moment $M. = {_i}\Sigma\, e\, l_i$ in the n and m states and is related to molecular (quantum) inductance, capacitance and resistance in the following way:

$$L = -1^2/e^2 \; m_e/f_{nm} \quad C_{nm} = 1/\omega^2_{nm} L_{nm.} \quad R_{nm} = \alpha_{nm} L\, n\, m \qquad (21)$$

where l is the dipole separation, m_e the electron mass, e is the elementary charge and α_{nm} the attenuation coefficient related to Einstein's spontaneous radiation probability:

$$A_{nm} = (n \longrightarrow m) \qquad (22)$$

All the above expressions may be related to energy transforming structures such as mitochondria, microtubules, microsomes, peroxisomes, melanosomes, chloroplast etc, in which a resonance tuning of the metabolic energy ΔG leads to "structural energy focusing" [33,34] $E = \phi\, \Delta G$ (ϕ - focusing coefficient dependent on geometry).

Taking values of ε, μ and χ typical for biological materials and $2\rho = 0.2 - 2$ μm, one can calculate the frequency of the resonance oscillations. It appears to cover the range from 0.3 to 3 µm or 10^{14} - 10^{15} Hz that is relevant to such EM phenomena as microwave, infrared, visible and UV light oscillations [32,35,36, 37]. A functional analog of a living system may be a closed circuit of a „representative EM oscillator" producing stationary carrying oscillations ω_s that correspond to a homeostasis. The information (In) content of this system is related to its geometric (l, d, ρ) and EM properties (ε,μ) and reflects only parameters of the carrying EM field (Fig. 5).

Different modes of information(Inf) coding (AM, FM, PM and others) imposed on the carrying field and realized e.g. by the Central Nervous System-processed information from the external world (Inf), may correspond to information accumulation, attaining the contents of Inf. A nonlinear, synergetic superposition of those „carrying" and „attained" fields, possessing an immense internal dynamism and self-integration, i. e. creating one integral entity, might be a model of an individual living being. A serious experimental support of this concept is provided by measurements of the electric potential between polarizing organs or structures in the human body. Metabolic activities in nearby tissues or adjacent organs create differences of potential, giving rise to a driving electromotive force through a biologically closed electric circuit [38].

The above proposed model accounts at least partially for the experimentally determined properties of a stationary and necrotic photon emission from intact and irreversibly perturbed (dying) organism, respectively. This agreement results from the analysis of the figure of merit or a quality factor Q characterizing the ability of a system to store energy E:

$$Q = 2\pi \; E \text{ stored in the oscillator} / E \text{ lost per period } T_0 \quad (23)$$

It describes the relative independence of an E-storing system from other systems or the quality of memory ability to the constant eigenfrequency ω_0:

$$Q = 2\pi / 1 - exp \, (-T_0/\tau) \; \omega_0 \, \tau \quad (24)$$

It is also related to the structural amplification (focusing) coefficient ϕ and to the halfwidth of the spectral distribution $I = f(\lambda)$ or $I = f(\nu)$:

$$\phi = W_{reson}/W_{max} = 1/RC\omega \quad \text{and} \quad Q = \omega_0/\Delta\omega \quad (25)$$

The closed circuit can dynamically store E and Inf (when the resonant oscillations are modulated), since its Q-factor has a high value (Fig. 6). Such a system loses (radiates out) only a small amount of its E. This feature agrees with a low intensity PE from intact non-perturbed organisms. A partial or total disintegration of the circuit diminishes its Q-value. The open disturbed system cannot store E and Inf in the form of cycling (alternating) E and M fields or virtual photons, i.e. associated with the electric and magnetic fields. It emits the E and Inf like a radio or TV antenna. Because of the malfunction of the E-supply systems, oscillations of the EM field are damped. Such a circuit functionally corresponds to the dying organism with enhanced necrotic PE.

5. PHOTOBIOLOGY AND BIOPHOTONICS

Photobiology and bioluminescence (also low-level biological luminescence) may be considered as the manifestation of the opposite or complementary aspects of the whole photons' involvement in life. This is evident from the equations describing basic photochemical and luminescent processes:

$$A + h\nu \begin{Bmatrix} \xrightarrow{k_a} A^* \xrightarrow{k_l} \\ \xleftarrow{k_f} A^* \xleftarrow{k_{ex}} \end{Bmatrix} B + C \qquad (26)$$

where $h\nu$ is either the "substrate" of a photoreaction (photobiology) or the „product" of a chemiluminescent reaction, and k_i are the rate constants of the corresponding absorption, photoreaction, chemiexcitation and emission processes, respectively (Fig. 7). The intensity of a luminescent process is related to these magnitudes in the following way:

$$I \sim dn/dt = n^\bullet = k_f \, (k_f + \Sigma \, k_q \, [Q] \,)^{-1} \, k_{ex} \, [B] \, [C] \qquad (27)$$

Here $n^\bullet$ stands for the number of counts per the sampling time Δt and k_q for the rate constant of quenching processes. The equation (27) is the expression of the microreversibility principle according to which chemiluminescence may be considered as the reversion of a photochemical reaction.

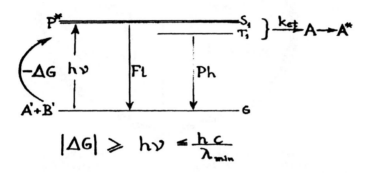

Figure 7: A scheme showing the relationship between photochemical and photon-generating processes. The production of an electronic excited state is a common element. Fl, Ph and k_{et} - fluorescence, phosphorescence and the rate constant of the excitation energy transfer to a acceptor molecule A (a secondary emitter), respectively.

The mechanisms and kinetics of "dark" chemical reactions and that of reversed photoreactions are the subject of different spectroscopic selection rules as they proceed *via* ground and excited states. This fact imposes an asymmetry on the photon circulation fluxes, i.e. it decreases the quantum efficiency Φ-value so that $n_{abs} > n_{emis}$. Teleological systems created by nature for the efficient utilization of the ΔG of photons for the energetic needs of the cell contribute predominantly to this asymmetry:

$$G \sim \mu (\lambda) \sim (hc/\lambda)(1 - T/\Theta) - k_B T \qquad (28)$$

Here μ is the photochemical potential, T and Θ are thermodynamical and nonequilibrium temperature, respectively. Electronic excited states are generated in certain biochemical (enzymatic) reactions for nonemissive important purposes of the cell's bioenergetics. This "dark photobiochemistry" [39] may significantly contribute to the asymmetry of photon circulation in living systems (Fig. 8). Photochemistry is a well developed scientific discipline focused mainly on the energetic aspects of radiation, but it neglects the theoretical possibility of regarding photons as conveyers or messengers of information. This aspect of biophotons: their possible informational/triggering role in the regulation in real time of the life-machine has to be exploited by biophotonics, as sketched in Fig. 8.

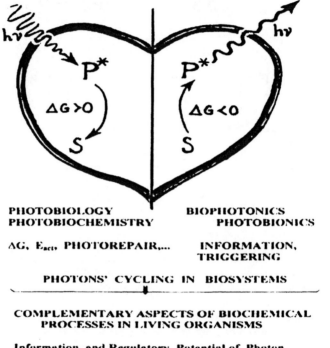

Figure 8: Photons hv as a driving force and "products" of metabolic processes and source of information in biosystems. Explanation in the text.

6. PHOTON CYCLING IN BIOSYSTEMS

Light has been always intrinsically related to the creative power that gives birth to life [40]. Indeed, the cell division, fertilization of an egg and the death of the cell are accompanied by PE [7,8, 9-11, 16, 21]. Light is considered as the unitary purposive principle which engenders the Universe and that has the nature of first cause [41]. It is an ultimate source of energy on Earth and a driving force of photosynthesis and other photobiological processes. This well established truism, accepted by the science community accounts barely for the energetic role of radiation. Recent findings suggest that biophotons carry hidden information coded in the spatiotemporal parameters of the electromagnetic field, such as a degree of coherence or possible squeezed states of the field. It cannot be excluded that the biophoton field conveying various kinds of inherent information is advantageous to take the squeezed states (photon number states) associated with antibunching and sub-Poissonian statistics. In such states the noise N can be "squeezed" to a separate phase of the entire cycle, improving significantly the S/N ratio $(S/N \longrightarrow 0)$ that enables extraction of the information from the UPE signals [42]. Therefore biophoton emission might deliver important information about the onset and the end of life processes, the influence of internal and external stress factors and the adaptation capability to adverse situations and environmental conditions.

Photons as the most perfect carrier of energy and information should be a precious "substantia prima". It has been experimentally well proven that a healthy intact organism emits fewer photons than the perturbed or injured one. It looks as if a photon field is protected or "arrested" within efficiently and coherently active biosystems with a high Q-value.

Despite fast progress in the development of advanced sensing photonics based on ultrasensitive single photoelectron counting (SPC), SPC-Imaging and optical heterodyne detection and photon counting methods [42], certain general aspects are still unrecognized and speculative. For example, the questions: is there a photon circulation in biosystems? Are photons involved in the global flow (circulation) of energy and matter as are all elements and their smallest entities, atoms? In the biosphere, atoms are immortal (excluding nuclear reactions) and Nature utilizes them to built up new life in a variety of cycles and processes. What about photons? As carriers of both energy and information, photons might be even more precious units of Nature than atoms. The next question: where do biophotons come from and where do photons of the necrotic radiation go to? (Fig. 9). Is there any Kingdom of Light, beyond the kingdoms of stars, planets, plant and animals? One can conclude by summarizing Lakhovsky's hypothesis in the form of a threefold principle [40]:

> Life is created by radiation;
> Life is maintained by radiation;
> Life is destroyed by oscillatory disequilibrium.

However, this thoughtful principle seems to be incomplete as it does not answer the aforementioned questions. These questions challenge us to enrich our understanding of biophotons and the mystery of life and death.

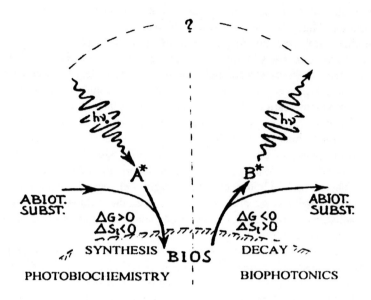

Figure 9: A simplified outline illustrating the framework of the photon cycling concept. For details see the text.

REFERENCES

1. Arnhold, J. (2001) Application of chemiluminescence methods in the investigation of redox regulation in cells. In *Chemiluminescence at the Turn of the Millenium,* Albrecht, S., Zimmermann,T. and Brandl, H. (eds.), Schweda-Werbedruck GmbH Dresden, pp. 85-94.

2. Devaraj, B., Usa, M. and Inaba, H. (1997) Biophotons: Ultraweak light emission from living systems. *Curr. Opin. Solid State Mat. Sci.*, **2**, 188-193.

3. Kobayashi, M. Takeda, M., Sato, T., Yamazaki, Y., Kaneko, K., Ito, K.-I., Kat, H. and Inaba, H. (1999) In vivo imaging of spontaneous ultraweak photon emission from a rats' brain correlated with cerebral energy metabolism and oxidative stress. *Neurosci. Res.*, **34**, 103-113.

4. Slawinski, J., Ezzahir, A., Godlewski, M., Kwiecinska, T., Rajfur, Z. Sitko, D. and Wierzuchowska, D. (1992) Stress-induced photon emission from perturbed organisms. *Experientia*, **48**, 1041-1058.

5. Slawinska, D., Polewski, K. and Slawinski, J. (1992) The stress-induced electromagnetic emission from biosystems: chemiluminescence response of plants to mechanical and chemical damage. *Bioelectrochem. Bioenerg*, **28**, 483-488.

6. Kochel,B. (1990, 1995) Perturbed living organisms: a cybernetic approach founded on photon emission stochastic processes. *Kybernetes* **19**, 16-25; Perturbed living organisms: a cybernetic approach to oscillatory luminescence. *Kybernetes*, **24**, 53-76.

7. Beloussov, L., Popp, F.A., Voeikov, V. and van Wijk, R. (eds) (2000) *Biophotonics and Coherent Systems*. Moscow University Press, Moscow.

8. Gurwitsch, A.A. and Livanova, T.T. (1980) Relationship between mitogenetic radiation and unbalanced molecular organization. *Bul. Exper. Biol. Med.* **89 (2)**, 179-180 (Russich).

9. Lepeschkin, W. (1934) Nekrobiotische Strahlung. *Protoplasma*, **21**, 594-597.

10. Perelygin, V.V. and Tarusov, B.N. (1966) Enhanced ultraweak radiation from injured tissues. *Biofizika*, **11**, 539-541.

11. Reiber, H.-O. (1989) Discrimination between different types of low-level luminescence in mammalian cells: the biophysical radiation. *J. Biolumin. Chemilumin.*, **4**, 245-248.

12. Ruth, B. (1978) Experimental investigation of low-level photon emissio., *Electromagnetic Bioinformation*, Popp, F.A., Becker, G., Konig, H.L. and Peschka, W. (eds.), Urban u. Schwarzenberg, Munchen, Baltimore, pp. 128-143.

13. Godlewski, M., Kwiecinska, T., Laszczka, A., Rajfur, Z., Slawinski, J., Szczesniak-Fabianczyk , B. and Wierzuchowska, D. (1997) Diagnostic value of ultraweak chemiluminescence of cells subjected to the oxidative stress. *Cur. Topics Biophys.*, **21(1)**, 96-101.

14. Rajfur, Z. (1994, 1993) Photon emission from chemically perturbed yeast cells. *J. Biolumin. Chemilumin.* **9**, 59-63; Photon emission from chemically perturbed normal and SOD-deficient yeast cells. *Bioluminescence and Chemiluminescence Status Report*, Szalay, A.A., Kricka, L.J., Stanley, P., John Wiley & Sons, Chichester, (eds.) New York pp.486-490.

15. Slawinski, J. (1988) Diagnosis of stress-induced perturbations of biohomeostasis evaluated by photon emission. *Zesz. Nauk. Akad. Roln. w Krakowie*, **233**,session 20, 53-66.

16. Slawinski, J. (1990) Necrotic photon emission in stress and lethal interactions. *Curr. Topics Biophys.*, **19**, 8-27.

17. Slawinski, J. (1991) Ultraweak luminescence and perturbations of biohomeostasis. Biological Luminescence, Jezowska-Trzebiatowska, B., Kochel, B., Slawinski, J. and Strek., W., *World Scientific*, Singapore 1990, pp.49-77; Stress-induced biological luminescence. Trends Photochem. Photobiol. 2, 289-308.

18. Godlewski, M., Rajfur, Z., Slawinski, J., Kobayashi, M. Usa, M. and Inaba, H. (1993) Spectra of the formaldehyde-induced ultraweak luminescence from yeast cells. *J. Photochem. Photobiol. B: Biol.*, **21**, 29-35.

19. Rajfur, Z., Kabyashi, M. and Slawinski, J. (submitted to Luminescence 2002) *Ultraweak luminescence from yeast-trichloracetic acid-interactions.*

20. Jaskowska, A., Milczarek, I., Borc, R. Zolnierczuk, R., Slawinski, J. and E. Spiewla (1996) Ascorbic acid and metabolic activity of plant cell. *Folia Histochem. Cytochem.*, **34** (S2), 12.

21. Jaskowska, A., Borc, R., Dudziak, A. and Spiewla, E. (1999) Oscillatory character of changes in ultraweak luminescence from Nitella cells. *Acta Soc. Bot. Polon.*, **68**, 281-285.

22. Jaskowska, A., Borc, R., Milczarek, I., Dudziak, A. and Spiewla, E. (2001) Kinetic studies of ultraweak luminescence induced by ascorbic acid in *Characeae* cells and their structures. *Luminescence*, **16** 51-56.

23. Jaskowska, A. and Gorski, Z. (2000) Electrical and luminescence responses of *Characeae* cells induced by local anaesthetics. *Molecular and Physiological Aspectsof Regulatory Processes of the Organism*, 9-th Int. Symp. Molec. Cell. Biol. UNESCO/PAS, Lach, H. (ed.) Wydawnictwo Nauk. AP, Krakow, pp.184-186.

24. Gorski, Z., Slawinski, J. and Lach, H. (submitted for publication 2002) Photon emmission imaging from mice subjected to stress and lethal factors. *Acta Bio. Cracov. Ser. Zool.*

25. Buttke, T.M. and Sandstrom, P.A. (1995) Redox regulation of programmed cell death in lymphocyten. *Free Radic. Res.*, **22**, 389-397.

26. Dirngal, U. Landauer, U. and Them, A. (1995) Global cerebral ischemia in the rat:online monitoring of oxygen free radical production using chemiluminescence in vivo. *J. Cerebral Blood Flow Metab.*, **15**, 929-940.

27. Isojima, Y., Isoshima, T., Nagai, K., Kikuchi, K. and Nakagawa, H. (1995) Ultraweak bioluminescence detected from rat hippocampal slices. *NeuroReport*, **6**, 658-660.

28. Slawinski, J. (1987, 1989, 2000) Electromagnetic radiation and the afterlife. *J. Near-Death Studies*, **6**, 69-136; *Energetic-informational aspects of necrotic aura: an electromagnetic model of the metaphysical transformation*. Proc. 2-nd Int. Conf. Paranormal Res., Colorado State University, June 1-4., pp.122-135, *Necrotic radiation - the Requiem for dying cells*. Int. Conf. Biophoton Emission & Coherence, Int. Inst. Biophysics, September 1-10, Neuss, Germany. Invited lecture.

29. Popp, F.A. (1989) *Coherent photon storage of biological systems, in: Electromagnetic Bio-Information*, Popp, F.A. et al.(eds.), Urban u. Schwarzenberg, II edit., Muniche, Wien, Baltimore, pp.144-167.

30. Popp, F.A., Li, K.H. and Gu, Q. (1992) *Recent Advances in Biophoton Research and its Applications*, World Scientific, Singapore, London.

31. Chang, J.J., Fisch, J. and Popp, F.A. (1998) *Biophotons*, Kluwer Academic Publishers, Dordrecht, Boston, London.

32. Chwirot, W.B., Dygdala, R.S. and Chwirot, S. (1985) Optical coherence of white-light-induced photon emission from microsporocytes of Larix europea Mill. *Cytobios*, **44**, 239-249.

33. Smith, C.W. and Best, S. (1989) Electromagnetic Man, Dent, J.M. and Sons (eds.) Ltd, London.

34. Smith, C.W. (1990) *Bioluminescence, Coherence and Biocommunication, in, Biological Luminescence*, Jezowska-Trzebiatowska, B. Kochel, B,, Slawinski, J. and Strek, W. (eds.), World Scientific, Singapore, pp. 3-18.

35. Frohlich, H. and Kramer, F. (1983) *Coherent excitation in biological systems.* Springer Verlag, Heidelberg.

36. Smith, C.W., Jafary-Asl, A.H., Choy, R.Y.S. and Monro, J.A. (1987) The emission of low intensity electromagnetic radiation from multiple allergy patients and other biological systems. *Photon Emission from Biological Systems*, Jezowska-Trzebiatowska, B. Kochel, B., Slawinski, J. and Strek, W. (eds.), World Scientific, Singapore, pp.110-126

37. Tilbury, R.N. (1992) The effect of stress factors on the spontaneous photon emission from microorganisms. *Experientia*, **48**, 1030-1041.

38. Nordenstroem, B.E. (1992) Biologically Closed Electric Circuits. Clinical, Experimental and Theoretical Evidence for an Additional Cirlulatory Systems. Nordic Medical Publishing, Stockholm, 1978; Link between external electromagnetic field and biological matter. *Int. J. Environmental Studies*, **41**, 233-250.

39. Cilento, G. (1988) Photobiochemstry without light. *Experientia*, **44**, 572-576.

40. Lakhovsky, G. (1939) *The Secret of Life*, trs. M. Clement, Heinemann, London.

41. Young, A.M. (1976) *The Reflexive Universe.* Delacorte Press/Seymour Lawrence, San Francisco.

42. Inaba, H. (1995) Ultraweak biophoton imaging and information characterization. *Ultrafast& ultraparallel Optoelectronics*, Sueta, T. and Okoshi, T., Ohmsha/John Wiley & Sons, Tokyo, pp.570-580.

Chapter 10

MITOGENETIC RADIATION, BIOPHOTONS AND NON-LINEAR OXIDATIVE PROCESSES IN AQUEOUS MEDIA

Vladimir L.Voeikov
Lomonosov Moscow State University,
Faculty of Biology,
Moscow, Russia

Abstract: Mitogenetic radiation (MGR), discovered by A. G. Gurwitsch, gave birth to the field of biophotonics that is now energetically developing. However, most facts of unique properties of MGR and of processes in which they originate, of the surprising discoveries made during several decades of intensive MGR research, are practically forgotten. Present-day biophotonics may gain a lot for its further development from discoveries and insights made at that time. In particular, it was discovered that usual enzymatic reactions are followed with MGR, that MGR emission from aqueous solutions of simple amino acids is correlated with spontaneous polypeptide synthesis, that substances possessing specific enzymatic activities may self-reproduce in such solutions. All these processes crucially depend on oxygen (and in some cases on illumination with visible light). An extremely sensitive analytical method – MGR spectral analysis – helped to show that branched chain reactions with the participation of reactive oxygen species and other free radicals serve as energy sources for the emergence of high energy mitogenetic photons. All these amazing phenomena are discussed, in particular here in relation to the growing understanding of the important role of reactive oxygen species and their reactions taking place in aqueous milieu for bioenergetics and bioinformatics.

Keywords: Mitogenetic radiation, reactive oxygen species, water, biophotons.

1. INTRODUCTION

Mitogenetic radiation (MGR), discovered by Alexander Gurwitsch, is met even now with scepticism by many biochemists and biophysicists. There are several reasons for such an attitude. First, classical bioenergetics deals with biochemical reactions with the energy output not exceeding 0,5 eV, equivalent to the middle infrared region of the spectrum of electromagnetic waves. Photon emission having the mitogenetic properties belongs to the UV-region of the spectrum and energy of mitogenetic photons reaches as high as >5 eV. Second, many properties of MGR seem very strange not only from the position of classical physical chemistry on which current biochemistry is based, but also from the position of classical quantum physics.

Beginning in the 1960s with the development of very sensitive single photon detectors it has been demonstrated that all living systems are indeed emitters of low level photon fluxes, though it turned out that its intensity in the visible and near IR-range of the spectrum is enormously much stronger than in the UV range. Still the majority of investigators dealing with low-level photon measurements acknowledge the priority of Gurwitsch as the discoverer of electromagnetic radiation of living objects. At the same time Gurwitsch's evidence that this emission plays any functional role is paradoxically rejected. Low-level photon emission phenomenon is generally looked upon in the frame of "Imperfection Theory". According to it photon emission originates from random metabolic aberrations due to uncontrolled free radical reactions [1, 2].

Nevertheless, there is another line of investigation of ultra-weak or low-level photon emission of living things, originated and persistently developed by F.-A. Popp, who bases his studies on the coherence theory. This theory allows explaining many peculiar features of photon emission by living things and predicts also that this emission may play a significant informational role in living matter [3]. Such photon emission got the name of "Biophotons", and Gurwitsch's MGR is considered to be a particular case of biophoton emission. However, biophoton research is based mainly on registration of photons using photomultipliers, while MGR was registered due to its specific ability to induce cell mitosis. Both methods have their advantages and drawbacks, as they reveal different "faces" of biophotons. Unfortunately, most of the information collected by Gurwitsch and his school is unknown to present-day scientists, and even to those who work in the field of biophotonics. But information obtained in these studies by no means is obsolete, on the contrary, it adds a new dimension to the fast developing field of biophotonics.

Here we'll present a digest of some forgotten discoveries in the field of MGR (biophotons) that were made as far back as 50-70 years ago, with our comments based on our own experience in biophoton research. We believe that taking into consideration old Gurwitsch's findings will provide a new impetus for the development of this frontier science.

2. PARADOXES OF MITOGENETIC RADIATION

2.1 Intensity and Spectral Properties of MGR.

A.G. Gurwitsch discovered MGR in 1923 in his famous and repeatedly described "onion experiments". Very soon studies performed in Gurwitsch's laboratory and multiple independent laboratories all over the world demonstrated that this radiation induces mitosis in all types of cells [4]. Major information of MGR properties was obtained in experiments with a more convenient and reliable biological detector – yeast culture grown on agar plates. A plate with a yeast layer is cut into two parts and one of them is "irradiated" with a supposed source of MGR, while another serves for a control. Some time after irradiation the number of cells entering mitosis are counted on both plates, and in the case of significant difference between experimental and control samples it is concluded that the tested object was indeed a source of MGR. Here we omit all other important experimental details and controls (including, e.g., double blind evaluation of results) carefully described in many of Gurwitsch's publications. Their neglect practically always resulted in failure to detect MGR followed with the statements that MGR does not exist at all. Gurwitsch and his collaborater repeatedly discussed this strange situation [5].

Though the notion of most properties of MGR was obtained based on its biological effects, in the1930s UV-radiation emitted by some biological objects and chemical reactions was successfully registered in several laboratories using physical detectors -- modified Geiger-Muller counters. Their photocathodes were made from materials having maximal light sensitivity in the range of 190-280 nm and practically insensitive to visible light – copper, magnesium, aluminium or their compounds. When such a material absorbs UV-photons it emits photoelectrons that trigger a gas discharge in a counter [6, 7, 8].

It was shown with such counters that intensity of UV-photon emission from developing frog eggs or a nerve-muscle preparation excited with an electrical current is very small -- 10-10000 photons ($10^{-10} - 10^{-8}$ erg) per 1 sec from 1 cm^2 of the emitting object. Figure 1 shows that counters detect UV-photons only in the moment of a nerve excitation and only if the nerve is alive. The same was demonstrated using biological detectors and nerve preparations as emitters of MGR.

The original conclusion that MGR is the flux of UV-photons was made based on experiments with glass and quartz filters: when a glass slide was put between an emitter and a detector, the mitogenetic effect disappeared, while a quartz plate did not eliminate it. More rigorous evidence in favour of this conclusion was obtained using spectral analysis of MGR with the help of precise spectrographs with quartz optics (Fig. 2). The emitting object was placed in front of the input slit and detectors – agar plates with yeast cultures – against the output slit.

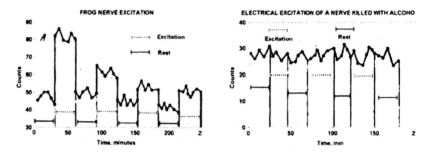

Figure 1: Registration of UV-radiation using modified Geiger-Muller counter with photocathode sensitive in the range of 215-240 nm from excited nerve of a frog leg. Left – alive nerve, right – nerve killed with alcohol.

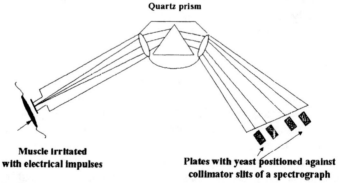

Figure 2: Outline of the MGR spectral analysis of a muscle preparation using a yeast culture as a detector.

Analysis of MGR spectra from different objects including such "simple ones" as enzymatic reactions performed *in vitro* (see below) revealed a lot of unusual facts. First, MGR spectra represented sets of spectral bands; some of them were as narrow as 1 nm wide. Second, sets of bands from different emitters differ very much in the number of lines, their width and position (Figure 3). Thus, MGR spectra are at variance from typical spectra of luminescence of aqueous solutions. In particular, spectra of fluorescence of aqueous solutions of luminofores at room temperature represent more or less wide "humps" with half-width usually exceeding 10 nm, which resolve into more narrow lines only at very low temperatures. Rather wide spectral bands (in visible range of the spectrum) are characteristic also for bioluminescent and chemiluminescent systems in water. But one needs to note that first, MGR spectra are obtained using a biological detector, while luminescence spectra are obtained using physical detectors, and second, MGR intensity is several orders of magnitude lower than that registered in most physical and chemical experiments. Taking into consideration such low intensity of MGR it is difficult to explain in terms of usual optic laws the very opportunity to register MGR spectra. If the total photon flux does not exceed several tens or thousands photons, while many spectra are represented by multiple bands, one needs to conclude that absorption of only a few photons is enough to induce multiple mitoses in a cell culture containing many tens of thousand cells. Thus there should exist a mechanism with a tremendous coefficient of amplification which is switched on by UV-photons.

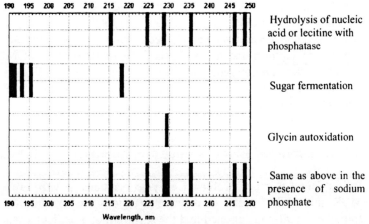

Figure 3: Mitogenetic spectra of different enzymatic reactions and of the reaction of glycin autoxidation induced with glycin solution irradiation with MGR (see below)

Mitogenetic effect may be induced also by physical sources of UV-light – arc discharges, hydrogen lamp strongly attenuated with grids or filters. But their efficiency in induction of mitoses was incomparably lower than that of "natural" sources of MGR. In other words, for effective commencing of mitogenetic effect emission should not only belong to a UV frequency range, but it should possess some other specific properties.

2.2 "Fractionation" of MGR

Efficiency of mitogenetic effect was measured based on duration of the irradiation period needed to trigger mitogenetic effect. Besides, too long irradiation from intense sources of MGR resulted in inhibition of mitotic activity. A biological detector may be exposed to an MGR source either continuously or intermittently. To obtain an intermittent mode of irradiation, a rotating disk with slits was installed between an emitter and a biological detector. It turned out that intermittence might significantly induce the efficiency of mitogenetic effect depending on width of slits, distance between them, and rate of disc rotation. For example, if two yeast cultures are positioned opposite to each other, they may induce mitogenetic effect in each other (mutual induction) only at a distance of less than 3 cm. Under these conditions threshold period at continuous irradiation was 6-8 min, while at the optimal parameters of intermittence it reduced 40-50-fold. At a larger distance without intermittence yeast cultures did not stimulate each other, but under the optimal parameters of intermittence they interacted even at a distance of 15 cm. This result seems extremely paradoxical: irradiation time might be reduced down to 1,5% of the total time of exposure and the effect was not lost but rather enhanced. Only due to intermittence it was possible to perform spectral analysis from many sources of MGR. Most important that intermittence increases efficiency of MGR only if it is performed in a rhythmic mode; at arrhythmical intermittence mitogenetic effects are usually lost.

2.3 Sensibilized Fluorescence and Selective MGR Scattering

It was found that practically all enzyme reactions *in vitro* are sources of MGR, and MGR spectra of different reactions are significantly different. Initially Gurwitsch suggested that the spectrum of a particular reaction, e.g., urea hydrolysis with urease is characteristic of this particular process, but later it turned out that addition of glucose to the urea-urease system resulted in appearance of the same bands as in glycolytic process, that addition of phosphate to a glycolytic system resulted in appearance of bands characteristic of a phosphatase reaction, etc. This indicated that MGR spectral bands result from fluorescence of simple compounds and even of their chemical residues excited by energy released from some highly exothermic process. This phenomenon was named "sensitized fluorescence" (SF). SF was characteristic to small molecules and their residues, such as a peptide bond, phosphoric acid residue of a nucleic acid or a phospholipid, amino group of an amino acid, and even to such ions as Na^+ and Cl^-. On the contrary such big molecules, as nucleic acids or proteins did not produce SF. Thus there are many differences between SF and usual fluorescence, but conditions for their origination and observation also differ. For usual fluorescent analysis of solutions of flourophores they are irradiated with rather intense light sources and its own intensity is rather high. SF is excited by energy coming from a reaction system itself – from reactions of free radicals (see below). Besides SF spectra were analysed using extremely sensitive biological test systems, and until now no physical methods allowing registering and analysing this effect have been devised.

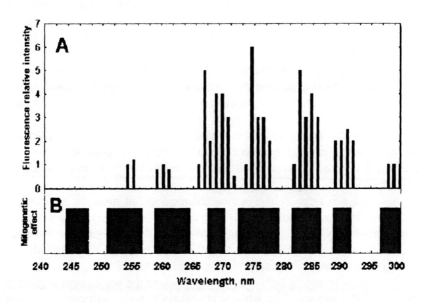

Figure 4: Comparison of fluorescence spectrum of benzol vapour (A) and MGR spectrum of glycin autoxidation reaction to which 1% of benzol water was added (see text)

Validity SF was directly demonstrated by comparison of benzol SF spectrum and usual fluorescence spectrum of benzol vapour. For this experiment a reaction system of glycin autoxidation in water (see below) having a very simple spectrum (only one narrow line) was taken. To it 1% of water, being in contact with benzene and in which only traces of benzol are present, was added. Fig. 4 compares MGR spectrum of this system with fluorescence spectrum of benzol vapour, which is resolved in many groups of narrow lines. It can be seen that all these groups are really present in MGR spectrum. This experiment shows an enormous sensitivity of SF that cannot be easily understood in the frame of our current knowledge. One needs to explain how energy from rare reactions of free energy recombination may be efficiently transferred to benzol present in a reaction system in an extremely low concentration and remitted by them. It should be realised that from the point of view of purely stochastic processes the observed phenomenon is improbable. Precise mechanisms of such a transfer is not known, but it is interesting to speculate that water due to its unique properties should play a key role in this process.

SF may arise in aqueous solutions which are not irradiated from outside only if energy is supplied by highly exothermic reactions running in these solutions. Quanta of energy providing excitation followed with fluorescence in UV-range of the spectrum might come only from reactions of very active free radicals recombination. For example, recombination of two bi-radicals: $\downarrow\downarrow C=O + O\uparrow\uparrow \rightarrow CO_2$ (arrows here symbolise unpaired electrons – see below) may supply a quantum of energy equivalent to a UV-photon with the wavelength of 170 nm sufficient to excite a certain fluorophore which remits an MGR photon with the shortest observed wavelength of 190 nm. Verification that free radicals with such high reactivity really arise in reaction systems is a very difficult task because of their extremely short lifetime. Nevertheless they were identified experimentally with the new method developed by Gurwitsch and collaborates – "selective scattering" (Figure 5).

A tested substance (either a dry chemical or an aqueous solution) is irradiated with a monochromatic line, cut out by a spectrograph from a spectrum of a hydrogen lamp. Radiation scattered by a sample is dispersed by a second spectrograph and its MGR spectrum is obtained. It was demonstrated that mitogenetic effect was observed only in case the wavelength of incoming light was exactly equal to the wavelength registered in the MGR spectrum. The latter was characteristic of particular chemical groups or free particles, like ions or electronically excited free radicals present in irradiated solutions.

This extremely sensitive and informative method allowed confirming quite independently that MGR bands registered in the course of different enzymatic and non-enzymatic reactions result from SF of the substances and their chemical radicals present in a solution. In particular such free radicals,

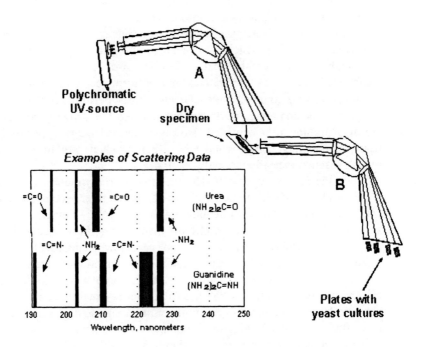

Figure 5: A set-up of an experiment for obtaining MGR spectra of a substance or a solution using a selective scattering phenomenon. (A) and (B) – two quartz monochromators.

as $\uparrow\uparrow C=O$, $\uparrow NH_2$, $\uparrow OH$, which emit photons with wavelengths of 202-204 nm, 253 nm, and 306-308 nm respectively could be identified. Selective scattering allowed also to detect particular fluorescent groups, such as R-OH, CH_2O, $R=C$—$N=R$, $R=C=N$—R and others. This precise method allowed understanding the intimate mechanisms of the processes accompanied with MGR.

2.4 Autocatalytic Formation of Enzyme-Like Substances and Multiplication of Low Molecular Weight Substances in Aqueous Solutions of Amino Acids.

One of the most "shocking" discoveries made using the methods of mitogenetic analysis is the discovery of spontaneous polycondensation of amino acids into high molecular weight proteinatious compounds in aqueous solutions. Polycondensation is formation of a polymer from monomers with release of water. This reaction provides for the synthesis of all biopolymers: proteins, nucleic acid, polysaccharides in a cell.

A high barrier of energy of activation forbids spontaneous condensation of two amino acids producing a di-peptide and releasing a water molecule. To synthesize a peptide chain chemists use specially activated and "defended" amino acid derivatives and perform reactions in non-aqueous solutions. A very complex molecular «machine» performs protein synthesis in living cells, and amino acids that are attached to a growing peptide chain are also preliminarily activated. However, initial energy expenditures are nearly completely compensated when final products are formed and the overall reaction is nearly thermally neutral.

Thus the discovery, that 10-15 min after a brief irradiation of a glycin (H_2N—CH_2—$COOH$) solution with MGR, the solution becomes a source of MGR, and compounds with peptide bonds (—$CONH$—) emerge in it was a surprise. The solution emits MGR for many hours, but concentration of a polypeptide stabilises at a certain and very low stationary level due to established equilibrium between its continuous formation and degradation.

The process could be initiated only with photons with $\lambda < 326$ nm. Its caloric energy equivalent is >87 kcal/mol, that is equal to energy of a bond between hydrogen and nitrogen atoms of an amino acid. It was also shown that after the reaction initiation ammonia was accumulating in a reaction system and that for its progress its illumination with visible or near IR-light ($\lambda < 1300$ nm) was needed. From this the supposed scheme of the reaction could be described as following. Energy of MGR photons is somehow used to cleave a hydrogen atom from an amino acid with the formation of an amino acid free radical:

$$H_2N\text{-}CHR\text{-}COOH + h\nu \rightarrow H\uparrow + (HN\text{-}CHR\text{-}COOH)\downarrow \qquad (1)$$

The radical condenses with an amino acid molecule, and a di-peptide plus a hydroxyl radical appear. This reaction needs energy of activation of 22 kcal/mol supplied by visible or near IR light:

$$H_2N\text{-}CHR\text{-}COOH + (HN\text{-}CHR\text{-}COOH)\downarrow + 22\ kcal/mol$$
$$\rightarrow H_2N\text{-}CHR\text{-}CO\text{-}HN\text{-}CHR\text{-}COOH + OH\downarrow \qquad (2)$$

Recombination of

$$H\uparrow + OH\downarrow \rightarrow H_2O \qquad (3)$$

gives 110 kcal/mol, and as 110 kcal/mol - 22 kcal/mol = 88 kcal/mol – enough energy to start a new cycle of reactions 1 → 3. But this simple chain reaction should rapidly decay and can not produce any emission. The emergence of MGR in the contact with air indicates that additional and very high portions of energy come from oxidative deamination of an amino acid, which may be catalyzed by the proteinatious polymer emerging at the first stage, e.g.:

$$2H_2N\text{-}CHR\text{-}COOH + O_2 \to 2O\text{=}CR\text{-}COOH + NH_3 + \Delta E \qquad (4)$$

Another peculiar property of this reaction is that it can be "multiplied". Transfer of an aliquot of the solution already emitting MGR to a fresh amino acid solution initiates the same reaction in the latter: it becomes an MGR source and a proteinatious polymer possessing oxidase-deaminase activity appears in it. The procedure could be repeated many times with the same effect. These amazing experiments did not attract attention of a scientific community and were completely forgotten. Only recently we partially reproduced them and confirmed that after irradiation of aqueous solutions of different amino acids with very low intensity sources of UV-light high molecular weight substances possessing deaminase arise there [9].

In glycin aqueous solutions there may develop even more startling processes – "self-reproduction" of enzymes, or to be more precise, of compounds possessing specific enzymatic activities – "fermentoids" according to Gurwitsch's terminology. If one makes a very dilute solution of, e.g. urease in glycin solution and incubate it in visible light with air access activity of urease measured by its ability to cleave urea increase after a lag period of 10-20 min. Same results were obtained with other tested enzymes, e.g., phosphatase and arginase. Enzyme activity increased not due to some form of added enzyme activation, but due to self-reproduction of its "active principle". Gurwitsch never claimed that the process was a genuine protein self-reproduction, because specific activity of "fermentoids" was much lower than that of template enzymes. It looked like it was reproduction of its rudiment with the properties of an active site. However, the same specific inhibitors that suppressed activity of the respective natural enzymes suppressed activity of fermentoids, and if specific metals (e.g., manganese) were needed for a template enzyme, no fermentoid activity was observed without this metal. Though activity of fermentoids was very low, it could be registered by conventional biochemical methods [10]. Multiple consecutive transfers of portions of active solutions into large volumes of fresh glycin solutions resulted in reinitiating of the same process there Besides performing their specific activities fermentoids catalyzed oxidative deamination of glycin, providing energy for their own reproduction and MGR emission from the solutions.

For the performance of specific catalytic functions fermentoids should contain different amino acids in addition to glycin, which could form an active center. Thus substances that are more complex than glycin should somehow arise in reaction systems. In fact, free radicals $=CO$, $—CH_2—$, $—NH_2$ u OH were detected there. These radicals arising from fragmentation of glycin molecules could serve as building blocks for substances having specific catalytic activities. Later Anna Gurwitsch, the daughter of A.G. Gurwitsch, demonstrated that complex specific substances might really quasi-spontaneously form from glycin [11]. After addition of tyrosine in a very small concentration (e.g., $5,5 \times 10^{-12}$ M) to glycin solution with diffuse visible light and being in contact with air tyrosine concentration increased 6 orders of magnitude. Selective scattering revealed the presence in this solution of the same free radicals as in solutions of fermentoids. Similar "self-reproduction" was shown under the same conditions for other natural aromatic and heterocyclic compounds (adenine, pirrol, indole). These revolutionary discoveries, demonstrating the new principle of matrix synthesis of complex compounds from the simplest one passed practically unnoticed though it was never disproved.

Nevertheless, recently there appeared some circumstantial evidence of unusual transformations of biomolecules in aqueous solutions without enzyme participation. For example, irradiation of an aqueous solution of phenylalanine and H_2O_2 with low intensity monochromatic UV-light ($\lambda=253,7$ nm) results in appearance in it of only four new compounds. All of them are amino acids: alanine, aspartate, lysine, and serine [12]. This set contains major functional amino acids to build different proteins. Notably that H_2O_2 is not consumed in this process, rather it behaves as a catalyst.

Besides, more and more attention is now attracted to the so-called Maillard or amino-carbonyl reaction. This reaction starts from condensation of simple sugars (ribose, glucose) or other carbonyl compounds and simple amino acids such as glycin. It proceeds in mild conditions (room temperature, slightly basic or even neutral pH), and in the course of it much more complex compounds rapidly arise. They may become precursors of nucleic bases, other cyclic and heterocyclic molecules, complex polymers. It is notable that in the course of all these reactions continuous generation and consumption of reactive oxygen species occur, and that the reactions are followed with very weak photon emission. Recently we have shown that such photon emission may proceed as a highly organized oscillatory process, indicating that self-organization is an essential feature of such reactions. Processes of this type may serve a contemporary adequate model of reactions discovered by Gurwitsch.

2.5 Role of Visible Light and Oxygen in the Development of Gurwitsch Reactions.

As mentioned, MGR accompanies many enzymatic hydrolytic reactions, for example hydrolysis of urea by urease. According to thermodynamic calculations, such reactions are practically thermally neutral: energy needed to break down the substrates is nearly completely compensated by quantity of energy released when the final products are formed. Thus the major reaction process can not provide energy for emergence of such high-energy quanta as UV-photons. From where can the additional energy come?

It was stressed above, that contact of a solution with air was a necessary condition for the emergence of MGR and in many cases illumination with weak visible light was needed. Gurwitsch suggested that oxygen and water actively participate in processes. They should disintegrate together with substrate molecules into atoms and simple free radicals such as $=CO$, $—CH_2—$, $—NH_2$, $=O$, $—OH$. Then additional energy may be supplied by reactions of radical recombination in which electron excited products arise. For example for urease reaction MGR spectral analysis and selective scattering spectra confirmed presence of the predicted radicals and products of their recombination in the reaction system. From calculation of energy balance it followed also that for dissociation of O_2 into atoms extra 118 kcal/mol were needed. Gurwitsch suggested that energy of two photons of visible light were added constructively to reach this energy quantum. Then the reaction system should emit MGR only if it was irradiated with photons of $\lambda<473$ nm (equivalent to energy of 60 kcal/mol x 2=120 kcal/mol). In fact when it was irradiated with monochromatic light with $\lambda>500$ nm no MGR from the reaction system could be registered, and it appeared only under illumination of light with $\lambda<(=)470$ nm. Similar predictions made based on the calculation of energy balance for other enzymatic reactions also were completely confirmed experimentally.

Thus, two crucial facts were found out. First, oxygen participates in all chemical reactions followed with MGR emission though it does not follow from chemical equations of these reactions derived from their analysis by less sensitive methods. Second, constructive adding up of two photons was discovered. This phenomenon seemed unbelievable until recently when it was demonstrated, that constructive two- and even three photons adding up may be observed in physical experiments. In particular, irradiation of a solution of a tyrosine derivative with a picosecond Ti-saphire laser ($\lambda=780$ and 855 nm) may excite it and provide its fluorescence in UV spectral range [13]. However, even now it is not easy to realize that similar processes may occur in systems where the processes accompanied with MGR emission proceed under illumination with non-coherent light sources.

The role of visible light in MGR emergence was impressively demonstrated in another type of experiment. A 10 cm long tube was filled with a solution of a highly diluted extract of cancer tissue or liver which contains a low molecular weight substance. When the tube was irradiated from one side with monochromatic light (λ=<480 nm) MGR photons of λ=218-220 nm were emitted from another side. No effect was observed if a tube was illuminated with light λ>500 nm. In this case energy of anMGR photon exceeds even the double energy of photons of visible light by 0,5 eV, indicating that some energy donating processes takes place in this solution.

3. BRANCHING CHAIN REACTIONS AND PUMPING OF ENERGY INTO NON-EQUILIBRIUM AQUEOUS SYSTEMS.

3.1 Basics of Branching Chain Reaction Chemistry

Gurwitsch noted that many features of reactions accompanied with MGR are characteristic for branching chain reactions (BCR), which were discovered at the end of the 1920s by Nikolai Semyonov [14] and Cyrill Hinshelwood [15]. BCR may be initiated by an introduction of only a few active particles (free radicals, atoms) into a reaction mixture or emergence of such species in it, for example, due to photodissociation. Under certain conditions each active centre may give birth to more than a single new one and the reaction rate accelerates in these processes exponentially. Accordingly free energy of a reaction system spontaneously increases. However, the active centres being very short-lived, intermediate products completely disappear from the reaction mixture when the process comes to its end. The final result of BCR is the same as for any other chemical and physical process proceeding in closed systems: absolute reduction of free energy and entropy growth. However, the essence of BCR is manifested in its dynamics rather than in its overall balance.

Chemical BCR in gaseous and organic liquid phase, physical BCR taking place in an atomic bomb or a nuclear reactor (known also as "run-away reactions") are extensively studied. However, Gurwitsch's claim that reactions of such type may proceed in aqueous systems under mild physiological conditions was neglected: it contradicted traditional views on the mechanisms of chemical processes running in water. Nevertheless, contemporary studies indicate that BCR may occur in aqueous systems though in a very peculiar form. Before we pass to discussing special properties of BCR in aqueous systems, general features of such reactions should be considered by the classical example: idealized BCR of oxidation of hydrogen with oxygen. We'll discuss it based on a "polycyclic" model for the first time suggested by us (Figure 6).

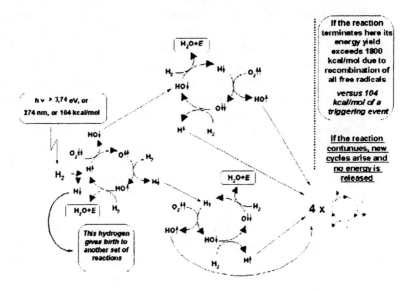

Figure 6: «Polycyclic» scheme of BCR of detonating gas ($H_2 + O_2 \rightarrow H_2O$). Arrows at atoms and molecules symbolize unpaired electrons.

A mixture of oxygen and hydrogen may stay for a long time without any visible changes until an active particle such as a hydrogen atom emerges in it. It may appear as a result of breaking down of H_2 by a UV-photon of the proper energy. Under the appropriate conditions a hydrogen atom may react with an oxygen molecule producing a hydroxyl radical and an oxygen atom. The former reacts with a hydrogen molecule producing a new hydroxyl radical, and the latter oxidizes a hydrogen molecule to water and regenerates a hydrogen atom, which starts a new turn of the cycle.

Thus two active particles – a hydroxyl radical and a hydrogen atom – are released when the first cycle makes one turn, and they may give birth to new cycles. Note that when all radicals either "rotate" in cycles or generate new ones practically no energy is released from the system. But it is well known that the reaction of hydrogen oxidation with oxygen is the explosive "detonating gas" reaction, releasing an enormous quantity of energy. This happens when chains break off due to radicals' recombination, which occur near the boundaries of the vessel or on other defects of the system. On the other hand, the system may hold a substantial quantity of free energy which many times exceeds the quantity of energy needed to initiate and to support the process after only three cycles arise.

What is the source of this energy? It was initially "locked" in oxygen molecule (O_2) which is unique among other molecules in the environment. O_2 has two electrons with parallel spins (unpaired electrons) on its valence molecular orbitals ($M\uparrow\uparrow$, where $\uparrow$ represents an electron with a certain spin). Such constitution of an outer electron shell is termed a triplet state. Particles in a triplet state are paramagnetic (attracted into the magnetic field) and possess an excess of energy over their singlet state [$M\uparrow\downarrow$], in which all their electrons are paired. Triplet O_2 is a vast energy store, able to release more than 180 kcal/mole upon its reduction to two water molecules after gaining 4 electrons (together with their carriers – protons). But it is rather inert by itself, because according to Wigner spin conservation rules it cannot directly interact with singlet state molecules. On the other hand, a free radical – a particle with an unpaired electron (in our case it is a hydrogen atom) rather easily reacts with oxygen. This reaction represents a chain branching because it produces two free radicals, which can easily react with other molecules generating two new chains.

Polycyclic scheme of BCR (Fig. 6) gives an idea why energy may accumulate in the reaction system before its dissipation. It can be also seen that energy from reaction systems where BCR runs may be released in the form of light, rather than heat (the so called "cold flame"), and often strong luminescence may be observed while the temperature of the reaction system is relatively low. Under certain conditions regular flashes (oscillations) of luminescence appear. From the presented sketch one can presume that the conditions for oscillatory release of photonic energy may be in principle provided by electron currents in closed circles of free radical transformations. It can not be excluded that these currents may interact with each other providing stability of the process as a whole, and beget oscillations with longer and longer relaxation times.

BCR similar to the reaction of detonating gas are termed "throughout branched chain reaction" and are schematically presented like this:

Here the arrows indicate reproduction of active centers (free radicals). But from Fig. 6 it can be seen, that the cycles generating free radicals themselves consist of free radicals ($O\uparrow\uparrow$, $H\uparrow$, and $OH\uparrow$). The cycles are sustained due to continuous consumption of hydrogen and oxygen molecules. It is easy to conceive that any perturbations in the reaction system hindering supply of "fuel" impede cycles "rotation", and in this case radicals creating the cycle may recombine. If triple recombination of radicals occurs, an H_2O_2 molecule is produced with the release of an energy quantum equivalent to a photon ~ 200 nm; if only oxygen and hydrogen atoms recombine, an excited hydroxyl radical together with another one can initiate new chains.

In other words, in a system where BCR flows, metastable objects may exist which retain a lot of labile high density energy. This is a most general property of BCR and when it is pronounced the reactions are termed "degenerate BCR", or more precisely – "chain reactions with delayed branching" (CRDB). Delayed branching is typical even of nuclear run-away reactions. After a nucleus of a splitting atom absorbs a neutron and breaks down releasing more than 1 new neutron, metastable fragments of the nucleus also appear. They decay later, when the major chain has already run away or has been terminated. But when these fragments desintegrate they release neutrons that initiate new chains. The property of delayed branching provides for a long life of the process as a whole, while its participants (active sites and individual chains) may be very short-lived.

BCR running in a liquid phase are practically always CRDB. They develop much more slowly than reactions in gaseous phase and do not look like explosions though kinetically they have all the features of explosions. CRDB are always oxygen-dependent oxidative reactions, and branching in them is provided basically due to an un-pairing of electrons of triplet molecular oxygen. Branching generally occurs due to the appearance of metastable by-products in the course of the development of major chains, e.g., peroxides.

For a delayed branching to come about, a peroxide should decompose in to free radicals that launch two new chains (Figure 7). Oxygen present in the system provides for production of an excess of peroxide free radicals. When they recombine, extremely unstable and energy rich compounds –tetroxides– appear. Tetroxides decompose generating electronically excited products which either relax, or their energy may be non-radiatively transferred to other substances present in the system. Such energy quanta either perform photochemical work or are remitted with the wavelength typical to a fluorophore.

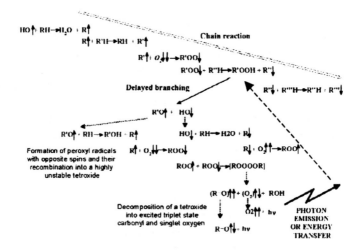

Figure 7: An outline of a chain reaction with delayed branching. A dashed arrow indicates that energy of electron excitation may be used for the decomposition of a metastable intermediate product into two radicals thus providing a chain branching.

This reasoning is true for the description of mechanisms of CRDB running in liquid organic phase [16]. But from the first sight it cannot be applied to the processes taking place in aqueous solutions at room temperature because energy of activation needed for peroxides to decompose into radicals reaches several tens of kilocalories per mole. If so, the probability of such reactions in water at usual temperatures seems to be very small. However, the results of many experiments performed by Gurwitsch and his colleagues (some of them have been described above) as well as the results of our studies of oxygen-dependent oxidative reactions in aqueous systems accompanied with photon emission [17, 18, 19, 20], as well as many properties of oxidative reactions occurring in biological systems, such as blood, need to admit that BCR may proceed in aqueous milieu [21].

The possibility of disintegration of peroxide and other metastable compounds in the course of the reactions under consideration may be provided by high non-equilibrium state of reaction systems which may be looked upon as the active medium. Its stable non-equilibrium state is sustained by the continuous inflow of energy due to oxidative processes (and in some cases – due to illumination with visible light) and by the special properties of the aqueous milieu, that provide energy circulation over the common electron levels of system components before its dissipation. The latter statement is based on the phenomenon of "degradative radiation" discovered by Gurwitsch and repeatedly observed by other authors.

3.2 Degradative Radiation and Non-Equilibrium Molecular Constellations

Properties of MGR discussed above had been established mainly in the studies of enzyme and other reactions going by in homogenous systems. Yeast and bacterial cultures, eggs at early stages of their development and other biological systems where intensive cell division takes place, blood, nervous and contracting muscle cells also spontaneously emitted MGR. It was a surprise that liver, kidney and most other tissues of adult animals do not emit MGR in a resting state, though intensity of oxidative metabolic processes in them is undoubtedly high. However, it turned out that all biological systems regardless of their capability to emit MGR spontaneously responded to different injuries or physiological irritants – rapid cooling, mechanical compression, passage of an electrical current even of very low power, narcotics, and even glucose with an intensive flash of MGR. The flash lasts for many minutes and until it completely relaxes irritants cannot induce a new one. Spectra of degradative MGR are very different from that of homogenous systems. For example, bands characteristic of specific fluorophores may be identified in spectra of spontaneous MGR from yeast cultures and they are relatively constant. Spectra of degradative MGR of yeast differ not only from that obtained in a resting state, but they differ significantly for different strains of yeast; they change depending on the nature of a factor inducing degradative MGR, on the physiological state of the living system, on the state of its development.

The latter is well illustrated by A.A. Gurwitsh [22]. She studied the properties of MGR of the naked baby rabbit muscle *in vivo* in different times of postnatal development. Spontaneous MGR was measured at a body temperature and degradative after the muscle was doused with cold physiological solution. Evolution of intensity of both kinds of MGR and of the spectrum of MGR are presented in Figure 8.

It can be seen that in the course of postnatal development intensity of spontaneous MGR decreases, while that of degradative MGR increases. During this period the processes related to muscle tissue functioning become progressively consolidated so that by the 15-th day intensity of degradative MGR reaches its maximum. Exactly by this time the posture and coordination of movements of an animal stabilizes indicating that the processes related to muscle tissue functioning reached maximal interrelation. Spectrum of MGR also changes in a specific way. At the early stage of development it looks like spectra of spontaneous MGR from systems containing many different fluorophores. Later the number of spectral lines decreases, they widen, until only one wide line is left which has no correlation with fluorophores known from selective scattering data.

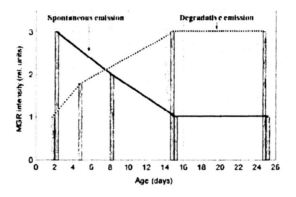

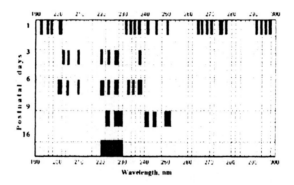

Figure 8: Upper part: evolution of intensity of spontaneous and degradative MGR of a baby rabbit muscle in vivo in different terms after birth (intensity was evaluated as a reverse of threshold period of irradiation by a muscle of yeast culture for obtaining a mitogenetic effect). Lower part: MGR spectra of a muscle (recorded at normal temperature).

All the properties of degradative MGR indicate that it differs in its origin from that of a spontaneous one from homogenous systems. As already mentioned the latter most probably originates due to remission of energy originating in reactions of radical recombination by fluorophores present in reaction systems. The properties of degradative MGR suggest that it arises due to disintegration of some preexisting objects retaining a lot of easily mobilizing energy, in particular of energy of electron excitation. Gurwitsch named these presumed objects "non-equilibrium molecular constellations". He supposed that they represent groups of excited macromolecules kept together due to constant energy circulation along their common energy levels. "Constellations" are fundamentally different from usual molecular associations and clusters. Components of the latter are kept together with different types of chemical bonds; energy is released when bonds lock, and sufficient energy inflow is needed for their dissociation. On the contrary, molecular constellations are sustained due to constant energy inflow, and any variations of energy supply let alone its blockade results in dispersion of constellations with release of energy retained by them. Lability of constellations precludes their revealing in fixed and even faulty biological material, and only the methods of studies of living cells and cellular systems similar to mitogenetic analysis may provide an insight into the existence of such dynamic structures.

Components of constellations – excited molecules – are prone to continuous exchange, because as soon as a molecule relaxes to a ground state it leaves a constellation. But at any given moment closely arranged excited molecules mutually orient, consequently blending in constellations as systems. In other words, there are weak orientation interactions between the elements of constellations, but free energy of the constellation as a whole is higher than that of its elements in time preceding a constellation formation and after its desintegration due to the reasons mentioned above.

It is natural that if a constellation is disturbed by any means and loses its potential energy, the next distortion would not bring about a flash of degradative MGR until new constellations are formed. From the biological point of view there is nothing extraordinary such in behavior of a constellation. Existence of refractory periods for excitable (more precise – irritable) tissues is well known. There are many ways to trigger a nervous impulse (discharge), and until the critical value of membrane potential is restored, the next irritation would not induce new impulse.

Gurwitsch applied to constellations the notion of Nobelist Albert Szent-Gyorgyi of migration of energy along the common electronic levels of protein molecules [23]. But he broadened it to the possibility of migration of excitation energy along the constellations consisting of different molecules and also considered the possibility of energy quanta summation in different localities to the levels enough to emit UV-photons. Developing this concept he stressed that because of univocacy of energetic and spatial parameters of constellations fluctuations in energy migration should result in spatial realignments of constellations. That is why he stated that at the molecular level of living systems "... it is wrong to oppose the notion of a structure to the notion of a process. The only correct approach to living systems is an approach to them as to the structured processes, flowing in molecular complexes widely different in the degree of their lability" [24].

The possibility of high density energy transfer to macroscopic distances has been experimentally demonstrated using mitogenetic analysis by A.A. Gurwitsch. She has shown that if one fills a narrow capillary with a dilute protein solution and exposes it to MGR from one end, no emission may be registered from the opposite end under usual conditions. However, if the capillary is placed in the longitudinal (parallel to its axis) electrical field (50 v/m), it becomes an MGR conductor. The same effect is observed if the protein solution flows in a capillary at a rate of 1 m/sec. This phenomenon may be explained by a thread-like form of protein molecules, and their alignment along the capillary axis under the action of anelectrical field or a fluid flow, that increases the probability of energy transfer from one molecule to the next one. It is interesting, that the rate of photon "diffusion" in this experimental system was around 30-32 m/sec, which is very close to the rate of a nerve impulse travelling along an axon.

It also follows from this experiment that pumping of constellations with energy is a necessary but not the sufficient condition of their emergence. As the elements of constellations can not mutually orient in them due to usual chemical bonding there should exist an external vectorial factor that imposes a certain spatial arrangement to the elements of constellations. In an example with protein solution able to conduct MGR the role of such a factor was played by an electrical field or a fluid flow.

As constellations are postulated to be the most fundamental necessary condition for the existence of living matter ("structured processes") the uninterrupted existence of such a vectorial factor of dynamic nature is also to be postulated. That is why Gurwitsch's theory of a biological and cellular fields – an imprescriptible property of all living systems – can not be considered without referring to his experimental work in "mitogenetic biology". However, Gurwitsch's theory of biological field cannot be considered here, and the reader may consult other sources for more information on it [25, 26].

4. GENERAL CONCLUSIONS

We depicted only a small fragment of Gurwitsch's heritage concerning the problems of "mitogenetic biology". We practically did not touch an enormous amount of data obtained in application of mitogenetic methods of analysis to cell biology, general physiology and physiology of the central nervous system, oncology, plant physiology, etc. Taken as a whole the results obtained in these studies and conclusions from them should have drastically changed the very foundations of biology. However, mitogenetic biology until now is practically unknown to the great majority of biologists, and most of those who have heard of it do not consider it a serious science. One of the major reasons for this is the apparent absence of serious physical-chemical background for phenomena related to MGR.

Here we concentrated on the most "simple" model systems, which are able to emit MGR – aqueous solutions of amino acids where very peculiar processes develop. It should be stressed again that the very term – mitogenetic radiation – means that these systems "act at a distance" upon populations of cells inducing in them a definite response, a burst of cell divisions. Thus, from the biological point of view MGR should be considered as a signaling (triggering) factor, and it may display biological activity only if it acts on a responsive biological system*.

On the other hand, these signals should have a very definite physical nature. Gurwitsch claimed that MGR are UV-photons. After a long debate it is currently accepted that living systems may emit photons in UV range (e.g., [27]). But why did it take so much time to accept this, though in the last 50 years it was indisputable that both living things and some chemical reactions (such as lipid peroxidation) may be sources of weak photon emission in the visible and near IR-range of electromagnetic spectrum? There are both technical and psychological reasons for this paradox.

Ultra-weak photon emission from living things is currently registered with sensitive photomultipliers (PMT), while Gurwitsch used biological test systems. Even if PMT sensitive to UV-range of EM-spectrum were used, researcher had a lot of problems to detect UV-photons, though it had been claimed in the 1930s that MGR intensity could reach tens-thousands photons per second per 1 cm^2. But a very important thing stressed above was neglected: in most cases the ability of an experimental system to emit MGR depends upon its illumination with visible light. In the darkness MGR is either drastically reduced or disappears and precisely under these conditions photon emission is registered with PMT.

Another reason for misunderstanding of many peculiarities of MGR is in a certain sense psychological. In order to get reliable information about the system a researcher tries to obtain a good "signal-to-noise ratio" and prefers to deal with more or less intense signals. So, *a priori* there is a tendency to consider that informative part of the signal should be "loud", while a

---

* In fact, in one of the papers with alleged refutation of the existence of MGR (Hollaender A., Klaus W. Bull. Nat. Res. Council 1937, 100:3-96), which is sited until now to claim that the whole problem of MGR is an example of "pathological science" (Langmuir I. Physics Today, 1989, 36: 47.) this very mistake was made: a yeast culture taken as a detector of MGR was used at a stage when it is unresponsive to it.

"whisper" is insignificant. As the intensity of radiation measured with PMTs in a visible and IR-part of the spectrum is usually much higher than that in the UV-region, and may be additionally amplified by addition to a natural system of chemiluminescent and fluorescent compounds (e.g., luminol, lucigenin, Rhodamin), most researcher judged the properties of systems under study taking into consideration only intensity of radiation and not its other properties. As a matter of fact, in our studies of photon emission from the chain reaction with delayed branching running in solution of glycin in the presence of H_2O_2, we could barely register any photons in UV-region, but on addition to the reaction mixture of tiny quantities of fluorophors that could be excited only by UV-photons (or equivalent energy quanta) and remit in the visible range, photon emission intensity increased enormously.

But is intensity of photon emission really so important? Let us return to a very significant Gurwitsch observation: supply of MGR (which is very weak by itself) to a biological detector in the intermittent mode, when the overall intensity of incident photons is reduced drastically significantly increases its biological efficiency. This does not happen when intermittence is chaotic. Therefore not intensity of radiation, but rather its orderliness is of primary importance for considering this radiation "biophotonic"! Besides, we should realize that radiation from a living system registered by us is energy, which is lost by it, and is very "expensive" high density energy. And the phenomenon of "degradative radiation" discovered by Gurwitsch shows, that a living system may store a vast amount of it which may be easily mobilised as photonic energy under perturbations of its stationary state.

And here we come to a controversial term of "biophotons". This term was first suggested by Friz-Albert Popp to designate photons belonging to a very low level of photon emission from living organisms (not to be mixed up with "bioluminescence" – high intensity radiation of many organisms due to specific enzymatic reactions) [28]. Popp discovered that statistical and temporal properties of this radiation are very different from those expected from stochastic emission of usual chemiluminescent reactions. He proved that at least the so-called delayed luminescence induced by irradiation of living organisms with a flash of light results from the relaxation of an intrinsic coherent electromagnetic field [3]. As far back as in 1983 Nagl and Popp suggested that there is a negative feedback-loop in living cells which couples together states of a coherent ultra-weak photon or biophoton field and the conformational state of the cellular DNA. They assume photon transfer or non-radiation chemical pumping from the metabolism which results in changes of the DNA conformation via exciplex/excimer formation [29]. Recently energy transfer to tremendous (in molecular scale) distances without its dissipation has been demonstrated for DNA molecules and for helical regions of proteins [30, 31].

It should be taken into consideration that when Szent-Gyorgyi and Gurwitsch formulated concepts of "energy circulation along the common electronic levels" no corresponding physical notions were yet formulated and no physical evidence for such phenomena existed. Apparently the time has come to consider the properties of biological systems starting from the molecular level from the position of soliton and exciton physics, from the notion of active medium in laser physics.

However, even if special properties of "biophotons" emitted by living organisms are acknowledged and explained due to special structural properties of macromolecules present in living matter, there remains a need to explain the origin of biophoton fields in living matter. We believe that processes developing in "simple" aqueous solutions of amino acids followed with emission of ultra-weak radiation of high energy photons are the foundation of all other processes related to generation and relaxation of electronic excitation, which may or may not follow with photon emission from living systems [32]. Very recently we have demonstrated that in such initially simple reaction systems real self-organisation takes place under the conditions when oxygen-dependent chain reactions with delayed branching develop: temporal, spatial and even chemical complexity of these solutions increase, and they behave as the active medium [33, 34]. Diverse external mild stimuli induce intense flashes of photons from these solutions, indicating that they accumulate a huge amount of easily mobilised high density energy – in close analogy to "degradative" radiation. Of special interest is that photons may be emitted from solutions where the named processes develop in an oscillatory manner, sometimes highly ordered and displaying a wide range of frequencies and amplitudes. We registered emission mainly in the region of 400-500 nm, but there is every reason to believe that these systems are radiating also in other spectral regions (though at much lower intensity) and that this radiation is also well-ordered.

Thus "biophotonic-like fields" may emerge even in initially very "plain" systems, such as aqueous solutions of simple organic molecules under soft but specified conditions which imply the development of oxidative chain reactions with delayed branching in aqueous milieu. Water should be regarded as an active participant of these processes rather than a passive medium for them [35] due to its unique dynamic organization. Thus, aqueous systems where there develop processes first discovered by Gurwitsch gain ability to emit "biophotons" – electromagnetic radiation of very low intensity, in a very wide spectral range, and possessing specific orderliness. That provides them capability to trigger realignment and modulate reorganization of their own biophotonic fields in living organisms accompanied with specifically patterned biophoton emission from them, and followed with an array of responses including cell division, differentiation, dedifferentiation, apoptosis, shifts in metabolic activity, locomotion and other manifestations of their vitality.

5. REFERENCES

1 Zhuravlev A.I. (1972) "Substrates and Mechanisms of Endogenous (Chemical) Generation of Excited Electron States and Superweak Luminescence in Tissues". In *Ultraweak luminescence in biology*. Transactions of the Moscow Society of Naturalists. **Vol. XXXIX**, 17-32.

2 Cadenas E.(1984) Biological chemiluminescence. *Photochem Photobiol.*, **40**, 823-830.

3 Popp, Fritz-Albert (1992) "Some Essential Questions of Biophoton Research and Probable Answers". In *Recent Advances in Biophoton Research and its Applications,* Fritz. A. Popp, Ke-hsueh Li and Qiao Gu (eds), Singapore, World Scientific, 1-46.

4 Bibliography on MGR problem from 1924 to 1948 includes more than 1000 publications from the laboratories of USSR, Germany, France, Italy, Japan, and other countries. Major information presented below is based on the analysis of the following sources:

Gurwitsch, A. (1932) *Mitogenetic Emission*. Gos. Med. Izdat., Moscow.
Gurwitsch, A. and Gurwitsch, L. (1934) *Mitogenetic Radiation*. The All-Union Institute of Experimental Medicine Publishing House, Leningrad.
Rahn, O. (1936) *Invisible Radiations of Organisms*. Verlag von Gebruder Borntraeger, Berlin.
Gurwitsch, A. and Gurwitsch, L. (1937) *Mitogenetic Analysis of Tumor Cells*. The All-Union Institute of Experimental Medicine Publishing House, Leningrad.
Gurwitsch, A. and Gurwitsch, L. (1937) *Mitogenetic Emission*. Medgiz, Moscow.
Gurwitsch, A. (1968) *Problem of |Mitogenetic Radiation as an Aspect of Molecular Biology*. Medicina, Leningrad.
Gurwitsch, A. (1991) *Principles of Analytical Biology and of the Theory of Cellular Fields*. Nauka, Moscow.

5 Gurwitsch A.G., Gurwitsch L.D. (1999) *Twenty Years of Mitogenetic Radiation: Emergence, Development, and Perspectives.* Uspekhi Sovremennoi Biologii. 1943; 16: 305-334. (English translation: 21st Century Science and Technology. Fall, 12, **No 3**: 41.)

6 Zalkind S.Ya. (1937) Contemporary state of the problem of mitogenetic biology. Part 1. Review of critical literature. *Uspekhi Sovremennoi Biologii*, **7**, 50-66. Part 2. Recent positive data. *Uspekhi Sovremennoi Biologii*, **7**, 216-233.

7 Audubert R. (1938) Emission from chemical reactions. Uspekhi chimii. 1938. 7: 1858-1883. Translated from *Angew. Chemie*, **51**, 153.

8 Rabinerson A.I., Filippov M.V. (1938) Emission of short UV-photons from chemical reactions. *Zhurnal Fizicheskoi Chimii*, **9**, 688-701.

9 Baskakov I. V., Voeikov V. L. (1996) Formation of a polymer with glycin deaminase activity upon UV irradiation of amino acid solutions. *Russ. J. of Bioorganic Chemistry.*, **22**, 77-82.

10 Kuzin A.M., Polyakova O.I. (1947) On enzyme activity of highly diluted enzyme solutions in the presence of amino acids. In *Collection of Works on Mitogenesis and Theory of Biological Fields*. Moscow: Izdatelstvo AMN, 54-64.

11 Gurwitsch A.G., Gurwitsch A.A. (1958) The problem of autocatalysis (autoreproduction) of some cyclic compounds from lower amino acids. *Enzymologia – Acta Biocatalytica.*, **30**, 1-16.

12 Ansari A.S., Tahib S., Ali R. (1976) Degradation of phenylalanine in the presence of hydrogen peroxide. *Experientia*, **15**, 573-574.

13 Gryczynski I., Malak H., Lakowicz J.R. (1999) Three-photon excitation of N-acetyl-L-tyrosinamide. *Biophys Chem.*, **79**, 25-32

14 Semyonov, Nikolai (1935) *Chemical Kinetics And Chain Reactions.* Oxford, Oxford University Press.

15 Hinshelwood, Cyrill (1940) *The Kinetics of Chemical Change.* Oxford., Claredon Press.

16 Emanuel N.M., Saikov G.E., Maizus Z.K. (1973) *Role of Medium in Radical Chain Reactions of Oxidation of Organic Compounds.* Moscow, Nauka.

17 Voeikov V.L., Baskakov I.V. (1995) Study of chemiluminescence kinetics in aqueous amino acid solutions in the presence of hydrogen peroxide and ethidium bromide. *Biofizika (Russ.)*, **40**, 1150-1157.

18 Voeikov V. L., Baskakov I. V., Kafkialias K., Naletov V. I. (1996) Initiation of degenerate-branched chain reaction of glycin deamination with ultraweak UV irradiation or hydrogen peroxide. *Russ J Bioorganic Chem*, **22**, 35-42.

19 Voeikov V.L., Naletov V.I. (1998) "Weak Photon Emission of Non-Linear Chemical Reactions of Amino Acids and Sugars in Aqueous Solutions" In *Biophotons*. Jiin-Ju Chang, Joachim Fisch, Fritz-Albert Popp (eds.), Dortrecht, Kluwer Academic Publishers, 93-108.

20 Voeikov V.L., Koldunov V.V., Kononov D.S. (2001) Sustained oscillations of chemiluminescence in course of amino-carbonyl reaction in aqueous solutions. *Russ J of Phys Chem*, **75**, 1579-1585.

21 Voeikov V L., Novikov C N., Vilenskaya N D. (1999) Low-level chemiluminescent analysis of nondiluted human blood reveals its dynamic system properties. *J Biomed Optics*, **4**, 54-60.

22 Gurvich, Anna. (1992) "Mitogenetic Radiation as an Evidence of Nonequilibrium Properties of Living Matter." In *Recent Advances in Biophoton Research and its Applications*, Fritz. A. Popp, Ke-hsueh Li and Qiao Gu, (eds.) Singapore, World Scientific, 457-468.

23 Szent-Gyorgyi A. (1941) Towards a new biochemistry. *Science*, **93**, 609.

24 Gurwitsch, Alexander, (1944) *Theory of Biological Field.* Moscow, Sovetskaya Nauka.

25 Lipkind, Michael. (1992) "Can the Vitalistic Enthelechia Principle be a Working Instrument?" In *Recent Advances in Biophoton Research and its Applications*, Fritz. A. Popp, Ke-hsueh Li and Qiao Gu, (eds). Singapore, World Scientific, 469-490.

26 Beloussov L.V. Developmental biology and physics of today. *This volume.*

27 Slawinski J. (1988) Luminescence research and its relation to ultraweak cell radiation. *Experientia*, **44**, 559-571.

28 Popp, F.A. (1979) "Photon Storage in Biological Systems." In *Electromagnetic Bio-Information*, F.A. Popp, G. Becker, H.L. Konig and W. Peschka, (eds.), Munich-Baltimore, Urban & Schwarzenberg.

29 Nagl W, Popp F.A. (1983) A physical (electromagnetic) model of differentiation. Basic considerations. *Cytobios*, **37**, 45-62.

30 Wan C., Fiebig T., Kelley S.O., Treadway C.R., Barton J.K,. Zewail A.H. (1999) Femtosecond dynamics of DNA-mediated electron transfer. *Proc Natl Acad Sci*, USA, **96**, 6014-6019.

31 Schlag E.W., Sheu S.Y., Yang D.Y., Selzle H.L., Lin S.H. (2000) Charge conductivity in peptides: dynamic simulations of a bifunctional model supporting experimental data. *Proc Natl Acad Sci*, USA, **97**, 1068-1072.

32 Voeikov V.L. (2000) Processes involving reactive oxygen species are the major source of structured energy for organismal biophotonic field pumping. In *Biophotonics and Coherent Systems. Proceedings of the 2nd Alexander Gurwitsch Conference and Additional Contributions.* Lev Beloussov, Fritz-Albert Popp, Vladimir Voeikov, and Roeland Van Wijk, (eds.), Moscow, Moscow University Press.

33 Voeikov V.L., Koldunov V.V., Kononov, D.S. (2001) New oscillatory process in aqueous solutions of compounds containing carbonyl and amino groups. *Kinetics and Catalysis* (Moscow), **42**, 670-672.

34 Voeikov V.L., Koldunov V.V., Kononov D.S. (2001). Sustained oscillations of chemiluminescence in course of amino-carbonyl reaction in aqueous solutions. *Zhurnal Phys. Chem*, (Russian Journal of Physical Chemistry), V. 75, N **9**, 1579-1585.

35 Voeikov V. (2001) Reactive oxygen species, water, photons, and life. *Rivista di Biologia/Biology Forum*, **94**, 193-214.

Chapter 11

BIOPHOTONS: ULTRAWEAK PHOTONS IN CELLS

Hugo J. Niggli[a,b]* and Lee Ann Applegate[b]
[a]BioFoton AG, rte. d'Essert 27,
CH-1733 Treyvaux, Switzerland,
Phone/Fax: ++41-26-4131445
[b]Department of Obstetrics,
Laboratory of Oxidative Stress and Ageing,
University Hospital, CHUV,
CH-1011 Lausanne, Switzerland
*Address for correspondence : BioFoton AG, P.O. Box 28,
CH-1731 Ependes, Switzerland
E-mail:biofoton@swissonline.ch

Abstract: Photons participate in many atomic and molecular interactions and changes. Recent biophysical research has detected ultraweak photons or biophotonic emission in biological tissue. It is now established that plants, animals and human cells emit a very weak radiation which can be readily detected with an appropriate photomultiplier system. Although the emission is extremely low in mammalian cells, it can be efficiently induced by ultraviolet light. Photons in the visible range are coupled with radical reactions while photons in the UV are linked with the DNA as the source. In recent years, we developed a cell culture model for biophotonic measurements using fibroblastic differentiation in order to test the growth stimulation efficiency of various bone growth factors. Since that fibroblasts play an essential role in skin aging, skin carcinogenesis and wound healing, the biophotonic model using the fibroblastic differentiation system provides a new and powerful non-invasive tool for the development of skin science. Its high sensitivity can be applied in all fields of skin research, in investigating skin abnormalities and in testing the effect of such products as regenerative, anti-aging and UV-light protective agents.

Cyclobutane-type pyrimidine dimers participate as important factors in ultraviolet-induced lethality, mutagenicity and tumorgenicity. Substantial efforts have been made in recent years to understand the induction of pyrimidine photodimers and their repair in human skin cells exposed to low fluences of UV-light. Dimers are efficently induced after UVC and UVB-irradiation, but these photoproducts are even produced in DNA of human skin in vivo after UVA irradiation. By developing a highly sensitive immunohistochemistry dimer detection assay, we confirm that UVA radiation induces substantial amounts of these DNA-changes in the epidermis and this enhanced immunohistochemistry technique detects them even down to the reticular dermis. A considerable number of these photodimers were also seen in non-irradiated control skin up to 2 cm away from the irradiation site which persist for at least two days postirradiation time.

These results sustain the hypothesis that pyrimidine dimer formation and excision repair as well as ultraweak photon emission seem be involved in a type of intra- and extracellular photobiostimulation and may be important triggers of UV-induced signal pathways expressing epidermal communication.

Keywords: Ultraweak, photon, fibroblast, skin, human, ultraviolet, growth factor

1. MATTER AND RADIATION

Photons, the smallest elements of light, necessarily play an important role in all atomic and molecular interactions and changes in the physical universe. Light was probably created about 20 billion years ago as a result of a massive explosion of infinite energy (Big Bang). The huge fireball expanded at unbelievable speed, during which it cooled and led to the formation of elementary particles. They are known in modern physics as fermions which contain the basic elements of the atoms: proton + neutrons (hadrons) as well as electrons (leptons), the building blocks of matter. The hadrons are composed in turn of quarks and, based on our most recent knowledge, of so-called preons. The photons, the quantum elements of light, are bosons. Their electromagnetic force holds the elementary particles together. Modern physics sees matter as changing into energy according to Einstein's formula $E = mx c^2$. Therefore mass is a form of bound energy. On the basis of modern quantum physics it became clear that fermions and bosons no longer represent separate, but complementary terms. Photons can be regarded as the information carriers of matter because they not only glue elementary particles together but can be exchanged within atomic and molecular interactions.

2. HISTORY

2.1. Overview

At the beginning of this century, Alexander Gurwitsch, born almost 130 years ago in Russia, suggested that ultraweak photons transmit information in living systems [1] as summarized recently [2]. Although the results of Gurwitsch were refuted by Hollaender and Klaus [3] and the interest in this subject declined in the following decades, the presence of biological radiation was re-examined with the development of photomultiplier tubes in the mid-1950s by Facchini and co-workers [4]. In the 1960s most of the work on ultraweak photon emission was performed by Russian scientists [5-7], while in western countries several pioneers, Quickenden in Australia [8], Popp in Germany [9] and Inaba in Japan [10], independently developed methods for ultraweak photon measurements in a variety of different cells by the use of an extremely low noise, highly sensitive photon counting system which allows maximal exploitation of the potential capabilities of a photomultiplier tube. In the meantime it is commonly agreed that plant, animal and human cells emit ultraweak photons often called biophotons [11-17]. From these and additional investigations different origins for this very weak radiation have been proposed.

2.2. Radical Reactions as a Source of Biophotons

Most investigators think that very weak radiation results from radical reactions which can be produced by biological events such as lipid peroxidation. In studies of microsomal lipid peroxidation [18,19], it has been shown that the amount of malonaldehyde production and the intensity of emitted light are related to each other. Based on these studies, Inaba and co-workers proposed [20] that oxygen dependent light emission in rat liver nuclei was caused most likely by lipid peroxidation in the nuclear membrane. As discussed in detail by Cadenas and Sies [21], free radical decomposition of lipid hydroperoxides leads to the formation of excited chemiluminescent species by the self-reaction of secondary lipids peroxyradicals, producing either singlet molecular oxygen or excited carbonyl groups.

2. 3. Does Chromatin Contribute to the Emission of Ultraweak Photons?

However, there also exists a very interesting model published in 1983 by Nagl and Popp [22] suggesting that there is a negative feedback-loop in living cells which couples together states of a coherent ultra-weak photon or biophoton field and the conformational state of the cellular DNA. The authors assume photon transfer or non-radiation chemical pumping from the cytoplasmic metabolism which results in changes of the DNA conformation via exciplex/excimer formation. Their hypothesis is based on experimental data reviewed by Birks [23] who also suggested these excimers as precursors of the pyrimidine photodimers which play a key role in the radiation damage of DNA and has been proposed as biomolecular signals in communication [24]. It was hypothesized that there exists an effective intracellular mechanism of photon trapping in normal human cells [25]. This type of light-trapping mechanism in UV-induced pyrimidine dimers could be responsible for influencing metabolic and cellular events by UV-induced pyrimidine dimers once they are excised as was recently shown for photodimer induced melanogenesis [26].

2. 4. Conclusions

Tilbury discussed in the multi-author review of van Wijk [27] that ultraweak photon emission has been detected in both the visible and ultraviolet region. Radiation in the visible region appears to be due to excited carbonyl groups and/or excited singlet oxygen dimers arising from lipid peroxidation, which in turn are associated with an increase in various reactive oxygen species such as the superoxide anion, hydrogen peroxide, hydroxyl radical and singlet oxygen. There is also substantial evidence for DNA playing a key role in these emissions as discussed above [9,28,29]. This macromolecule may especially be involved in the emission of ultraweak photons by the cell in the UV-region of the spectrum [27].

3. BIOPHOTONIC EMISSION IN CELL CULTURES

3. 1. Ultraweak Photon Emission in Human Skin Fibroblasts

First experiments with cultured human cells were reported [25] in which normal and DNA excision repair deficient Xeorderma pigmentosum (XP) cells were UV-irradiated in medium and balanced salt solution (EBSS) and assessed for ultraweak photon emission. There was some evidence of induced photon emission from normal cells in balanced salt solution, but clear evidence of a fluence-dependent emission in XP cells in medium and in EBSS were seen. Overall, it was clear that an important difference between normal and XP cells was present and it is possible that XP cells are unable to store ultraweak photons which are efficiently trapped by normal cells.

3.2. The Fibroblast Differentiation System

Since Hayflick's pioneering work in the early 1960s, human diploid fibroblasts have become a widely accepted *in vitro* model system. Recently, Bayreuther and co-workers extended this experimental approach showing that fibroblasts in culture resemble, in their design, the hemopoietic stem-cell differentiation system [30]. They reported morphological and biochemical evidence for the fibroblast stem-cell differentiation system in vitro. They showed that normal human skin fibroblasts in culture spontaneously differentiate along the cell lineage of mitotic fibroblasts (MF) MFI-MFII-MFIII and post-mitotic fibroblasts (PMF) PMF IV-PMFV-PMVI-PMFVII. Additionally, they developed methods to shorten the transition period and to increase the frequency of distinct post-mitotic cell types using physical or chemical agents such as mytomycin C, 5-fluorouracil, and 5-bromo-2-deoxyuridine, that have been reported to induce differentiation in a variety of cell systems, as we have recently summarized [30-32,35-37]. Mitomycin C is an effective chemotherapeutic agent for several cancers in man; this effect is probably related to the interaction with DNA leading to DNA-DNA crosslinks and DNA-protein crosslinks. We have previously demonstrated that the mitomycin C-treatment of three different normal human fibroblast strains (CRL 1221, GM 38 and GM 1717), frequently used in mutation, transformation, and aging research, induces characteristic morphological changes in the fibroblasts and brings about specific shifts in the [^{35}S]methionine polypeptide pattern of total cellular proteins. These results support the notion that mitomycin C accelerates the differentiation pathway from mitotic to postmitotic fibroblasts. Using this system, we were also able to demonstrate that no significant difference exists in the rate and the extent of the excision-repair response to thymine-containing pyrimidine dimers following UV-irradiation shortly after mitomycin C treatment of distinct strains of human skin fibroblasts and in the mitomycin C-induced PMF stage of these cells [31]. In addition, aphidicolin inhibits excision repair of UV-induced pyrimidine photodimers in low serum cultures of mitotic and mitomycin C-induced postmitotic fibroblasts of human skin [32]. Since that fibroblasts play an essential role in skin aging, and wound healing, our results imply that the fibroblast differentiation system is a very useful tool to unravel the complex mechanism of skin aging, skin carcinogenesis and wound healing.

3.2.1. Bone Factor Induced Proliferation in Fibroblastic Differentiation

Proliferation of fibroblasts was determined by counting cultured cells following a one to two week treatment with 5 ng of each lot of bone growth factor. The percentage of increase of cell proliferation was between 0 and 60% as shown in Table 1.

Table 1: Percentage of cell counts in immediately trypsinzed monolayer fibroblasts with and without treatment with growth factors (100 % correspondend to 3.0×10^6 mitotic fibroblasts (controls and MMC-treated) and 0.9×10^5 postmitotic fibroblasts)).

Differentiation stage of cells	Control	BP 2	BP 3	BP 5
Mitotic fibroblasts	100%	100%	140%	120%
Mitotic fibroblasts (MMC)	100%	100%	160%	120%
Postmitotic fibroblasts	100%	133%	133%	155%

Growth Factor Efficiency: BP3>BP5>BP2

3.3. Ultraweak Photon Determinations in Human Skin Fibroblasts

3.3.1. Cell Culturing

Skin fibroblasts from a Xeroderma Pigmentosum from a 9 year old patient of complementation group A (XPA), XP12BE (GM05509A) were obtained from the Human Genetic Mutant Cell Repository (Camden, NJ, USA). XPA cells derived from a 10-year -old female (CRL 1223) were purchased from the American Type culture collection (Rockeville, USA). Cells were cultured in tissue culture plastic flasks (surface, $75 cm^2$, Gibco Basel, Switzerland) in 15 ml Dulbecco's modified Eagle's medium (DMEM; Gibco, Basel Switzerland) supplemented with 10% fetal calf serum and 100 units (U) ml^{-1} penicillin-streptomycin as previously described [37]. XPA cells were plated 1:3 and passages 15-17 were used. Postmitotic fibroblasts were prepared as previously described [32,34]. For the determination of ultraweak photon emission in mitotic fibroblasts, 3×10^5 cells were seeded into 10 cm diameter tissue culture dishes and grown for 1 week until confluency with and without growth factors at a density of 5ng/ml. UVA-irradiated fibroblasts and control cells were frozen gently in liquid nitrogen in DMEM from which phenol red had been omitted. For the determination of ultraweak photons in postmitotic fibroblasts, 1.5×10^5 cells were seeded into 10 cm diameter tissue culture dishes. 48 h later the medium was removed and a new complete medium containing 2×10^{-7} M mitomycin C with and without growth factors was added for 24 h. The medium was then removed and new complete DMEM with and without growth factors was added for 4 days. Cells were additionally treated with 2×10^{-7} mitomycin C essentially as described elsewhere [32,35]. The medium was removed and new complete DMEM with and without growth factors was added for 7 days until they were in a postmitotic stage. Control cells which had immediately treated with mitomycin C were also used. Cells were counted in a haemocytometer from Neubauer (Flow Laboratories, Baar, Switzerland) and cells were counted in triplicates ($\pm 10\%$). Cells were frozen in liquid nitrogen using the cryopreservation apparatus from Biotech Research Laboratories (Rockeville, MD, USA). Controlled gradual temperature reduction during cryopreservation was critical for the maintenance of cell life and viability. This apparatus preserves cells at the rate of 1°C per minute when placed in a -70° C freezer. After 5 hours or overnight, frozen samples were transferred to the liquid nitrogen transport storage system. DMSO can interfere with ultraweak photon emission and therefore no DMSO was added.

3.3.2. Biophotonic Measurements in Human Skin Cells

For ultraweak photon emission measurements different fibroblasts samples were transported in liquid nitrogen. Some cells were transported fresh in T-75 flasks and used for ultraweak photon emission. The cells were thawed and were diluted in a volume of 10 ml Dulbecco's modified Eagle's medium, from which phenol red was omitted, and were transferred to a quartz sample glass ($2.2 \times 2.2 \times 3.8$ cm^3; thickness 0.15 cm). Detection and registration of spontaneous and light-induced emitted photons was accomplished as described before [25]. The test samples were kept in a dark chamber in front of a single photon counting device equipped with an EMI 9558 QA photomultiplier tube (diameter of the cathode 48mm, cooled to -25° C). This high-sensitivity photon counting device is described in detail by Popp and co-workers [9,38] and measures photon intensities as low as 10^{-17} W in the range between 220 and 850 nm. Signal amplification is normally 10^6-10^7 and dark count ranges between 10-15 counts per second (cps). Maximum efficiency of the S20-cathode is 20-30% at 200-350nm, decaying almost linearly down to 0% at a wavelength of 870 nm and mean quantum efficiency in the entire spectral range is about 10%. The integral intensity values within each interval of 40 ms were stored and processed by an interfaced computer. For light induced emission experiments, irradiation of the test sample was performed perpendicular to the detecting direction with focused white light from a 75-W Xenon lamp. Each measuring cycle was started by irradiating the sample for 5s. The measurements for the monochromatic light induction were performed by changing the spectrum of the inducing light through a monochromomator (PTI, Hamburg, Germany) from 300 to 450 nm (25nm interval).

3.4. Ultraweak Photons in Mitotic and Postmitotic Human Skin Fibroblasts

In normal cells, we have found that white light-induced photon re-emission relaxation dynamics are identical after successive irradiations of approximately 1 min intervals, and even several cycles of illumination and measurement do not quantitatively change the re-emission intensity for several hours. The re-emission curves are hyperbolic as we have reported most recently with mouse melanoma cells [25]. These results confirm several previous investigations in plant and mammalian cells [11,15,16,25,26]. As interpreted by Li and Popp [27], the hyperbolic decay kinetics found in living systems after pre-illumination with white light may indicate coherent re-scattering of ultraweak photons due to collective excitation of nucleic acids within the DNA of the investigated fibroblasts.

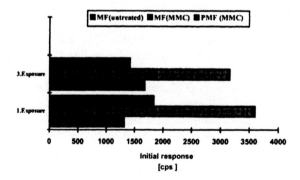

Figure 1: Total counts of photoinduced ultraweak photon emission registered in the first second after illumination with white light during 30 s of normal human skin fibroblasts (GM38).

As described above, Bayreuther and co-workers [30-32,35-37] showed biochemical and morphological evidence for the fibroblast differentiation system in vitro. They showed that normal human skin fibroblasts in culture spontaneously differentiate along the cell lineage of mitotic (MF) and postmitotic fibroblasts (PMF). Additionally, they developed methods to shorten the transition period and to increase the frequency of distinct postmitotic cell types using physical agents such as ultraviolet light (UV) and mitomycin C (MMC). Figure 1 depicts total light-induced photon emission in distinct differentiation stages of normal fibroblasts (GM38). There is no discernible difference between untreated mitotic fibroblasts, MMC treated mitotic cells and MMC-induced postmitotic cultures three weeks following MMC-treatment.
Similiar results (not shown) were obtained for other normal cell lines (CRL 1221, GM 1717).

We further performed ultraweak photon experiments in the fibroblast differentiation system with human skin fibroblast from Xeroderma pigmentosum patients (GM 05509A; CRL1223). Bayreuther and co-workers [30-32; 35-37] showed biochemical and morphological evidence for the fibroblast differentiation system in vitro. They showed that normal human skin fibroblasts in culture spontaneously differentiate along the cell lineage of mitotic (MF) and postmitotic fibroblasts (PMF). Additionally, they developed methods to shorten the transition period and to increase the frequency of distinct postmitotic cell types using physical agents such as ultraviolet light (UV) and mitomycin C (MMC). To differentiate the human skin fibroblasts we have used mitomycin C. Mitotic skin fibroblasts are small in size in contrast to postmitotic cells which are enlarged by a factor of up to ten [35]. As shown in Figure 2 there is no discernible difference between untreated mitotic fibroblasts, miotomycin C treated cells and MMC-induced postmitotic cultures one week following MMC-treatment in age-matched XPA cells (GM05509 A; CRL 1223; 3×10^7 cells/10 ml).

We have previously found that postmitotic fibroblasts are significantly reduced in UV-induced ornithine-decarboxylase responses [35]. In skin fibroblasts from a patient with Xeroderma pigmentosum (XPA) we detected a slight increase in the rate of thymine dimer induction formed after UV-irradiation of the MMC-induced postmitotic stage three weeks after MMC-treatment [31]. For ultraweak photon emission, our experiments with cultured human cells in which normal and XPA cells were UV- irradiated showed an important difference between normal and XP cells indicating that XP cells are unable to store ultraweak photons which are efficiently trapped in normal cells [25].

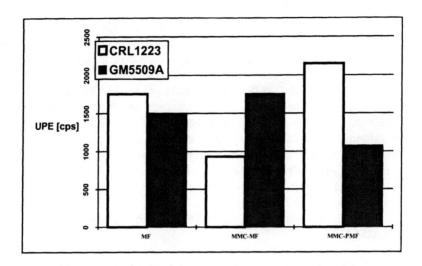

Figure 2: Total counts of photoinduced ultraweak photon emission registered in the first second after illumination with white light during 30 s of XPA human skin fibroblasts. Background values obtained with medium alone are subtracted. Average of two independent experiments in triplicates. Standard deviation is ± 10%.

In the first experiments with human XPA skin fibroblasts it was found that measurable amounts of ultraweak photons can be achieved with cell densities as low as $2-3 \times 10^4$ cells /ml (data not shown). At these cell densities, control values after irradiation with a white light source for induction of ultraweak photons were determined in mitotic and postmitotic fibroblasts from fresh cells and compared to those of frozen cells. The results are shown in Table 2.

Table 2: White light induced ultraweak photon emission (cps/25) in immediately trypsinzed monoayer fibroblasts (fresh) compared to fibroblasts transported in liquid nitrogen (frozen)*

Differentiation stage of cells	Induced emission (fresh)	Induced emission (frozen)
Mitotic fibroblasts	72	54.2 ± 5.4 (n=6)
Mitotic fibroblasts (MMC)	74	40.0 ± 5.0 (n=6)
Postmitotic fibroblasts	54	33.8 ± 3.4 (n=6)

*10 ml DMEM

Although there was a small increase in the induced photon emission from fresh cells, most probably due to the stress condition of the transport, it can be concluded that the frozen cells showed the same type of emission. Similiar results were obtained for monochromatic induction of the fibroblasts in the range of 300-450nm as shown for MMC- mitotic cells in Figure 3.

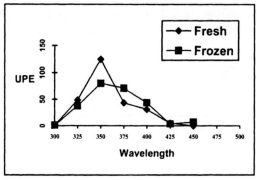

Figure 3: Monochromatic light (300-450 nm) induced ultraweak photon emission (UPE; cps/100) in immediately trypsinzed monolayer fibroblasts (fresh) compared to fibroblasts transported in liquid nitrogen (frozen). Average of triplicate determinations are given. Standard deviation is less than 10%.

In a previous report [25], we found a slight increase in the dark count rates in postmitotic fibroblasts in XPA cells in contrast to normal fibroblasts. Therefore, the dark count rate at a cell density of 3.6×10^4 cells/ml in untreated and growth factor treated postmitotic fibroblasts was determined (data not shown) and this increase of the dark count rate in postmitotic cells was confirmed.

3.5 Ultraweak Photon Emission in Assessing Growth Factor Efficiency

As shown in Figure 3 the most significant increase in ultraweak photon emission after monochromatic irradiation of fibroblasts between 300-450 nm was found at 350 nm. Therefore, the induced photon emssion after 350 nm light induction at a cell density of 3.6×10^4 cells /ml in untreated and growth factor treated MMC-mitotic fibroblasts was determined. In addition, the cells were treated with 400 kJ/m^2 UVA as described elsewhere [37]. For mitotic cells, the results are shown in Table 3 showing an improvement of light-trapping ability for the growth factors in the efficiency range of BP5>BP3>BP2. For MMC-treated mitotic fibroblasts similiar results have been obtained as shown in Table 4 as well as for MMC induced-postmitotic fibroblasts (data not shown).

Table 3: Induced ultraweak photon emission (%) at 350nm in mitotic fibroblasts with and without treatment with growth factors in controls and UVA-treated cells. Average of triplicate determinations are given. Standard deviation is less than 10%.

Differentiation stage of cells	Control	BP 2	BP 3	BP 5
Mitotic fibroblasts; control	100	50	80	185
Mitotic fibroblasts; UVA	100	47	0	10

Growth Factor Efficiency: BP5>BP3>BP2

Table 4: Induced ultraweak photon emission (%) at 350 nm in MMC-treated mitotic fibroblasts with and without treatment with growth factors in controls and UVA-treated cells. Average of triplicate determinations are given. Standard deviation is less than 10%.

Differentiation stage of cells	control	BP 2	BP 3	BP 5
Mitotic fibroblasts; control	100	161	320	106
Mitotic fibroblasts; UVA	100	0	0	33

Growth Factor Efficiency: BP3>BP2=BP5; All very efficient

We have previously found that after white light induction there is a small decrease from mitotic to postmitotic fibroblasts from XPA-cells while there is in MMC-treated normal fibroblast a significant increase [38]. Additionally we detected that normal cells trap UV-photons [25]. As shown in Table 4, growth factor treatment indeed increases the monochromatic induction in these cells similar to normal cells. Most interestingly, UVA irradiation in mitotic and postmitotic cells significantly reduce the monochromatic induction of ultraweak photons. It can be assumed, therefore, that the decrease in mitotic untreated and MMC treated cells which has been exposed to growth factors is due to trapping of photons, emphasizing a return to normal basal status.

4. PYRIMIDINE DIMERS IN MITOTIC AND POSTMITOTIC HUMAN SKIN FIBROBLASTS

4.1. The Importance of Pyrimidine Dimers in Human Cells

Cyclobutane-type pyrimidine photodimers are important factors in ultraviolet-induced lethality, mutagenicity and tumorgenicity. In the early eighties, a highly sensitive method was developed for the resolution and determination of pyrimidine dimers using high pressure liquid chromatography on reversed-phase columns and the rates of formation and excision of UV-light induced pyrimidine photodimers were determined in cultures of human skin fibroblasts in mitotitc and postmitotic fibroblasts .[32]. Recently, beneficial aspects of biological pathways involving pyrimidine dimers have been delineated. Formation of sunburn cells produced via pyrimidine dimers in skin, are important in preventing skin cancer [33]. Sunburn cells are dependent on the gene p53, considered to be the "guardian-of-the-tissue". This gene prevents mitosis induction prior to accomplished DNA repair and aborts cells damaged beyond repair possibilities. It therefore plays a major role in the prevention of skin tumours.

In another recent study [26] the repair of pyrimidine dimers and/or the accumulation of DNA fragments (topical application of dinucleotides) was shown to be responsible for increased, possibly protective melanogenesis in cultured human skin cells and in mammalian skin. Recent biophysical research has shown the existence of photons in biological tissue. Plant, animal and human cells emit a very weak radiation which can be readily detected with an appropriate photomultiplier system [27]. Although the emission of this radiation is extremely low in mammalian cells, it can be efficiently induced by ultraviolet light. We have reported that postmitotic XP-fibroblasts lose the storing capacity of ultraweak photons which are efficiently trapped in normal cells. Thus, it is evident that there exists an important difference between normal and XP cells and this suggests that there is an effective intracellular mechanism of photon trapping in normal human cells [25]. This type of light-trapping mechanism in DNA could be responsible for influencing metabolic and cellular events by UV-induced pyrimidine dimers once they are excised as was recently shown for photodimer induced melanogenesis [26].

These recent observations implicating pyrimidine dimers in important signal pathways for possible protection of human skin add new dimensions to the studies where they are shown to trigger mainly deleterious effects as depicted in Figure 4.

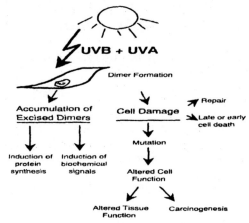

Figure 4: Various agents induce pyrimidine dimer pathways.

4.2. UV-Irradiation of Human Skin and Detection of Pyrimidine Dimers with Monoclonal Antibodies [24]

For UVA radiation a UVASUN 3000 lamp (Mutzhas, Munich, Germany) was used at a dose rate of 300-600 W/m^2 at a distance of 60-90 cm. Irradiation periods were on the average of 10-45 min maximum. The spectral output of the lamp was analyzed with a calibrated Optronic model 742 spectroradiometer (Optronics Laboratories Inc., Orlando, FL, USA). Before and after each experiment radiation fluences were monitored by an International Light Radiometer (IL 1700, calibrated against the spectroradiometer). The emission spectrum was between 340 and 450 nm with a broad peak between 360 and 410 nm and, therefore, this will be termed UVA I herein even though there is a very small amount of visible light present.

The solar simulator used was a 15S single port Solar UV Simulator filtered with Schott WG320/1 mm and UG 11/1 mm filters (290-400 nm) emitting energy similar to that of the overhead sun up to 400 nm (Solar Light Company, Philadelphia, Penn. USA). The dose rate used was between 13-16 W/cm^2 and irradiation periods averaged between 2 and 4 min. Radiation exposure (intensity and dose) was continuously controlled by a Dose Control

System (Solar Light Company) directly attached to the 15S Solar Simulator which automatically operates a shutter system that closes for a preset dose.

Healthy volunteers between the ages of 24 and 36 with skin type II or III were used for this study under informed consent and the nature of the study, procedures and possible side effects were fully explained (Table 2). Separate areas of the buttocks were subjected to increasing doses of either UVA I (25, 50, 75 and 100 J/cm^2 or solar simulating radiation (Solar Light 15 S; 30, 40, 50, 60, 70, 80, 100, 120 mJ/cm^2). Biopsies of 4 mm were taken under local anesthesia (1% lidocaine) at 24 hours following the UV exposure to ½, 1 and 2 minimal erythema doses (MED) and from adjacent non-irradiated control sites at various distances from the irradiation sites.

Immunohistochemistry for pyrimidine dimers was performed on frozen sections. The frozen sections were subjected to a mild alkaline hydrolysis (0.7 N NaOH for 4 min) followed by digestion with proteinase K (1 µg/ml for 10 min). The sections were then fixed with 4% paraformaldehyde for 30 min at 25° C and washed with PBS 3 times for 10 min each. Thereafter, all incubations were done in a humidified chamber in the dark unless otherwise specified. Tissue sections were incubated with 0.1% phenylhydrazine in PBS for 60 min at 37°C to block endogenous peroxidases and washed 2 times for 5 min each. Non-specific binding was blocked by an incubation for 2 hr at 25°C with a solution of PBS containing 5% fetal calf serum (FCS), 7% normal goat serum (NGS) and 0.1% Triton X 100. They were then incubated overnight at 4° C with H3 monoclonal specific antibodies against thymine dimers (generously offered by Dr. Len Roza, Netherlands) at a 1:50 dilution in PBS containing 5% FCS, 5% NGS and 0.1% Triton X 100. The following morning, tissue sections were washed 3 times for 10 min each in PBS and the sections treated with biotinylated goat anti-mouse at 1:200 in a solution of PBS with 5% FCS, 1% NGS and 0.1% Triton X 100 for 3 hr at 25°C. Tissue sections were washed 4 times for 5 min each in PBS and then treated with Vectastain ABC® (Vector, Burlingame,CA) as indica- ted by the company but diluted in 1% Triton X 100 for 3 hr at 25° C. After this incubation, tissue sections were washed 3 times for 10 min each in PBS and treated with 0.5 mg/ml 3,3'-diaminobenzidine with 0.32 µl 30% H_2O_2 added just before an incubation of 1-2 min. The samples were washed for 5 min under running water. They were counterstained with Papanicolaou (Harris' Hematocylin solution), dehydrated and mounted with Merckoglas® (Merck, Geneva,

Switzerland). All samples were treated at the same time. The antibody stained thymine dimers were visible as a brown coloration.

Table 5: VOLUNTEERS FOR IMMUNOHISTOCHEMICAL DETECTION PYRIMIDINE DIMERS IN UV-IRRADIATED SKIN IN SITU

	Volunteer	Age	Sex	Skin Type
UVA Radiations:				
UVA I				
	V1	26	M	III
	V2	36	M	II
	V3	25	M	II
	V4	28	M	II
	V5	28	F	II
	V6	27	F	II
	V7	30	M	III
Solar UV Simulators:				
Solar Light Co. 15S				
	V8	25	M	II
	V9	24	M	III

Control experiments to assure specificity for H3 antibodies were performed with T4-endonuclease V. Tissue sections were treated in 1 M KCL, 20 mM Tris, pH 7, 0.3% Triton X-100 to remove proteins attached to DNA strands. The sections were then incubated with 10 µg/ml of T4 endonuclease V in PBS in order to block dimer antibody binding sites before continuing the immunochemical method for dimer detection.

4.3. Induction of Pyrimidine Dimers Following UVAI and Solar Simulating Radiation

With avidin-biotin immunohistochemistry techniques using a monoclonal antibody specific for thymine dimers we explored the specific regions of human skin where these dimers localize and to what extent UVAI radiation (320-400 nm) modifies their formation in comparison to solar simulating radiation. Already following sub-erythemal UVB (solar simulating) doses, approximately 10-25% of the epidermal cells showed staining for pyrimidine dimers in any given individual tested as shown (data not shown). When comparable erythema doses were given, both UVA I and solar simulating radiation induced a similar quantity of pyrimidine dimers (Table six).

Table 6: Expression of pyrimidine dimers 24 hr post-UV irradiation of human skin in situ

		UVA I	Solar Simulator
Pyrimidine dimers			
MED:	0	-	-
	0.5	ND	++
	1	+++	+++
	2	++++±	++++
Time Post-UV:	-		
	0	-	-
	6	++++	ND
	24	++++±	ND
	48	+++±	ND

Multiple stained sections were used to analyze each skin biopsy semi-quantitatively. The staining was interpreted as very strong ++++, strong +++, moderate ++, weak + or no staining -. These observations were noted by two to three investigators independently.

Furthermore, the distribution of pyrimidine dimer induction throughout all layers of the epidermis was equivalent. In contrast, UVA radiation was shown to efficiently induce pyrimidine dimers well into the reticular dermis, a process not seen with solar simulating radiation. To assure specificity of the H3 antibodies, experiments were performed on the UV-irradiated tissue sections with T4 endonuclease V. This treatment removed all H3 antibody binding sites on UV-irradiated sections indicating a specificity towards pyrimidine dimers.

The sensitive *in situ* immunohistochemical assay for thymine dimers has several advantages over the previously performed quantitative measurements. Firstly, one is able to localize the dimers to the different layers of human skin and to assess the extent of their presence already before UV treatment. Previous detection methods using the action of *Micrococcus luteus* UV-endonuclease on isolated DNA, which induces a specific break at dimer sites, does not allow for a precise assessment of initial numbers of lesions. Secondly, dimers in dermal interstitial cells are readily seen with immunohistochemistry, but can not be detected in dermal tissue subjected to DNA extraction procedures (unpublished observation, this laboratory).Unfortunately, quantitation of pyrimidine dimers is rendered difficult due to the tissue processing. As the frozen sections used were subjected to alkaline hydrolysis, there exist obvious holes in the tissue sections where nuclei have disappeared. It was therefore preferable only to semi-quantitate our data.

Many of the UVA induced dimers persist for at least two days. Following UVB exposures, other studies have seen persistence for several days to weeks (Dr. Antony Young, personal communication).

4.4. Persistence of Pyrimidine Dimers Following UVA I Radiation

Pyrimidine dimer induction by UVB has been shown to be efficiently repaired in normal human skin cells by excision repair. In one report it was shown that 50% of pyrimidine dimers were excised one hr post-UVB [39], but other extensive studies have shown 50% dimer removal anywhere from 6 to 21 hr following UVB radiation depending on the individual [40]. We have seen in both the dermis and epidermis, the persistence of the induced dimers up to 24 hr post-UVA as shown in Table II. In the 48 hr biopsies, there was decreased antibody staining for thymine dimers in the epidermal keratinocytes and throughout the dermis although a remarkable number of dimers could still be detected 2 days following UV exposure.

4.5. Presence of Pyrimidine Dimers at Various Distances from UV-Irradiated Sites

Investigating pyrimidine dimer induction following UVA irradiation, we found that a considerable number of pyrimidine dimers were also present in the non-UVA-irradiated control skin taken from regions 1-2 cm away from the irradiation sites of the buttock skin as depicted in Fig. 5.

The dimers were seen throughout the epidermis; fewer dimers were visible in the rete ridge zone of the dermis. Biopsies analyzed from regions 1-2 cm away from UVB irradiated sites did not show the presence of pyrimidine dimers as seen following UVA radiation (data not shown). At 10 cm from the UVA irradiated sites there were no dimers detected.

The presence of pyrimidine dimers at 1-2 cm distances from the actual UVA-irradiated site is an unexpected finding, which can not yet be satisfactorily explained. The most obvious possibility is that of light scattering in the tissue although 2 cm does seem to be extreme for UVA wavelengths. This possibility would be compatible with the observation that solar simulator (primarily UVB) radiation does not produce the same effect. Another possibility for dimers being induced at distances from the irradiated site is that infrared radiation is capable of inducing the effect. Such a phenomen, with red to infrared was shown in a study by Albrecht-Buehler where cells were capable of communication at distances by a "rudimentary form of cellular vision" [41]. Another possibility that we can propose for dimers being present at distances from the irradiated site is the induction of pyrimidine dimers indirectly by the secretion of a factor or cytokine. If a chemical such as trimethyl-1,2 dioxetame can induce pyrimidine dimers in DNA *in vitro* [42], it can not be completely ruled out that photosensibilsating substances may also permit the same reaction to occur *in vivo*. In this respect in the UVB, dimer yields are influenced by the irradiation temperature [43], and the nucleosomal linker DNA is enriched in dimer content by a factor of 2-4 relative to bulk DNA [44]. In contrast, neither temperature [45] nor nucleosomal structure affect dimerization in the UVC-region.

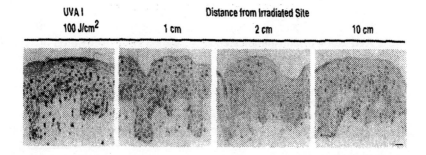

Figure 5: Antibody staining of dimers in human skin in situ at various distances from UVA I irradiation sites.

Dimer antibody staining in irradiated human skin (1 minimal erythema dose) and at various distances away from the irradiated site a) 1 cm, b) 2 cm, c) 10 cm. UV-treatment and staining was as described in Material and Methods. Bar is equal to 5 μm.

4.6. Pyrimidine Dimers as Photon Storing Vehicles in DNA in Order to Convert Light Energy into Biochemical Signals

A new hypothesis is proposed on the basis that light itself is responsible for this effect (see scheme 1).

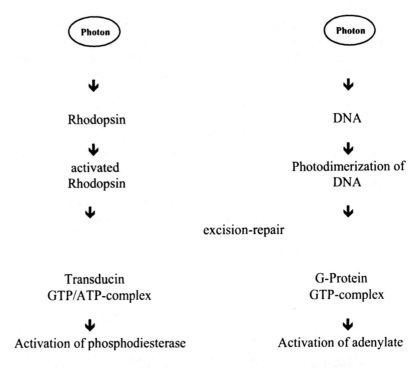

Scheme 1: Pyrimidine dimers as a powerful light trapping system in DNA in order to convert light,energy into biochemical signals .
Proposed mechanism of pyrimidine dimers as a powerful light trapping system in DNA in order to convert light energy into biochemical signals in comparison to the mechanism of the light-driven photoisomerization of the chromophore retinal discovered 30 years ago in bacteriorhodopsin [34].

In this respect, spontaneous ultraweak photon emission (PE) has been extensively described in yeast, plant and animal cells [15-17,27-29]. In a recent report, experiments with cultured human cells in which normal and DNA-excision-repair-deficient XP cells were UV-irradiated in medium and balanced salt solution (BSS) were assessed for ultraweak photon emission [25]. There was evidence of induced-photon emission from normal cells in BSS but clear evidence of a UV fluence-dependent emission in XP cells in medium and in BSS. Overall, these results revealed an important difference between normal and XP cells and it was proposed that XP cells are unable to store ultraweak photons which are efficently trapped in normal cells and perhaps used to regulate metabolic activity [25]. In the same study, we found in defined stages of the fibroblast differentiation system, which has been recently described in detail by Bayreuther and co-workers [32,35], that

UV-light elevates photon emission in MMC-induced postmitotic XP-fibroblasts at least by a factor of 2 compared to mitotic XP-cells. As depicted in Fig. 3, these ultraweak photons in human skin fibroblasts from XPA-cells are efficiently induced with light in the UVA region (325-420 nm). As shown by Kay and co-workers [46] circadian oscillators are present throughout the body of drosophila which are switched on most probably by light. This study confirms the recent results of Cohen and Popp [47] which found specific biological rhythms for the ultraweak photon emission in the hands and forehead of the human body. In this respect Sancar discovered that plant and mammalian cells have blue-light sensitive proteins that have nothing to do with DNA repair [48]. It is very likely, however, that not only proteins but also nucleic acids, as shown in Scheme 1, act as chromophores in the genetic material. Photon release within the cell may be via free radical reactions as proposed by Voeikov [49] and an important part of this energy may be pumped via excimer reaction in a kind of light storage process into DNA [22,50,51]. As shown by Birks [23] excimers are the precursor of pyrimidine photodimers which are known to be biochemical signals for activation of chemical reaction within the cell [24,26]. A similiar mechanism for activation of biochemical reactions is found as the light-driven photoisomerization of the chromophore retinal discovered 30 years ago in bacteriorhodopsin [34] and subsequently confirmed in the visualisation pathway of the eye. The last hypothesis would support that dimer formation and excision could be an expression of epidermal communication and not only a sign of UV damage as shown recently by Applegate et al. [24]. As discovered by Gilchrest and co-workers [26], UV-induced pyrimidine photodimers are stimulating melanogenesis confirming therefore that photodimers initiate biochemical processes and serve indeed as molecular signals. In conclusion, our data show experimental evidence that the genetic material is involved in the ultraweak photon emission process and therefore in intra- and extracellular biocommunication. Furthermore, our results imply that the complex mechanism of UV-induced DNA damage and repair may be a powerful light trapping system in DNA in order to convert light energy into biochemical signals and are therefore involved in an intra- and extracellular photobiostimulation.

ACKNOWLEDGEMENTS

This study was funded in part by grants from the Swiss League Against Cancer (KFS 695-7-1998) and the Erwin Braun Foundation. HN was generously supported by the H +M Kunz Stiftung (Urdorf, Switzerland).

REFERENCES

[1] Gurwitsch, A.G., Grabje, S. and Salkind, S. (1923) Die Natur des spezifischen Erregers der Zellteilung. *Arch. Entw. Mech.*, **100**, 11-40.

[2] Niggli, H.J. (1992) Ultraweak photons emitted by cells: biophotons. *J. Photochem. Photobiol. B: Biol.*, **14**, 144-146.

[3] Hollaender, A. and Klaus, W. (1937) An experimental study of the problem of mitogenetic radiation. *Bull. Nat. Res. Council* **100**, 3-96.

[4] Colli, L., Facchini, U., Guidotti, G., Dugnani Lonati, R., Arsenigo, M. and Sommariva, O. (1955) Further measurements on the bioluminesence of the seedlings. *Experientia,* **11**, 479-481.

[5] Konev, S.V., Lyskova, T.I. and Nisenbaum, G.D. (1966) Very weak bioluminescence of cells in the ultraviolet region of the spectrum and its biological role. *Biophysics,* **11**, 410-413.

[6] Popov, G.A. and Tarusov, B.N. (1963) Nature of spontaneous luminescence of animal tissues. *Biophysics,* **8**, 372.

[7] Zhuravlev, A.I., Tsvylev, O.P. and Zubkova, S.M. (1973) Spontaneous endogeneous ultraweak luminescence of the mitochondria of the rat liver in conditions of normal metabolism. *Biophysics*, **18**,1101.

[8] Quickenden, T.I. and Que-Hee, S.S. (1976) The spectral distribution of the luminescence emitted during growth of the yeast Saccharomyces cerevisiae and its relationship to mitogenetic radiation. *Photochem. Photobiol.*, **23**, 201-204.

[9] Chang, J.J., Fisch, J. and Popp, F.A. (1998) *Biophotons*, Kluwer Academic Publishers, Boston.

[10] Inaba, H., Shimizu, Y., Tsuji, Y. and Yamagishi, A. (1979) Photon counting spectral analyzing system of extra-weak chemi- and bioluminescence for biochemical applications. *Photochem. Photobiol.*, **30**, 169-175.

[11] Chwirot, W.B., Cilento, G., Gurwitsch, A.A., Inaba, H., Nagl, W., Popp, F.A., Li, K.H., Mei, W.P., Galle, M., Neurohr, R., Slawinski, J., Van Wijk, R.V. and Schamhart, D.H.J. (1988) Multi-author review on Biophoton emission. *Experientia*, **44**, 543-600.

[12] Hideg, E. and Inaba, H. (1991) Biophoton emission (ultraweak photon emission) from dark adapted spinach chloroplasts. *Photochem. Photobiol.,* **53**, 137-142.

[13] Slawinska, D. and Slawinski, J. (1983) Biological chemiluminescence, *Photochem. Photobiol.*, **37**, 709-715.

[14] Panagopoulos, I., Bornman, J.F. and Björn, L.O. (1990) Effects of ultraviolet radiation and visible light on growth, fluorescence induction, ultraweak luminescence and peroxidase activity in sugar beet plants. *J. Photochem. Photobiol. B: Biol.,* **8**, 73-87.

[15] Van Wijk, R. and Van Aken, H. (1991) Spontaneous and light-induced photon emission by rat and by hepatoma cells. *Cell Biophys.,* **18**, 15-29.

[16] Scholz, W., Staszkiewicz, U., Popp, F.A. and Nagl, W. (1988) Light-stimulated ultraweak photon reemission of human amnion cells and wish cells. *Cell Biophys.,* **13**, 55-63.

[17] Grasso, F., Grillo, C., Musumeci, F., Triglia, A., Rodolico, G., Cammisuli, F., Rinzivillo, C., Fragati, G. Santuccio, A. and Rodolic, M. (1992) Photon emission from normal and tumor human tissues. *Experientia,* **48**, 10-13.

[18] Cadenas, E. (1984) Biological chemiluminescence. Photochem. *Photobiol.,* **40**, 823-830.

[19] Wright, J. R., Rumbaugh, R. C., Colby, H.D. and Miles, P. R. (1979) The relationship between chemiluminescence and lipid peroxidation in rat hepatic microsomes. *Arch. Biochem. Biophys.,* **192**, 344-351.

[20] Devaraj, B., Scott, R.Q., Roschger, P. and Inaba, H. (1991) Ultraweak light emission from rat liver nuclei. *Photochem. Photobiol.,* **54**, 289-293.

[21] Cadenas, E. and Sies, H. (1982) Low level chemiluminescence of liver microsomal fractions initiated by tertbutyl hydroperoxide, *Eur. J. Biochem.*, **124**, 349-356.

[22] Nagl, W. and Popp, F.A. (1983) A physical (electromagnetic) model of differentiation. Basic considerations. *Cytobios,* **37**, 45-62.

[23] Birks, J. B. (1975) Excimers, *Rep. Progr. Phys.,* **38**, 903-974.

[24] Applegate, L. A., Scaletta, C., Panizzon, R., Niggli, H. and Frenk, E. (1999) In vivo induction of pyrimidine dimers in human skin by UVA radiaition: Initiation of cell damage and/or intercellular communication? *International J. of Mol. Medicine,* **3**, 467-472.

[25] Niggli, H. J. (1993) Artificial sunlight irradiation induces ultraweak photon emission in human skin fibroblasts. *J. Photochem. Photobiol. B: Biol.,* **18**,281-285.

[26] Eller, M. S., Yaar, M. and B.A. Gilchrest (1994) DNA damage and melanogenesis. *Nature,* **372**, 413-414.

[27] Popp, F.A., Li, K.-H. and Gu, Q. (1992) *Recent Advances in Biophoton Research and its Application,* World-Scientific, Singapore.

[28] Niggli, H.J. (1998) UV-induced DNA damage and repair: A powerful light trapping system in DNA in order to convert light energy into biochemical signals. In *Biophotons,* Chang, J.J., Fisch, J. and Popp, F.A. (eds.), Kluwer Academic Publishers, Dordrecht, Netherland, 79-86.

[29] Niggli, H. J. (1996) The cell nucleus of cultured melanoma cells as a source of ultraweak photon emission. *Naturwissenschaften,* **83**, 41-44.

[30] Bayreuther, K., Rodemann, H.P., Hommel, R., Dittman, K., Albiez, M. and Francz, P.I. (1988) Human skin fibroblasts in vitro differentiate along a terminal cell lineage. *Proc. Natl. Acad. Sci., USA*, **85**, 5112-1516.

[31] Niggli, H.J., Bayreuther, K., Rodemann, H.P., Röthlisberger, R. and Francz, P.I. (1989) Mitomycin C-induced postmitotic fibroblasts retain the capacity to repair pyrimidine photodimers formed after UV-irradiation. *Mutation Res.*, **219**, 231-240.

[32] Niggli, H.J. (1993) Aphidicolin inhibits excision repair of UV-induced pyrimidine photodimers in low serum cultures of mitotic and mitomycin C-induced postmitotic human skin fibroblasts. *Mutation Res.*, **295**, 125-133.

[33] Ziegler, A., Jonason, A. S., Leffell, D. J., Simon, J. A., Sharma, H. W., Kimmelman, J., Remington, L., Jacks, T. and Brash, D. E. (1994) Sunburn and p53 in the onset of skin cancer. *Nature*, **372**, 773-776.

[34] Oesterhelt, D. and Stoeckenius, W. (1971) Rhodopsin-like protein from the purple membrane of Halobacterium halobium. *Nature*, **233**, 149-151.

[35] Niggli, H.J. and Francz, P.I. (1992) May ultraviolet light-induced ornithine decarboxylase response in mitotic and postmitotic human skin fibroblasts serve as a marker of aging and differentiation? *Age*, **15**, 55-60.

[36] Rodemann., H.P., Bayreuther, K., Francz, P.I., Dittmann, K. and Albiez, M. (1989) Selective enrichment and biochemical characterization of seven fibroblast cell types of human skin fibroblast population in vitro, *Exp. Cell Res.*, **180**, 84-92.

[37] Niggli, H.J. and Applegate, L. A. (1997) Glutathione response after UVA irradiation in mitotic and postmitotic human skin fibroblasts and keratinocytes, *Photochem. Photobiol.*, **65**, 680-684.

[38] Popp, F.A., Li, K.-H. and Gu, Q. (1992) *Recent Advances in Biophoton Research and its Application*, World-Scientific, Singapore.

[39] D' Ambrosio, S.M., Slazinski, L., Whetstone, J.W. and Lowney, E. (1981) Photorepair of pyrimidine dimers in DNA in human skin in vivo. *J. Invest. Dermatol.*, **77**, 311-313.

[40] Niggli, H.J. and Cerutti, P.A. (1983) Cyclobutane-type pyrimidine photodimer formation and excision in human skin fibroblasts after irradiation with 313-nm ultraviolet light. *Biochemistry*, **22**, 1290-1295.

[41] Albrecht-Buehler, G. (1992) Rudimentary form of cellular 'vision'. *Proc. Natl. Acad. Sci. USA*, **89**, 8288-8292.

[42] Lamola, A.A. (1971) Production of pyrimidine dimers in the dark. *Biochem. Biophys. Res. Commun.*, **43**, 893-898

[43] Niggli, H.J. and Cerutti, P.A. (1983) Temperature dependence of induction of cyclobutane-type pyrimidine photodimers in human skin fibroblasts after irradiation with 313-nm ultraviolet light. *Photochem. Photobiol.*, **37**, 467-469.

[44] Niggli, H.J. and Cerutti, P.A. (1983) Nucleosomal distribution of thymine photodimers following far and near-ultraviolet irradiation. *Biochem. Biophy. Res. Commun.*, **105**, 1215-1223.

[45] Niggli, H.J. (1986) A new sensitive assay for detecting thymine containing pyrimidine dimers using HPLC. In *Proceedings of the first International meeting on cosmetic dermatology in Rome, Italy, March 7-9, 1985*, Morganti, P. and Montagna, W. (eds.) Ediemme, Rome, 335-342.

[46] Plautz, J.D., Kaneko, M., Hall, J.C. and Kay, S.A. (1997) Independent photoreceptive circadian clocks throughout Drosophila. *Science*, **278**, 1632-1635.

[47] Cohen, S. and Popp, F.A. (1997) Biophoton emission of the human body. *Photochem. Photobiol. B: Biol.*, **40**, 187-189.

[48] Miyamato, Y. and Sancar, A. (1999) Vitamine B2-based blue-light photoreceptors in the retinohypothalamic tract as the photoactive pigments for setting the circadian clock in mammals. *Proc. Natl. Acad. Sci. U.S.A*, **95**, 6097-6102.

[49] Voeikov, V. (2001) Reactive oxygen species, water, photons and life. *Rivista di Biologia/Biology Forum*, **94**, 193-214.

[50] Niggli, H.J., Scaletta, C., Yan, Y., Popp, F.A. and Applegate, L. A. (2001) UV-induced DNA damage and ultraweak photon emission in human fibroblastic skin cells: parameters to trigger intra- and extra-cellular photobiostimulation. *Trends in Photochem. & Photobiol.*, **8**, 53-65.

[51] Niggli, H.J., Scaletta, C., Yan, Y., Popp, F.A. and Applegate, L. A. (2001) Ultraweak photon emission in assessing bone growth factor efficiency using fibroblastic differentiation. *Photochem. Photobiol. B: Biol.*, **64**, 62-68.

Chapter 12

BIOPHOTONS – BACKGROUND, EXPERIMENTAL RESULTS, THEORETICAL APPROACH AND APPLICATIONS

Fritz-Albert Popp
International Institute of Biophysics,
Ehem.Raketenstation, Kapellener Straße o.N.,
D-41472 Neuss (Germany)
E-mail: iib@lifescientists.de

1. HISTORICAL BACKGROUND

As an outstanding developmental biologist of the third decade of the 20th century, the Russian scientist Alexander Gurwitsch [1, 2] tried to solve one of the most crucial problems of the life sciences, i.e. the "Gestaltbildungs" – problem, which is the question of how living tissues transform and transfer information about the size and shape of different organs. Since chemical reactions do not contain spatial or temporal patterns a priori, Gurwitsch claimed that a "morphogenetic field" is responsible for the regulation of cell growth. In particular, in his so-called "Grundversuch" ("basic experiment"), he found ample indication for the involvement of photons in the stimulation of cell division. Fig. 1 displays this famous "Grundversuch" of A. Gurwitsch.

He used the stem of an onion root as a "detector" and the tip of another one, very near to the detector but not actually touching it, as an "inductor". The subject of observation was the cell division rate at just the region of the stem where the tip pointed onto it. It turned out that the cell growth on this region of the stem did not change in the case of normal "window glass" being squeezed between inductor and detector. However, as soon as the window glass was substituted with a quartz glass plate (which is transparent for UV light of about 260 nm), the cell division rate (number of mitoses) increased significantly. Gurwitsch interpreted this effect as the mitotic activity of single photons of about 260 nm, triggering cell divisions. He called this photon

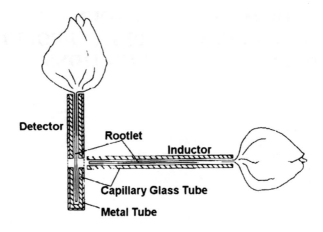

Figure 1: Arrangement of Gurwitsch's experiment with onion roots.

emission from biological systems "mitogenetic radiation" and repeated the experiments successfully also with other biological systems, e.g., yeast.

However, despite confirmation of his results, also shown in a paper by the later Nobel laureate D. Gabor [3], the scientific community forgot Gurwitsch's work in view of (i) some (unessential) objections that came up, (ii) the rather difficult experimental work in this field involving the lack of appropriate photon counting systems, and (iii) a fast developing biochemistry which tried to explain cell growth in terms of hormones and similar biomolecules. "Mitogenetic radiation" was therefore considered some kind of artifact.

After World War II, technical devices for measuring single photons improved through the development of photomultipliers. Russian biophysicists, and others, too, confirmed the existence of a "dark luminescence" of all living systems in the visible range, which could not be explained in terms of heat radiation. The rather viable work of the Russian groups who published mainly in the Russian journal *Biophysics* (translated in the USA) has been reviewed by Ruth [4]. However, about 1970 the Russians stopped their activities in this field and turned to more practical questions about photosynthesis. Apart from biochemists in Poland around the researcher Slawinski [5], the Russian work received almost no attention.

In the Western world "low-level luminescence" of living systems did not ever become a serious subject of fashionable science. With the exception of groups around Inaba (Japan) [6], Boveris (USA)[7], and Quickenden (Australia)[8], this phenomenon of single photons from active biological tissues was completely disregarded or even disreputed. In cases where this non-thermal photon emission has been accepted at all, the common opinion reflected the statements of the Russian biophysicist Zhuravlev [9] and the American chemist Seliger [10], i.e. their hypothesis that "weak bioluminescence" originates from "imperfections" in metabolic activity. This means that occasionally photons should be emitted since the living system is in the situation of a permanent excited state subject to falling back into thermal equilibrium. Under these conditions, some scientists considered to be obvious that highly reactive compounds such as radicals and oxidation reactants are the most likely candidates for photon sources.

Around 1970, a new idea came up in our group at the university in Marburg (Germany) when we found significant correlations between some optical properties of biomolecules (including polycyclic hydrocarbons) and their biological efficacy (including carcinogenic activity)[11, 12, 4, 13, 14]. The basic question was whether the excited states of biomolecules are responsible for the light emission or whether the photon field in living systems is the regulator for the excitation of the biological matter. This problem is similar to the question "Which came first? The chicken or the egg?"

In contrast to the purely biochemical point of view, this search for the original regulator could be approached in terms of information transfer in biological systems supported by increasing understanding of quantum optics, in particular in the non-classical range. First, I will confine myself to the most essential experimental results that have been obtained by careful investigation of "low-level luminescence" in the last 25 years. Then, I will show that the more physical basis of interpretation provides a rather consistent picture of this universal phenomenon of weak photon emission from living systems. Last, some practical applications of "biophotonics" will be discussed.

2. MEASUREMENTS OF ESSENTIAL PROPERTIES OF BIOPHOTONS

Biophotons are measured by detectors based on photomultiplier techniques. These instruments provide both high sensitivity and resolution. Our single photon counting system functions at a sensitivity of about 10^{-17} W and a signal-to-noise ratio of at least 10. The cathode of an EMI 9558 QA photomultiplier is sensitive within the range of 200 to 800 nm. The noise is reduced by inserting the multiplier into a cooling jacket, where copper wool provides thermal contact. In addition, a grounding metal cylinder protects the multiplier from electric and magnetic fields. In order to prevent the multiplier from freezing, the whole tube together with the cooling jacket is kept in a vacuum. Thus, the quartz glass in front of the multiplier tube has no thermal contact with the cooled cathode and cannot become covered with moisture. An optimal cooling temperature is produced at about -30° C. With the use of a chopper, the equipment is able to register a real current density of 2 photons/(s cm^2) at a significance level of 99.9% within 6 hours. A detailed description of the method has been presented elsewhere [4]. Fig. 2 displays an implementation of the equipment.

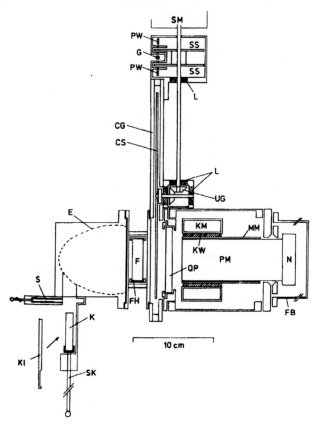

Figure 2: The measuring equipment. PM –photomultiplier; CS-disc of the chopper; CG – housing of the chopper-disc; E – ellipsoid; F – filters; FB – faraday cup; FH – filter holder; G – lamp; K– test quartz glass; Kl – flap: L– ball bearing; KM – cooling jacket; KW – Copper wool; MM – metal cylinder; N – network; PW – photosensitive resistor; QP – quartz glass; S – slide to close the ellipsoid; SK – rod to move up the test glass; SM, UG – geared motor; SS – sector-discs.

We report here only results that have been reproduced several times, and that have been confirmed by different groups. Thus, the essential characteristics of biophoton emission may be summarized as follows[15, 16, 17]:

- The total intensity i from a few up to some hundred photons/(s cm^2) indicates that the phenomenon is quantum physical, since fewer than about 100 photons are ever present in the photon field under investigation.

- The spectral intensity $i(v)$ never displays small peaks around definite frequencies v. Rather, the quite flat spectral distribution has to be assigned to a non-equilibrium system whose excitation temperature $\vartheta(v)$ increases linearly with the frequency v. This means that the occupation probability $f(v)$ of the responsible excited states does not follow a Boltzmann distribution $f(v) = \exp(-hv/kT)$ but the rule $f(v) =$ constant (Fig.3).

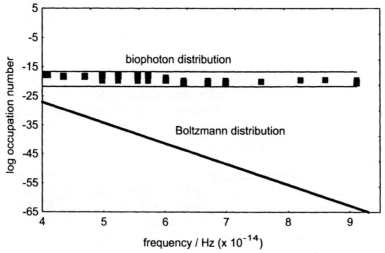

Figure 3: In the case of average occupation numbers we obtain an $f =$ const.-distribution that with increasing frequency displays increasing deviation from the Boltzmann-distribution.

- The probability $p(n,\Delta t)$ of registering n biophotons (n=0,1,2,....) in a preset time interval Δt follows under ergodic conditions, surprisingly accurately, a Poissonian distribution $(\exp(-<n>) <n>^n/n!$, where $<n>$ is the mean value of n over Δt. This holds true at least for time intervals Δt down to 10^{-5} s. For lower time intervals Δt there are no results known up to now [18]. (Fig.4)

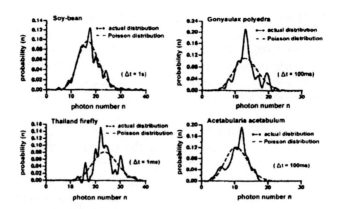

Figure 4: The photocount statistics (= probability $p(n, \Delta t)$) of registering n counts in a preset time interval Δt, where $n= 0, 1, 2, ...$) is very similar for all biological systems. If Δt is sufficiently small such that the mean number of photons in the field becomes lower than about 100, $p(n, \Delta t)$ displays a Poissonian (and sometimes even a sub-Poissonian) photocount distribution. There are 4 different examples with different Δt. 100 measurement values have been used always for evaluation, where the biological state was kept quasi-stationary

- After excitation by monochromatic or white light, the "delayed luminescence" of every biological system relaxes quite slowly and continuously down to "spontaneous" biophoton emission, not according to an exponential function, but with a strong relationship to a hyperbolic-like $(1/t)$ function, where t is the time after excitation (Fig.5)

- The optical extinction coefficient of biophotons passing through thin layers of sea sand and soya cells of various thickness can have values of at least one order of magnitude lower than that of artificial light with comparable intensity and spectral distribution, indicating that this difference cannot be explained in terms of wavelength dependence on extinction [19].

- The biophoton emission displays the typical temperature dependence of physiological functions, such as membrane permeability, glycolysis, and many others. This means that with increasing temperature one gets overshoot reactions, while with decreasing temperature an undershoot response may take place. The resulting „temperature hysteresis loops" of biophoton emission (Fig.6) can be described by a Curie-Weiss law dependence [20].

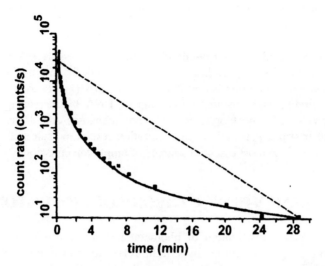

Figure 5: Instead of an exponential decay (dashed line), living cell populations (here tissue of Bryophyllum daigremontanum) exhibit a hyperbolic relaxation of photon intensity after exposure to white-light-illumination. This holds for total as well as for spectral observation (here at 676 ± 10 nm). Under ergodic conditions, hyperbolic decay is a sufficient condition of perfect coherence.

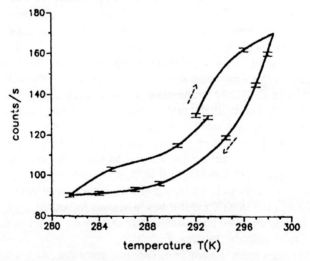

Figure 6: The biophoton intensity of living tissues shows a hysteresis-like dependence on the temperature T, if T is cyclically varied. The example shows etiolated barley, 4 days of germination. The variation of temperature starts at T=292K with the rate $\partial T/\partial t$=0.5K/min. At T=298.5K the rate of temperature change is reversed to $\partial T/\partial$ t= -0.5K/min, and again at T=281.5K with $\partial T/\partial$ t=0.5K/min. This hysteresis-like behaviour of biophoton intensity can truthfully be described as a Curie-Weiss-law-dependence.

- Reactions to stress are frequently indicated by an increase in biophoton emission.

- There is evidence that the conformational states of DNA influence biophoton emission. This has been demonstrated, for instance, by the intercalation of ethidium bromide (EB) into DNA (Fig.7). According to the upwinding and renewed unwinding of DNA by increasing concentrations of EB, the biophoton emission intensifies and drops down in a rather strong correlation. This and other results indicate that chromatine is one of the most essential sources of biophoton emission [21, 22].

3. ELEMENTS OF A THEORY OF BIOPHOTONS

The Poissonian distribution of photocount statistics $p(n,\Delta t)$ under ergodic conditions together with the hyperbolic relaxation function of delayed luminescence is a sufficient condition of a fully coherent photon field [23].

Thus, we can conclude that biophotons originate from a coherent field. It is necessary to note that it is *quantum* coherence and *not classical* coherence, since in the case of classical coherence one would expect a geometrical distribution instead of a Poissonian one, as long as the coherence time τ is not smaller than the preset time interval Δt [24]. This is an important point in understanding biophotons. Take the case of a chaotic field which has typical relaxation time of 10^{-9} s, corresponding to the lifetime of allowable optical transitions. In this time interval a light wave travels over a distance (10 cm), which is large compared to the diameter of a single cell (10^{-3} cm). In other words, it is rather unlikely that biophotons lose their phase information within and also around small cell populations, whether we look at it from the point of view of classical or quantum physics. However, quantum coherence is rather important since it excludes photon bunching and since it points to possible properties of non-classical light, such as antibunching and squeezed states [25]. A theory of biophoton emission is therefore linked to the mechanism of evolution in so far that it has to answer the question of how a single cell could develop from an unstable state of classical coherence to the rather stable state of quantum coherence.

There is agreement among scientists that some necessary conditions have to be fulfilled in order to arrive at this stage. First, the system has to escape thermal equilibrium. The most ingenious way of solving this problem is to *increase* entropy by exposing the cell to an external pump (e.g., sun rays). Instead of the Boltzmann distribution $f(v) = \exp(-hv/kT)$, this open system

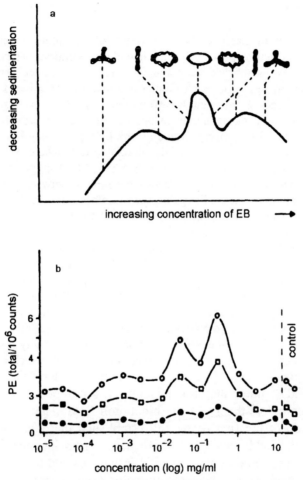

Figure 7: (a) The more ethidiumbromide (EB) is added, the more EB-molecules are inserted between the base pairs of DNA. This intercalation leads to the unfolding of the helix structures of the DNA. The degree of this unfolding is experimentally determined by the sedimentation of the DNA. After complete unfolding further insertion of EB leads to a new upwinding of the DNA helix in the opposite direction. (b) The observation of the biophoton emission after adding EB shows (lower curve: after 1 hour) that the intensity displays the same dependence on the concentration as in (a). This typical profile becomes even more evident after longer measurement time (middle curve: after 3 hours; upper curve: after 5 hours) which indicates a dependence of the biophoton emission on the spatial structure of DNA.

becomes the subject of $f(v) = 1/N$, where N is the number of degrees of freedom [17].The entropy $NlnN$ may then be higher than under thermal equilibrium conditions. The next step is the reduction of N by mode coupling. For $N \to 1$, entropy may approach even 0 without violating the maximum entropy principle.

In order to stabilize $f(v)$ = constant and quantum coherence, suitable matter has to be inserted to fix the boundary conditions. Thus, the f = const. rule can be established by polymers where the distance between the molecular units has to be much smaller than the wavelength of light and where Casimir forces are relevant. It is likely that this role is taken by the DNA. On the other hand, the reduction of degrees of freedom provides some Bose-condensation-like process. Actually, the living cell may have developed to just the degree of coherence of the biophoton field where Bose-condensation and thermal dissipation balance each other [20]. This provides a relatively broad zone of "homeostatic" feedback coupling between matter and radiation, explaining qualitatively "delayed luminescence" and the necessity for permanent biophoton emission and quantitatively the temperature dependence of biophoton emission.

From the more "classical" point of view, this development from low (classical) to higher and higher orders of quantum coherence by spreading out the position- and momentum space corresponds to the expansion of coherent states in terms of more and more efficient storage of external light (and other electromagnetic waves). Consequently, one expects that biological structures are based on cavity resonators and wave guides. Fig.8 displays a striking example. Actually, the whole of mitotic activity can be described in terms of superpositions of cavity resonator waves, which represent at the same time examples of coherent states of the biophoton field.

It should be noted that it is difficult – if not impossible – to substantiate a complete theory of the biophoton field. The reason is that the initiation of the first living cell established the coherent state of the biophoton field, which provided what are now incomprehensible conditions, i.e. certain environmental influences and the presence of certain substances over an indeterminable time, where the probability of the initiation process itself was certainly rather small. However, as soon as one accepts the coherence of the biophoton field as the initial state of the living cell, there is no fundamental problem in understanding the physical properties of biophoton emission as well as its biological significance. The substances of a living cell and – in particular – the maintenance of the special non-equilibrium state may well reflect the capacity for maintaining the coherence of the biophoton field under physiological conditions. Models have been developed on the basis of exciplex excitation [17], chaos models of two-level systems excited under the constraint of the entropy maximum principle [27] and the van der Waals forces of stacked base pair interactions [28], models of optical phase conjugation [29] and of a Hamiltonian that keeps coherent states coherent [30]. All models are consistent with at least a part of the known physical properties of biophoton emission, including delayed luminescence.

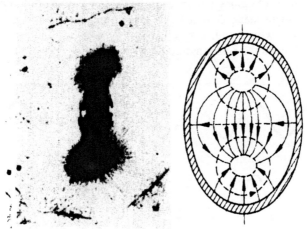

Figure 8: Left side. Completely developed spindle apparatus of a fish (Corregonus) in mitosis. (From: Darlington, C.D., Lacour, L.F.: The Handling of Chromosomes. Allen and Unwin, London, 1960).
Right side. Electric field of TM_{11} cavity modes in a right circular cylindrical cavity. Comparison with Fig.8 left side shows that mitotic figures are striking examples of long-lasting photon storage and coherent fields within biological systems (From: Popp, F.A.: Photon Storage in Biological Systems, In: Electromagnetic Bio-Information, Urban & Schwarzenberg, Muenchen-Wien-Baltimore 1979).

At present we may formulate the role of biophotons in a picture rather than by means of a quantitative theory. Biophotons originate from a coherent field which keeps its coherence through an unknown "homeostatic" coupling between radiation and matter. The electromagnetic field pattern is the driving force for the movement and interaction of the molecules (including the spatio-temporal regulation of biochemical reactivity by suitable excitation of transition state complexes), while matter constitutes the boundary conditions of the field. Field and matter in biological systems are somehow "married" in a way that they control each other by mutual coupling of dynamical interference patterns of the field with the rather flexible material boundaries.

4. BIOLOGICAL IMPACTS

Once the coherence of biophotons is accepted, it is not difficult to predict a variety of biological phenomena, which deviate considerably from the "conventional" point of view, thus providing a reliable basis for examining the theory and for obtaining a more profound understanding of biology.

It is evident that coherent fields give rise to destructive and constructive interference, where with respect to energy conservation law zones of destruction have to be compensated for by zones of construction (Fig.9).

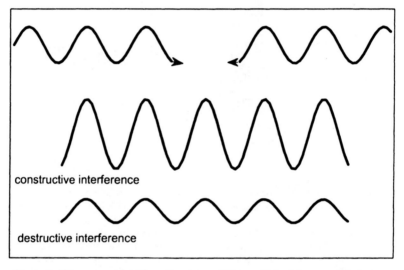

Figure 9: If two waves interfere, the phase relations will lead in general to zones where they amplify mutually ("constructive interference") or alternate ("destructive interference"). For coherent fields, these processes provide a basis of regulation and communication.

According to the theory of R. Dicke [31], there is a preference for constructive interference ("super-radiance") in the initial phase of the interaction between radiation and non-randomly oriented matter of suitable size, while destructive interference ("sub-radiance") dominates after longer periods of time. Consequently, there is always a considerable probability of destructive interference of the biophoton emission of living systems in the space between the living cells.

This means that the biophoton intensity of living matter cannot increase linearly with the number of units, but has to follow the effective amplitudes of the interference patterns of the biophoton field between living systems.

A striking example is the measurements on daphnia [32, 33]. *Daphnia magna* Strauss were put in darkness into water at 18 ^{0}C within the quartz cuvette of the biophoton measuring equipment. We altered the numbers n of daphnia from 1 to 250, always selecting animals of about equal size. After each alteration the intensity of the biophoton emission was registered. Since every one of the inbred animals emits almost the same intensity, one expects a dependence of biophoton intensity on the number of animals like that displayed in Fig.10a. After correction for self absorption, it should not significantly deviate from that of Fig.10a. However, careful measurements showed evidence of the results displayed in Fig.10b. The results from interference patterns of biophotons between the animals under investigation were as expected. There is a tendency for destructive interference resulting in a lower intensity than expected from the linear increase. The most efficient destruction of the biophoton field outside of the animals is obtained at about 110 animals, corresponding to the population density of daphnia in free nature. This zone of most efficient destruction according to the energy conservation law is at the same time the zone of highest efficacy in "storing" light *within* the animals.

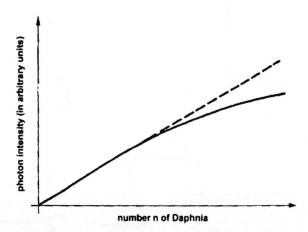

Figure 10a: In case that photon emission from daphnia is dependent on mutual interactions of the animals one expects a linear increase with increasing number of daphnia. This linearity will show a small decline for a large number of animals as soon as self-absorbence has to be taken into account.

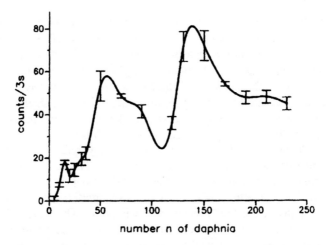

Figure 10b: Mean values of the photon intensity of adolescent daphnia in 15 ml volume with the weighted standard deviation.

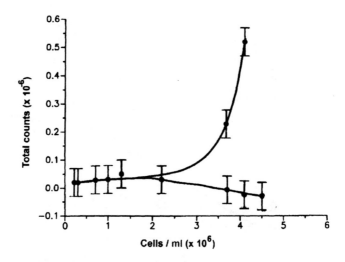

Figure 11: "Delayed luminescence" from tumor cells (upper curve) and normal cells (lower curve), as measured by Schamhart. The different curves can be approximated by a non-linear (cubic) dependence of intensity from cell-number n.

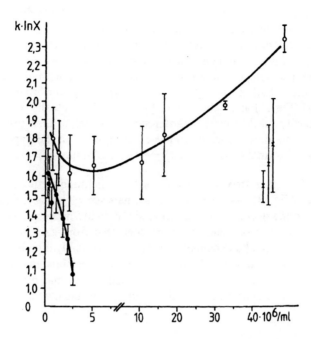

Figure 12: The decay parameter of the hyperbolic approximation that is adjusted to the relaxation dynamics of the afterglow of different cell suspensions after exposure to weak white light illumination is shown versus cell density. The lower curve displays the improvement of hyperbolic relaxation of normal amnion cells with increasing cell density. The upper curve shows the opposite dependence exhibited by malignant Wish cells. The three measurements at the right side of the figure correspond to the nutritive medium alone.

To some extent one is justified in saying that living systems "suck" the light away in order to establish the most sensitive platform of communication. A more detailed description of this phenomenon has been presented elsewhere [29]. Actually, this biocommunication by means of mutual interference of the biophoton field provides necessary information about the equality or difference of species, since similar animals have similar wave patterns. The signal/noise ratio becomes optimized as soon as the wave patterns interfere under maximum destruction between the communicating systems, since every perturbation leads then to an increase (signal) that the connected systems have to become aware of.

This rather ingenious means of biocommunication provides the basis for orientation, swarming, formation, growth, differentiation, and

"Gestaltbildung" in every biological system [17]. On the other hand, as soon as this capacity for coherent superposition of modes of the biophoton field (where longer wavelengths may also be included) breaks down, in the first stage of destruction one expects consequently an increase in biophoton emission (or delayed luminescence) with increasing numbers of living units within a biological population. This was first confirmed by D. Schamhart [34] (Fig.11) and Scholz et al. [35] (Fig.12). Actually, tumor cells loose the capacity for destructive interference according to their loss of coherence. At the same time, delayed luminescence turns from the hyperbolic-like relaxation of normal cells to the exponential one of tumor cells.

A further striking example is the synchronous flickering of dinoflagellates (Fig.13). As soon as these animals see each other, their bioluminescent flickering decreases and displays significantly more synchronous light pulses than in the case when they are separated from each other [36]. This phenomenon can be explained in terms of chemically amplified biophoton emission (which is called "bioluminescence"), establishing destructive interference as soon as the animals "see each other" and displaying synchronous pulses as a consequence of the disruption of the destructive interference patterns.

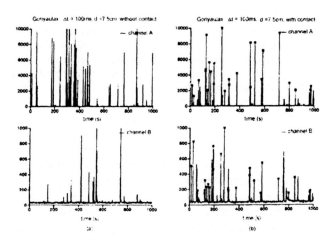

Figure 13: If one separates two cultures of dinoflagellates, their bioluminescence flickering is completely unsychronous (left side). As soon as they are in optical contact, a big amount of flickering is synchronous (right side). The stars indicate the synchronous flashes.

Even bacteria seem to use this kind of "communication" within their nutrition media [37]. Fig. 14 displays one of the measurements on

Enterococcus faecalis. Growing bacteria emit such low biophoton intensity that it cannot be registered, in contrast to the permanent photon emission of their nutrition media. (It is impossible, by the way, to produce nutrition media *without* spontaneous photon emission, originating from oxygenation processes). At a definite number of bacteria, the total intensity of the system drops down as a consequence of active photon absorption of the bacteria within the medium. It may happen then that at higher numbers of bacteria this absorbance disappears.

Again, the destructive interference of bacteria within the coherence volume of the light-emitting nutrient molecules provides an explanation for this obviously rather universal process in living nature.

It should be noted that growth regulation through biophoton emission has to follow a law where in addition to linear stimulation ($\dot{n} \propto n$) a nonlinear inhibition ($\dot{n} \propto n^2$) has to take place. Consequently, the correlation between growth rate and biophoton emission should be based on such a relationship. Fig. 15, as a result of measurements, confirms this connection.

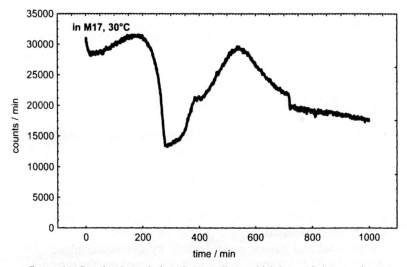

Figure 14: Growing bacteria in culture medium, which by oxidative reactions always emit light, absorb from a definite density on the light of the medium. For higher densities this absorbance may decrease.

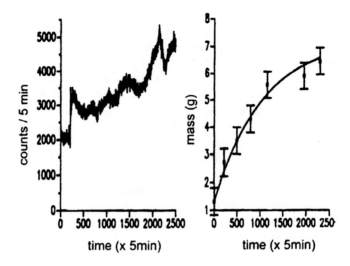

Figure 15: The biophoton intensity from 9 soy-seeds increases during germination in darkness as shown in the left figure. The mass m during germination increases according to the curve in the right figure, where the measured values (points with error bars) have been approximated by the growth curve $m = A/(B+(1-B)\exp(-Ct))$, with $A= 1.3$g, corresponding to the mass of 9 soy-seeds, $B=0.178$ $C=2.6\ 10^{-4}(\text{min}^{-1})$, t representing the time.

5. APPLICATIONS

Since biophoton emission displays a rather reliable and sensitive fingerprint of a living system, the non-invasive measurements of the spectral intensities provide a powerful tool for identifying biological systems as well as for characterizing their response to external influences. This bio-indication by means of biophoton emission (and/or delayed luminescence) is at present probably the most sensitive test for examining the influence of external factors. It has demonstrated, for instance, the differences between the tolerance for artificial and natural plant extracts by unicellular algae (*Acetabularia Acetabulum*) [38]. The usual test methods, based on biochemistry, could not resolve these differences between natural and artificial extracts which, on the other hand, had been long observed in tests on human skin. At the same time, the tolerability of toxic agents can be examined down to lowest concentrations of their limit of chemical detectability.

The advantage over chemical analysis becomes evident in the case of the germination capacity of seeds. While it is impossible to determine their viability by means of content analysis, biophoton emission is strongly correlated to germination capacity. After calibration of the equipment, it is possible to determine the germination capacity of seeds with an uncertainty of only a few percent by registering their biophoton emission [39].

A further striking case that demonstrates the superiority of biophotonics compared to conventional analysis has been documented by investigations of battery farming eggs compared with free-range eggs [40].

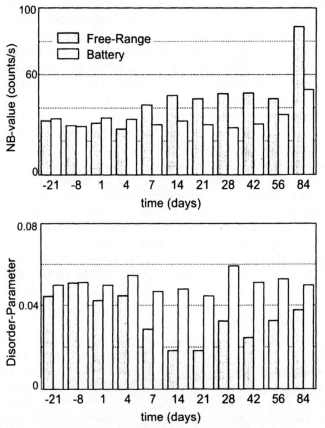

Figure 16: The hens were together in battery farming during the first 3 weeks (-21 up to 1). The NB-values as well as the disorder parameter remain similar during this time. After 3 weeks half of the hens was let free. The NB-values increased then significantly (gray bars, upper curve). The disorder-parameters changed also in the direction of a better adaptation to a hyperbolic decay curve. This adaptation was the best during time of sunshine.

The governmental "Tierversuchsanstalt Celle", responsible for the keeping of animals, sent us blind samples of eggs over some months for measurement of biophoton emission. One group originated from chickens kept in battery farming, the other group from the same species of chicken, fed on the same food but free to go outside onto sand and to return to the coop. For the first three weeks, both – battery farming and free-range chickens – were kept in the same conditions in the coop such that a difference in the quality of eggs would not be expected. And, in fact, as shown in Fig.16, it turned out that the biophoton intensity (delayed luminescence) of the yolk was the same for all chickens. After the three weeks, when one group was released, the eggs from indoor chickens kept their same properties in terms of biophoton emissions, while the free-range chickens' eggs displayed an increase in the intensity of delayed luminescence with increasing time. The disagreement with the hyperbolic relaxation function decreased for the free-range eggs and took the lowest value at a period of sunshine and favorable climate. Later, the weather got dreary, and, in correlation, the disagreement to a hyperbolic relaxation function increased.

Parallel trials conducted to look for differences in the contents of the eggs failed. This example shows that biophoton emission reflects the organizational properties of biological matter (phase relations of electromagnetic interactions between the biomolecules) rather than the molecular contents. They provide a measure of the "order" of the system which may well reflect important parameters of the "quality" of the system under consideration.

Thus, it is not surprising that biophoton parameters correlate significantly with ripening and freshness, and with a corresponding shelf life for fruits, vegetables, and meat. It is simply a question of technical development to make use of all these – not always linear – dependencies.

The method is then useful for optimizing the quality of products by investigating the relationships between the choice of the product, fertilization, harvesting, storage, processing, and the quick and reliable measurements of biophotonic parameters.

In its most refined application, biophotonics through the use of factorial analysis defines the dimensions of the quality of comparable products (e.g., color and smell are two different dimensions) and assigns to the different dimensions a quantitative description of quality.

A rather new field is the application of biophotonics for whole-body measurements, in order to investigate therapeutic effects or environmental influences on human health. (41)

6. IMPLICATIONS FOR AN INTEGRATIVE BIOPHYSICS

The most basic macroscopic laws of physics are energy- (and momentum-) conservation and the Second Law of Thermodynamics. Both are valid for closed systems, and they can also be applied to open ones when one surrounds the open system with a heat bath and applies these laws to the total system. Consequently, essential statements about the nature of living tissues should become available through the application of common macroscopic physics.

On the other hand, the elementary units of life, cells and biomolecules in cells, cannot be taken simply as subjects of macroscopic physics, for several reasons. Among them are:

(1) Essential sub-dimensions of a living system are in the range from some Angström-units, corresponding to the distance of neighbored base pairs in the DNA, up to the total length of some living cells or organisms on the order of meters, able to resonate from X-rays down to radio waves, respectively. Statistical mechanics has limited significance for understanding resonating systems with relaxation times longer than the time intervals of observation, which vary between at least microseconds to years.

(2) Many essential biological relaxation times are either too small or too large compared with time intervals where random phase approximations are useful. This means that in addition to the frequency of interactions, phase may play a dominant role for understanding biological functions, in contrast to ordinary "dead" systems.

(3) Biology, in its material compounds as well as in its radiation fields, is subject to quantum theory, i.e. the dielectric constant of biomolecules and the photocount statistics of biophotons cannot be calculated by classical electrodynamics. Actually, living matter is subject to permanent non-equilibrium excitation, which doesn't allow a simple assigning of a definite temperature to its contents. The number of photons in the field is too small, and their coherence length compared to the size of a cell is too high for getting reliable results from classical statistics.

This does not mean that a classical approach is senseless. Just the opposite--one may expect that biological systems become much more understandable if one compares the results of classical models on the one hand and quantum theoretical ones, on the other hand, to get better insight into the completely unknown mechanisms of their extraordinary regulatory capacities.

Erwin Schrödinger [42] was one of the first scientists to point out that living systems do not violate the energy conservation law and also do not violate the Second Law of Thermodynamics. Actually, food, as one of the most important energy and "negentropy" sources of the living state, provides energy balance.

$$EF + EL = EF^* + EL^*, \qquad (1)$$

where EF and EL are the energy contents of the food and the living system, respectively, before consumption, and EF^* and EL^* are the corresponding energy contents after consumption. One of the purposes of food is $EL^* > EL$, which means $EF^* < EF$.

A similar balance equation holds for entropy SF, SL, SF^*, and SL^*

$$SF^* + SL^* \geq SF + SL \qquad (2)$$

The > sign reflects the Second Law of Thermodynamics. In the case that living systems take „orderliness" from food, we have $SL^* < SL$. It is obvious from (2) that consequently $SF^* > SF$, and that the difference $SF^* - SF$ is a measure of the quality of the food. Food works by organizing the living state. Therefore, Schrödinger called living systems "orderliness- robbers".

One of the most basic and hitherto unsolved problems of the life sciences is the question of where living systems got this capacity of "sucking" orderliness from. There is no reason to believe that they do not follow macroscopically, and after sufficiently long time intervals of observation, the Second Law of Thermodynamics, and I would like to show that--just the opposite--they exhaust the possibilities that are provided by making use of this law.

But let me first return to some historical remarks. It is now generally accepted that in accordance with Schrödinger's point of view, the living system has to turn to an open system, since a necessary condition of establishing order is the uptake of non-random (non-chaotic) energy far removed from the equilibrium conditions of a heat bath. Actually, food provides this kind of energy, which has to be considered for this reason as a message rather than as a "calorie bomb". Well-known examples of this kind of information feeding are the Laser and related regulatory engines, in the most simple case even a refrigerator. All these engines have in common that they are pumped by non-chaotic (non-thermal) energy or under non-chaotic boundary conditions in order to develop some kind of "orderliness".

A decisive step forward in understanding this mechanism was the discovery of "dissipative structures" by Ilya Prigogine[43]. He focused his attention on "chemical clocks"(Beloussov-Zhabotinski-reactions, "Brusselators"), but was at the same time the pioneer of developments such as "deterministic chaos", limit-cycle-models, catastrophe theory, and--last but not least-- Herbert Fröhlich's claim for "long-range coherence" in biological systems[44]. However, the terms "order out of chaos" or "self-organization" were very misleading since they evoked the wrong impression that it is actually possible to violate the Second Law of Thermodynamics. Rather, all these new insights into the development of order in non-living and living matter are based on the idea of autocatalytic creation of order by means of external feeding. The autocatalytic behavior is expressed in terms of non-linear re-couplings within the systems under study. They die out sooner or later if they are not exposed to external pumping. The story of Münchhausen's pigtail does not work in physics, neither for energy nor for entropy. And there is no reason to believe that living systems are exceptions.

However, the question is justified, whether there are basic differences between a refrigerator, a Laser, dissipative structures, and the living state. From the phenomenological point of view there are certainly differences, among them the following:

1. Living systems display from the beginning a remarkable degree of orderliness, if we omit for the moment the question of how the first cells got stable in the first place.
2. Living systems evolve by their own "creativity" in transforming the external pumping into some kind of "order" such as rhythms, cell differentiation, communication inside and with the outside world, and the evolution of consciousness.
3. Living systems are able to select from their surroundings just the necessary tools for stabilizing and improving their coherence.

It seems that in some respects biological systems actually work for at least limited time intervals against the driving force of the Second Law of Thermodynamics, which, of course, does not mean that they actually do violate it.

As long as there is no necessity for acceptance of a principal difference between systems that follow the known laws of physics, and the living state, there is only one possibility of assigning a specific feature to living systems, e.g. an optimization principle for using the known laws of physics in order to establish their survival. This goal reminds us of Neo-Darwinism, but is different in approach. It may remove the character of randomness and allow for the question of meaning or sense, whatever "sense" means. The development of an "ethics" is not excluded, and may even become one of the most decisive consequences of this optimization procedure.

In the case of Schrödinger's "orderliness-robbership", this principle may be satisfied if the energy EF^* is as small as possible, and SF^* is as high as possible, where this "possibility" is again a factor of survival. This means for instance that under almost isolated conditions EF^* should approach 0 and SF^* should increase up to $SL+SF$, while in case of a surrounding population of consumers some balance between the available EF, SF, EL, SL and EF^*, SF^*, SL^* and EF^* should be achieved. As a consequence, this optimization principle requires collective interactions and cannot be based on the properties of individual parts of the system under investigation alone. However, what we can provide is the statement that an optimized system of this kind is not only an absorber of energy but at the same time an absorber and emitter of information. Only under these conditions is optimization possible at all. We may go even a step forward and postulate that a necessary and sufficient condition for this principle is the development of an optimized basis of information under the constraint of the most efficient consumption of energy. This means that a living system has to find an optimal compromise between the two antagonistic properties of stability, on the one hand, and sensitivity, on the other hand. Usually, stable systems are not very sensitive, and sensitive systems are not very stable. However, living systems arrive at least partially at the highest possible sensitivity, limited only by Planck's constant, and at the same time they expose a remarkable stability, not only in the sense of survival of an individual, but even more so as expressed in the survival of the species.

It is clear that optimization principles are subjects of physics, since all physical laws (macroscopic as well as quantum laws) are results of optimization procedures. Therefore, we have to find suitable models, where the application of extremal principles may lead to a better insight into the specific optimization of which biological system make use.

There are two simple models in macroscopic physics that may help us to understand the principle. These are cavity resonators as paradigms of high sensitivity and the entropy concept as the macroscopic basis of the high stability factor of living systems.

A powerful tool for investigating and understanding life is biophotonics, a discipline that is based on the analysis of photons, which are emitted by biological systems. At least in the infrared range (the body's "heat radiation") and in the visible range ("biophotons"), there is ample evidence of a permanent photon current from living tissues. The infrared part is assigned to "black body radiation", originating from the fact that its excitation temperature (which is $\vartheta(\lambda) = hc/(k\lambda) \cdot 1/(\ln(8\pi/\lambda^4 \, cF/I_\lambda +1))$, where h, c, k, λ, F, I_λ denote Planck's constant, the velocity of light, the Boltzmann constant, the wavelength, the area of the subject under investigation, and the spectral photon intensity per unit of time and per unit of wavelength, respectively (see Appendix AII.15), correspond to about 37^0 Celsius. This interpretation is certainly wrong since the body's heat radiation is subject to homeostatic regulation. However, we confine this paper to biophoton emission within the spectral range from about 200 to 800 nm, where we will show that the excitation temperature is far above the thermal temperatures of the surroundings. Thus, the question of whether the radiation is thermal or not does not play a role for biophotons, which are certainly non-thermal.

The important results of biophoton emission analysis are the physical properties of biophotons and so-called „delayed luminescence" (DL) of living and non-living biological tissues, as well as correlations between biophotons, DL (delayed luminescence), and biological functions such as cell growth and differentiation, transformation into tumor cells, response to temperature changes, cell density, and other internal and external influences.

Most important for understanding biophotons in terms of physics, besides the total intensity of the spectral distribution of biophotons, are their photocount statistics (PCS), the relaxation dynamics of spectral delayed luminescence (DL), and the temperature dependence of both biophotons and DL.

7. THE SPECTRAL DISTRIBUTION REVISITED

The most careful measurements on spectral distribution of biophotons have been performed by Ruth [4, 13], who used interference filters and took cucumber seedlings as subjects of investigation. For months he repeated daily measurements of biophoton emission from the same kind of cucumber seedlings (Chinesische Schlangengurken) in the same development stage, standardizing the spectral intensity by always relating it to the total intensity of the system under study. Before starting the measurements, the system had relaxed to a stationary state such that during the measurements no change of the mean intensity took place. Fifteen interference filters of definite wavelengths with only a few nanometer bandwiths, covering the range from 300 to 730 nm, were used. The equipment was calibrated by a standardized thermal lamp, where the spectral intensity had been definitely weakened to the intensity range of biophotons. One count per second corresponds to about 1 photon/(s cm^2) of the radiation source under investigation. The results are displayed in Fig.3.

The reader who is not familiar with Thermodynamics should skip to Appendix II: The Entropy Concept and then come back.

The results of this work agree essentially with all the other known results that have been published on this topic, such that I can confine this discussion to the most characteristic feature of the spectrum of biophoton emission. It turned out that the spectral biophoton intensity – in contrast to heat radiation of living systems – certainly does *not* follow black-body radiation according to (AII.16). Rather, instead of getting

$f = 1/(\exp(h\nu/(kT))-1)$, one obtains with considerably high accuracy the result

$$f = \exp(-v \, ln2) \qquad (3)$$

where we introduced the factor $ln2$ for purposes that become clearer in the following discussion. v takes a value between about 50 and 75, corresponding (1) to the biological variance and (2) to the measurement errors. Fig.3 displays this typical $f \approx$ constant-rule of the spectral distribution of biophotons compared to the f-values of the black-body radiation of a closed system according to AII.16.

There are different possibilities for explaining this important experimental result.

A. The first approach is the introduction of a chemical potential $\mu = h\nu - vkTln2$ which, for $v \gg 1$ would result in (3). This means that biophotons have a chemical potential influencing the order of the system under study in such a way that their energy is transformed into "orderliness"-work. The uptake of photons by the living systems would then reduce the entropy according to

$$(\partial S/\partial N)_{E,V} = (\partial S/\partial E)_{V,N} (vkTln2 - \varepsilon) \qquad (4)$$

This is a remarkable result, since it postulates that photons that are taken up by a living system reduce its entropy, and biophotons that are emitted lead to an increase in entropy. The decrease after uptake requires some "feeding" functions of photons for living systems according to the ideas of Schrödinger, and the increase after emission assigns to biophotons some signal character. However, this interpretation does not take account of the fact that our measurements are performed *outside* of the system, and we do not know much about the intensity inside of the cells. v can get a rather variable value depending strongly on the thickness between the real absorber within the cell and the outside world. Since v can only increase from the inside to the outside, we look at the two limiting cases $v \approx \varepsilon/(kTln2)$ at the virtual outside and $v=1$ at the source and sink of biophotons. In the first case we would come back to black-body radiation, since for $\mu \approx 0$ equ.(AII.16) is valid. (This may happen, for instance, for heat radiation from the body, since $\varepsilon \approx kT$ in the infrared range). However, for $v=1$, we have to go back to the original Bose-Einstein distribution

$$f = 1/[\exp(v ln2)-1] \qquad (5)$$

to arrive for $v=1$ at the value

$$f = 1 \qquad (6)$$

for the source and sink of biophotons.

This is again a remarkable result since it invites the idea that biophotons are originating from a source (and sink) that works around the laser threshold ($f=1$), where all the available states are occupied with the same probability $f = 1$. At the same time, the entropy loss (by absorption) or gain (by emission) of a photon $\Delta S = k ln\ 2$ is then just in agreement with the Brillouin theorem. By scattering processes in the cytoplasm (including random and non-random processes) and absorbance, v increases from $v_0=1$ to maximally about $v\infty = hv/(kT)$, leaving the system with just this overshoot entropy ($v\infty - v_0$) k $ln2$.

B. The second model can explain (3) in terms of an „ideal open system" that does not work under the constraint of energy – and possibly also particle conservation such as is found in closed or isolated systems. Actually, going back to the maximum entropy principle, an ideal open system is defined by (7). It has its highest value of entropy according to (AII.12) for

$$dS - (\partial S/\partial X)_E (\partial X/\partial N)\, dN = 0 \qquad (7)$$

where $(\partial S/\partial X)(\partial X/\partial N) = \sum_i (\partial S/\partial x_i)(\partial x_i/\partial N) = v\ ln2$ (with the variable v) defines the internal boundary conditions of the system according to variables x_i. By definition they shall not explicitly depend on the quantum energy ε of the bosons under consideration. Actually, the biophoton emission is not subject to the energy conservation law like heat radiation is, so that means that the term $(\partial S/\partial E)$ has to be cancelled. In addition, the measured value f is obviously independent of ε, which requires a maximum entropy according to (7).

We arrive at the solution

$$N = C/[\exp(v\ ln\ 2)-1] \qquad (8)$$

in complete agreement with the experimental results.

Let us note here that S according to (8) can get a much higher absolute value than the S of a closed system. This can be seen by the following consideration. Take

$$S = -k \sum_i f_i \ln (f_i), \qquad (9)$$

where i denotes the i-th of the possible energy states of the system. Then it is obvious that for a normalized function S with $\sum_i f_i = 1$, the maximum S is obtained if and only if $f_i = f = 1/M$, and the entropy takes its highest value $S = k \ln M$, where M is the number of modes. Comparison with (8) results in

$$v = \ln(M+1)/\ln 2 \qquad (10)$$

For $v = 1$ we get $M = 1$, for $v = 50$ we have $M \approx 10^{15}$, for $v = 75$ we get $M = 3.8 \ 10^{22}$. This means that if a biophoton is released from a boson condensate in a phase space cell with $f = 1$, it splits on average into a huge amount of different modes with about equal occupation. These modes cover the range from about 10^{15} Hz down to 10^{-7} Hz.

But how is it possible to condensate photons from a thermal equilibrium into a single phase space cell? It seems impossible since one may expect a contradiction to the maximum entropy principle. However, if one connects a Bose-condensating system that sucks up photons with the probability of about $(1-\exp(-\gamma N!))$, where γ is a parameter describing the probability of the collision of a photon with the Bose-condensate of N stored photons, the thermodynamical probability of the combination of a thermal system and the Bose-condensate becomes

$$W = \exp(-\gamma N!)(C+N-1)!/((C-1)!N!) + (1-\exp(-\gamma N!)) \qquad (11)$$

where we provide that the thermodynamical probability W_B of the Bose-condensate equals 1 and that there is no gradient in the photon number between the closed system and the Bose-condensate. It is surprising that the maximum of W is neither at $N = 0$, nor at $N = 1$, but can take in principle every value $N<\infty$. After some straightforward calculations we get the interesting result

$$N_{max} \geq (1/\gamma)(C-1)/(N+1)! \ (W_T/(W_T-1)) \qquad (12)$$

where W_T is the thermodynamical probability of the Bose-Einstein statistics according to (AII.16) or (AII.6). For $W_T \gg 1$, N_{max} is independent of W_T, but only dependent on C and γ. In the case of $\gamma=1$, and inserting the results of spectral intensities of biological systems, we get the relation

$$ln\,(N+1)! \approx v\,ln2 \text{ with } 50 < v < 75 \qquad (13)$$

This corresponds to not more than 20 photons in the field, indicating that the whole subject of biophotons can be treated only by quantum optics. In the case of $v=1$ we have, as expected, $N=1$.

There is not very much known about the dependence of biophoton emission on external temperature. In general, one can expect that the intensity increases at least in the physiological range as soon as the temperature increases. However, this increase does not simply follow an Arrhenius law ($n \propto \exp(-h\nu/(kT))$). Rather, we found a strong dependence on the temporal temperature gradient $\partial T/\partial t$, giving rise to temperature hysteresis loops [45]. The intensity of biophotons, expressed in terms of the excitation temperature

$$\theta(\varepsilon) = \varepsilon/(\varepsilon-\mu)\,T \qquad (14)$$

follows the rule

$$\theta(t) = \theta(t_0)/(1 + \chi_1 ln(1-\chi_2 \dot{T}\,(t-t_0))), \qquad (15)$$

where χ_1 and χ_2 are system-specific constants.
It has been shown that this law is based on a Curie-Weiss-law-like hysteresis. It can be interpreted as a balance between Bose-condensation effects on the one hand and thermal dissipation on the other hand.

There are indications that with the change of temperature there are also temporal changes in the spectral distribution. However, systematic investigations are still lacking.

8. PHOTOCOUNT STATISTICS (PCS) REVISITED [46, 47]

A further characteristic physical property of biophotons is the Poissonian distribution of the probability $p(n, \Delta t)$ of registering n photons ($n = 0,1,2,....$) in a preset time interval Δt. Fig. 4 displays an example. It seems strange at first glance when one has to conclude from this result that biophotons are either coherent or chaotic. How is it possible that two opposite physical properties display the same photocount statistics?

Take first the case of a coherent field. Coherent states are defined as eigenstates of the annihilation operator:

$$a|\alpha\rangle = \alpha|\alpha\rangle, \qquad (16)$$

where $a+$, a, representing the normalized creation and annihilation operator, respectively, satisfy the commutation relations

$$[a,a] = [a+, a+] = 0, [a, a+] = 1, \qquad (17)$$

α is the wave amplitude that is annihilated by a in the coherent state $|\alpha\rangle$.

The product of the uncertainties Δp in the momentum and Δq in the position of a coherent state is a minimum

$$(\Delta p \Delta q) = h/2, \qquad (18)$$

which means that they provide the utmost powerful carriers of information. Conjugated observables such as momentum and position or amplitude and phase of these waves have at the same time the lowest possible uncertainties. In terms of number states

$$a^+ a|n\rangle = n|n\rangle, \qquad (19)$$

where n is a definite number of particles (like photons) in the field, coherent states are expanded as

$$|\alpha\rangle = \sum |n\rangle\langle n|\alpha\rangle = \langle 0|\alpha\rangle \sum_n \alpha^n/(n!)^{1/2}|n\rangle \qquad (20)$$

which means that the probability $p(n)$ of finding n photons in the coherent state α is

$$p(n) = |\langle n|\alpha\rangle|^2 = |\alpha|^{2n} \exp(-|\alpha|^2)/n! \qquad (21)$$

For a fully coherent field, the PCS always follows a Poissonian distribution independent of the preset time interval Δt in which the counts are always collected during the measurement time.

A chaotic field, on the other hand, is composed of number states $|n\rangle$ photons, according to

$$a^+a|n\rangle \geq n|n\rangle \qquad (22)$$

which are randomly fluctuating around their mean value $\langle n \rangle$. These fluctuations of phase can be expressed in terms of a Gaussian distribution of the probability $W(\alpha)$ of detecting $\langle n \rangle$ photons in the case that the field takes the amplitude $|\alpha|$:

$$W(\alpha) = exp(-|\alpha|^2/\pi\langle n \rangle)/\langle n \rangle \qquad (23)$$

$p(n, \Delta t)$ is then the result of the process of photodetection described by the folding of the sensor function $K(|\alpha|^2, \Delta t, n)$ that accounts for the probability of measuring n photons in the time interval Δt for an amplitude $|\alpha|$ of the field and the amplitude distribution function $W(\alpha)$:

$$p(n, \Delta t) = \int d^2\alpha \, K(|\alpha|^2, \Delta t, n) W(\alpha) \qquad (24)$$

In the case of a fully coherent field $W(\alpha)$ follows a Delta function $W(\alpha) = \delta^2 (\alpha-\alpha_0)$ such that

$$p(n, \Delta t) = K(|\alpha|^2, \Delta t, n) \qquad (25)$$

Consequently, if one measures a Poissonian distribution for a fully coherent field according to (21), one can evaluate $p(n, \Delta t)$ by use of (23) to (25), in order to get the PCS of a chaotic field:

$$p(n, \Delta t) = \langle n \rangle^n / (1 + \langle n \rangle)^{n+1} \qquad (26)$$

The PCS of a chaotic field follows a geometrical distribution.

However, this is only valid for $\Delta t \ll \tau$, where τ is the coherence time of the chaotic field. It is clear that for increasing $\Delta t/\tau$, the variance of n gets lower and lower, since the fluctuations of $|\alpha|$ smooth out more and more. For $\langle (\Delta n)^2 \rangle$ one arrives at

$$\langle(\Delta n)^2\rangle = \langle n\rangle(1 + \langle n\rangle/M) \quad (27)$$

where M is the "number of degrees of freedom" of the fluctuating field depending on the number of independent modes as well on the ratio $\Delta t/\tau$. While $M = 1$ for the photocount distribution (26), we get

$$M = \Delta t/\tau \quad (28)$$

for a chaotic field, if the preset time interval is $\Delta t > \tau$.
A careful analysis shows then that

$$p(n, \Delta t) = \Gamma(n + M)/(n!\,\Gamma(M))(1 + M/\langle n\rangle)^{-n}(1 + \langle n\rangle/M)^{-M} \quad (29)$$

This means that with increasing ratio $M = \Delta t/\tau$ the photocount statistics of a chaotic field approaches more and more that of a fully coherent field. A difference can no longer be registered.

In order to find differences between a chaotic and a fully coherent field, one has to provide for $\Delta t \ll \tau$ under the conditions of the measurement.

There are some realistic ways of satisfying this important provision. The first is to go down to rather low time intervals $\Delta t \ll \tau$ in the case that the value of τ is known. For unknown τ one has to investigate „delayed luminescence" after excitation of the field under ergodic conditions and to examine either the relaxation function or/and the photocount statistics for $\Delta t \ll T$, where T is the typical decay time of the system under study. Both have been done for biophotons. It turned out that the relaxation function follows a hyperbolic law, $\langle \dot{n}\rangle = A/(t + t_0)$, where A and t_0 are constant values and t is the running time, rather than an exponential one, and that the PCS remains Poissonian over the whole relaxation time with $\Delta t \ll T$. Both results are necessary and sufficient for a fully coherent ergodic field.

9. DELAYED LUMINESCENCE REVISITED

As soon as an atom or a molecule is excited by a photon of suitable energy, after a definite time τ corresponding to the classical coherence time of the atom or molecule, the photon is reemitted. This process, known as "luminescence" (fluorescence for allowed transitions and phosphorescence for forbidden ones) can be considered also as the absorption and reemission process of a resonator (see APPENDIX I) with a resonator value $Q = \tau$. However, in contrast to common luminescence and in contrast to a linear resonator, biological systems even after death do not display exponential decay of the reemitted photon intensity, but instead a hyperbolic-like relaxation function. This is a crucial point in understanding biophotons, since the deviation from an exponential function indicates clearly that the interactions of light with matter are not based on random rescattering but tell us something about the organization of the „biophoton field" within the subject under investigation. In order to understand this correspondence, let us compare two rather limiting cases of interactions, i.e. the completely random rescattering with the perfectly coherent one.

Random rescattering is characterized by „random phase approximation" of electromagnetic waves originating from the different rescattering atoms and molecules within a system. It is well known that under this condition there is no interference pattern that contains information about the interaction of the particles in terms of their absorption and reemission of electromagnetic waves. The superposition of N waves then always results in the simple superposition of an average amplitude of all of them where the total intensity (corresponding to the square of all the amplitudes) is always proportionate to the number of particles (or photons) that take part in the interaction. Consequently, this process is subject to

$$\dot{n} = -n/\tau \tag{30}$$

with the solution

$$n = n(0) \exp(-t/\tau) \tag{31}$$

In the opposite case of an interference pattern (where the amplitudes and phases are fixed, such as in Fig.9), the dominating term of the superposition of waves gets the interference term that increases with n^2 rather than with n. Consequently we get

$$\dot{n} = -n^2/\tau \tag{32}$$

with the solution

$$n = \tau/t \tag{33}$$

representing a hyperbolic relaxation function.

This simple approach can be expressed even more elegantly in terms of Quantum Theory[48]. A sufficient condition of a coherent field is the capacity of displaying "destructive interference" of overlapping field amplitudes. In terms of the displacement operator

$$D(t) = \exp\left(-\int H(t')dt'\right), \tag{34}$$

where H is the Hamiltonian of the electromagnetic field, this means that

$$\tfrac{1}{2}(D(t)+D(-t)) = D(0) \tag{35}$$

for a fully coherent field, at least in the zones of destructive interference. Since $(D(t)D(-t))^{1/2} = D(0)$, this means at the same time that for a coherent field the geometrical mean is equal to the arithmetic mean of the operators.

Now take the case of a chaotic field which is subjected to

$$D(t_1)D(t_2) = D(t_1 + t_2) \tag{36}$$

with the consequence of an exponential relaxation, since only

$$\exp(t_1)\exp(t_2) = \exp(t_1 + t_2) \tag{37}$$

In the case of a fully coherent field we use equation (36) and split $D(t_1)D(t_2)$ in two parts, one following the semigroup-law, and the second part satisfying destructive interference according to

$$D(\tfrac{1}{2}(t_1 + t_2)) = \tfrac{1}{2}(D(t_1)+D(t_2)) \tag{38}$$

in order to obtain

$$D(t_1)D(t_2) = D(\tfrac{1}{2}(t_1+t_2)) D(\tfrac{1}{2}(t_1+t_2)) = D(\tfrac{1}{2}(t_1+t_2))(\tfrac{1}{2}(D(t_1)+D(t_2)))^*, \tag{39}$$

where the equality of geometrical mean and arithmetic mean of the displacement operators for a fully coherent field has been partially used.

From (39) it is evident then that only a hyperbolic-like relaxation satisfies the physics of this field since

$$(1/t_1)(1/t_2) = 1/(\tfrac{1}{2}(t_1+t_2)) \tfrac{1}{2}(1/t_1 + 1/t_2) \tag{40}$$

Consequently, a hyperbolic-like relaxation function of delayed luminescence is sufficient for a fully coherent (ergodic) field.

It is now well known that the hyperbolic relaxation of living systems also displays "hyperbolic" oscillation of the type

$$\dot{n} = A/((t+t_{01})(t+t_{02}))\sin(b\ln((t+t_{02})(t+t_{01}))+\phi) \qquad (41)$$

where A, t_{01}, t_{02}, b and ϕ are constant values.

For more detailed information the reader is pointed to the references [49, 50, 51].

10 BIOLOGICAL PHENOMENA REVISITED

10.1 Cell Growth and Differentiation

Let us try now to understand fundamental biological phenomena such as growth and differentiation in terms of the physical properties of biophotons, where the wavelength range can be extended principally over the whole spectral range of the electromagnetic field.

Let us start with a rather basic phenomenological equation

$$\dot{n} = \alpha N - \beta N^2 \qquad (42)$$

which explains sufficiently accurately the growth rate of cells in any cell population.

According to Alexander Gurwitsch, we postulate that regulatory interactions are provided by biophotons ("mitogenetic radiation"), which follows also from Fig. 15.

Consequently, α and β are parameters which can be traced back to the properties of the biophoton field within the system under study.

The solution of (42) takes the form:

$$N = N(0)/(B + (1-B)\exp(-t/\tau)), \qquad (43)$$

where $N(0)$ is the number of cells at time 0, and $N_f = N(0)/B$ is the final number of cells in the tissue, while $1/\tau$ is the growth-frequency. Comparison to (42) shows that

$$\alpha = 1/\tau \text{ and } \beta = B/(N(0)\tau) = 1/(N_f \tau) = \alpha/N_f \qquad (44)$$

For photonic control of growth, it becomes evident that coherence, expressed here in terms of coherence time τ, is the determining factor of cell growth (and, of course, also differentiation). Note that $1/\tau$ can be substituted by $1/Q$, where Q is the Q-value. In an even more general sense, this factor may represent the degree of coherence of biophotons in the tissues under investigation, including the possibility of turning over into chaotic states or squeezed states. For a human body one gets with $N_f = 10^{13}$ values of α of the order 10^{-7} Hz, $\beta = 10^{-20}$ Hz, corresponding to $\tau = 10^7$ s and $\nu\tau = 10^{22}$ bits in the optical range. For soy beans (Fig.15), we obtain $\alpha = 4.3 \ 10^{-6}$ Hz and $\beta = 5.9 \ 10^{-17}$ Hz, corresponding to a lower N_f of about 10^{10} cells as well as to lower potential information of about 10^{20} bit. While α accounts for stimulation of the cell division rate, β is responsible for cell inhibition. As soon as the ratio α/β exceeds a definite threshold, instead of normal growth, tumor development will take place. However, this ratio is certainly not a permanent constant but will change during development of the organism and will even take different values in different organs. In order to get a deeper understanding of cell growth, one has to know what τ and N_f mean and what mechanism is behind them. The increased biophoton intensity of tumor cells and their loss of coherence invites the following hypothesis: α and β are to some extent antagonistic factors, where the growth-stimulating influence, described by α, is due to a loss of coherence, while the growth-inhibiting factor, described by β, results from a gain of coherence of the biophoton field within the tissue under consideration. β is at the same time also responsible for the differentiation of the tissue. The parameters τ and $N_f(\tau)$ formally describe the relations between these quantities α and β, where the antagonistic property is related to $N_f(\tau)$. In order to distinguish between normal growth and tumor development, let us look at $\ddot{N}$ under variations of α and β, where we use

$\alpha = N_f \beta$
and

$$\ddot{N} = N \cdot (\dot{N}_f \cdot \beta + N_f \cdot \dot{\beta}) - N\dot{\beta} \qquad (45)$$

Necessary and sufficient for tumor growth is

$$\ddot{N} > 0 \qquad (46)$$

for all future times around $N = N_f$.

This means that normal growth develops into tumor growth as soon as

$$\dot{N}_f > (N - N_f) \frac{\dot{\beta}}{\beta} \qquad (47)$$

$\dot{N}_f$ can be considered as the cell loss rate in the tissue that takes, for instance, for the human body, values of about 10^7 Hz. This means that the cell loss rate protects against cancer as long as $\dot{\beta} > 0$, which means that an increase of coherence during development does not allow cancer development at all. However, as soon as $\dot{\beta} < 0$, from a definite size of the tumor on ($N > N_f$), the cell loss rate may not be high enough to protect against cancer development. This criterium accords with experience, since it describes the increase of malignancy with increasing tumor mass, and it explains the strange observation that some rapidly growing tumors, such as cysts, cannot develop harmful cancer even in view of their high cell growth rate due to their high cell loss rate.

However, it should be noted that just the opposite problem occurs as soon as α decreases to such an amount that $\dot{N}$ is permanently < 0. This case takes place as soon as either τ or cell loss or both become large compared with $N(1-N/N_f)$, which is >0. A straightforward calculation shows that

$$-\tau \dot{N}_f \rangle N_f \left(1 - N/N_f\right) \tag{48}$$

leads to the kind of catastrophe that is a disease just the opposite of cancer. It can be reduced by reduction of cell loss and/or by diminishing the coherence time of the biophoton field.

It seems that natural ways of increasing the coherence of the biophoton field and/or the cell loss rate are the right therapy concepts for cancer, and that decreasing the coherence time of the biophoton field and/or the cell loss rate are the right steps to improve, for instance, therapy for multiple sclerosis. In any case, however, the interference of cell death rate and coherence has to be taken into account.

From the molecular point of view, the origin of this relation is likely to be the physical property of the biophoton field within the DNA. As long as it displays quantum coherence, the factor β is dominating insofar as it inhibits cell growth and stimulates cell differentiation. As soon as quantum coherence gets lost in a local zone, the field decays into a chaotic field with two remarkable properties:
(1) The coherence time of this chaotic field is rather high such that the reactions (i.e. translation and transcription of the DNA) take place within the coherence time of the chaotic field.
(2) The reactivity is two times higher than that for the fully coherent field since photon bunching (which takes place only in a chaotic field) increases the probability of finding a second photon after a first one has been recorded, up to a factor 2.

Consequently, after local decay of the fully coherent field into the chaotic one, factor α dominates. It is clear that α and β of formula (42) are then rather rough average values of refined spatio-temporal molecular events of the DNA-biophoton field[52].

10.2 Cell Communication

A further basic biological phenomenon concerns cell communication by biophotons. The fundamental experiment is displayed in Fig.10 (Daphnia). In general, it demonstrates that between cells and cell populations coherent photons are exchanged in such a way that preferentially destructive interference takes place in the intercellular space and – for reasons of energy conservation – constructive interference within the intracellular space. Modulations in the spectral distribution may change this rule such that there are in generally three possibilities, e.g. destructive, constructive, or negligible interference. In any case, this effect takes place under the following conditions:
(1) The photons have to be at least partially coherent,
(2) The coherence time has to be long compared to the relaxation time in the movement of the biological systems under study,
(3) The „biological material" has to display non-linearly polarizable double layers in order to provide phase conjugation (see reference)
(4) It is rather likely that the biological matter has to be in an optically active state in order to respond to the incoming signals.

This kind of communication [51] seems to be the most perfect one that can be imagined. It is the most sensitive and accurate tool for identification of the communication partner, since the signals may contain any kind of species-specific pattern, printed in terms of the compositions of frequencies, phases and – to some extent – also the amplitudes, polarization directions, and propagation directions of the interference fringes. They work with the highest possible signal velocity. The signal-to-noise-ratio is optimized. For squeezed light it can even theoretically approach an infinite value. At the same time, the communication is rather flexible. It allows all imaginable variations, from complete "silence" to the highest possible information rate at all. It is strongly connected to attractive and repulsive forces between the systems under study, strong enough to explain the formation of cell aggregation and weak enough to allow the highest flexibility.

10.3 Consciousness

Let us try to construct from these basic insights into the phenomena of biophotons the simplest model of what we call "consciousness". At first we postulate that the results of the daphnia experiment (Fig.10) can be generalized to all kinds of communicative interactions between a subject (observer) and its surrounding objects. It is evident that this interaction is not based on passive absorption of signals, but induces a change of the biophoton field within the body of the observer, where we extend the name "biophoton field" to the whole frequency spectrum from ELF up to the UV-range. It is not necessary to reduce this interaction to certain "receptors". Rather, we may say that the zone of interaction, even in the case that rather local "receptors" work for receiving the signals, and the whole body may be influenced by destructive and constructive interference processes stimulated by the impinging signals in just the same way as with daphnia. This basic process changes at the same time the resonator values of the network of all cavities such as organelles, single cells, cell populations, organs, ... up to the whole body. The resonator values, on the other hand, are identical to the potential information of the interacting system. It is very likely that what we call "consciousness" has to be assigned to this concert of resonator values. It is the only "sensory" system that can bridge the gap between syntactic (structural) and semantic (meaning) information, e.g. physics on the one hand and the state of an "informational detector system" called "awareness". It is obvious that dead material cannot achieve this state of translating physical signals into "feelings". However, the necessity of transforming the syntactic information into semantic information invites us to take the resonator concept of the daphnia as the most useful model of understanding the process of consciousness, since all the conditions of translating physics into the language of "sense" can be satisfied. By inclusion of the Hamiltonian principle one can even go a step further, in order to find a necessary and sufficient definition of consciousness. We define consciousness as a process where actual information (the world of spatio-temporal events) is transformed into potential information (the world of possibilities) in a way that the image of the potential information contains relevant parts of the actual information. By taking up this definition, we immediately see four essential elements of this process.

(1) The self-interaction of a subject S-S: I would like to call this interaction the "point of Descartes" (Cogito, ergo sum). This self-reflection of a subject combines actual and potential information of the system under study and defines the reference point of consciousness. Psychologically it is responsible for the self-confidence of the observer.
(2) The interaction of the observer with an object outside of the observer O-S: This interaction is the basis of becoming aware about the outside world. It is not the absorption of light as in a camera, but an active process where the object changes the interference pattern of the subject in a way that it got not only "seen" but also "felt" in its interaction with the present information imprinted in the interference pattern of the subject.
(3) The repeated interaction of the observer with the object: O-S-O-S-O-S-....
    (a) In case this process leads to an infinite loop where nothing is changed anymore, we get the state of "confirmation", where the subject gets

the feeling that the object follows certain laws (which he calls physical laws). For instance, by repeatedly observing the sunrise, at the end of this process the potential information is able to predict precisely the dates of sunrise every day.

(b) In case this process is unlimited, either "forgetting" or "creating new possibilities" are the possible results. Both processes have evolutionary significance, since "forgetting" as well as "creativity" provide the openness for extending consciousness to more and more freedom and responsibility at the same time.

APPENDIX I: THE Q- VALUE

Fig.8 displays an example that demonstrates that cells establish dielectric cavity resonators with rather high resonator values. Actually, the whole dynamics of mitotic figures can be described in terms of superpositions of electromagnetic field patterns of electric and magnetic modes that satisfy the boundary conditions of the cell [54]. Consequently, cell division has to be looked upon as the dynamics of the storage of electromagnetic modes in the optical frequency range. As stable solutions to Maxwell equations, they are to some extent fixed by the dimensions and the dielectricity constant and permeability of the cell walls, but they not only steer the movement of the biomolecules, they influence, in turn, the electromagnetic properties of matter, including the geometry of the cell. In this mutual resonance interaction between radiation and matter, the specific pattern of mitotic figures and its spatio-temporal dynamics are determined. This is a holistic phenomenon where every part of the electromagnetic field and every molecule influences the whole system and is, in turn, controlled by the whole. It is evident that this kind of mutual interaction between field and matter is not confined to mitotic figures. It is the rather extraordinary principle of spatio-dynamical organization of living cells. I do not see another reasonable mechanism. It explains at the same time the extremely low error rate in distributing biomolecules to the daughter cells, which is orders of magnitude lower than expected by the laws of random statistics.

In order to understand the basic physics behind this phenomenon, let us first confine ourselves to the simplest case of a cavity resonator. It is a closed box, consisting of reflecting walls of certain physical properties of definite dielectric constant and permeability. Suitable electromagnetic waves, fitted to the boundaries of this box, can be stored over a definite time T before the thermal loss considerably exceeds the initially stored energy. The storage capacity of a resonator is called resonator value Q and expressed in terms of the ratio of stored energy U to the power loss $P = -dU/dt$ of the cavity:

$$Q \equiv U/P \tag{AI.1}$$

Q has the dimension of a time, reflecting the lifetime τ of the stored electromagnetic wave within the cavity.

It is evident that this definition of Q can be used for every kind of energy storage, including the excitation of molecules, the detection of microwaves or radiowaves (cavitities or antennae), or the number of reflections of a Laser mode between the mirrors, before the mode is emitted. Allowed optical transitions have a Q-value of nano-seconds, forbidden optical transitions may reach values between 10 micro-seconds to milliseconds, and Lasers display maximum Q-values as high as seconds. As a rule one may keep in mind that the Q-value can (but must not) increase the higher the complexity and the „orderliness" of the system under investigation is developed. Living systems should reach at least a Q-value of one day in order to use with sufficient efficiency the photon storage of sunlight [54, 55].

Of basic importance is the fact that the Q-value links energetical and informational aspects of the resonator. This makes it the simplest and most powerful model of bridging the gap between energy and information and taking this connection as the subject of optimization procedures of the living system.

Actually, by multiplication of the Q-value with the frequency of the mode under consideration, one gets the potential information (in units of bit) that can be transferred from this cavity under consideration. This relation is simple and basic. Ask for the highest number of „yes-no" modulations that can be imprinted onto a wave of frequency ν during the time $Q = \tau$. The shortest time interval within which the wave can get switched off or on, corresponding to one bit of information, is just $1/\nu$. The highest possible number of bits during time τ is consequently

$$I_{max} \text{ (bit)} = \nu \, Q = \nu \tau \tag{AI.2}$$

This makes it understandable that in order to create systems with high informational capacity, one has to find ways of getting high Q-value at high frequencies. Note that the intensity is of less importance.

It is evident that the potential information I is time-independent only under stationary conditions. However, this is certainly not the case in biology, where I is subject to evolutionary development. The appearance of what we call consciousness is based on the creation and optimization of I, since otherwise there is no way of bridging the gap between energetical and informational aspects of development. This invites us to take I as a Lagrange function of the Hamilton principle that is the basis of macroscopic (and also microscopic) physics. Consequently, we may write

$$J(t) = \int_0^t v \, Q(x) d(x) = Extremum \qquad (AI.3)$$

There are two basically different solutions to this problem. One is the case of chaotic resonators $Q = Q(0) = const.$, and the other is the case of coherent resonators $Q = Q(0)/t$.

The explicit solutions are

$$J(t) = v\tau \, t \qquad (AI.4a)$$

for chaotic resonators, where τ is the coherence time of the classically chaotic field within the resonator, defined by

$$dU/dt = -(1/\tau) \, U \qquad (AI.4b),$$

and

$$J(t) = \tfrac{1}{2} v \, t^2 \qquad (AI.4c)$$

for coherent resonators, where

$$dU/dt = -U/t \qquad (AI.4d)$$

Since the information obtained during time t is $I = <dJ(t)/dt>$, we obtain the rather profound results for the potential information in chaotic and coherent resonators, e.g.

$$I(t) = v\tau \qquad (AI.5a)$$

for chaotic resonators, and

$$I(t) = v \, t \qquad (AI.5b)$$

for coherent resonators.

This difference is obviously rather important for understanding the nature of biological systems. Chaotic resonators are not able to register a time t. They have no memory. Coherent resonators, on the other hand, register the time t, therefore displaying a memory, and correspond to a much higher extremum of the Hamilton principle than chaotic ones as soon as $t > \tau$.

We can go even a step further. Since the time T is a preset quantity in physics and not explainable in physical terms, we have the possibility to *define* it now as

$$t \equiv (1/v) \, I(t), \qquad (AI.6)$$

where we provide that $I(t)$ is known and subject to the Hamilton principle (5). In other words, biological evolution may maximize the memory time of biological systems, which can be looked upon as the basis of the evolution of consciousness.

In order to demonstrate a simple example: Expose a resonator to the sunrise for ten days. It has a frequency 1 day^{-1} and a potential information of 1 bit per day (sunrise, yes or no?). A chaotic system will register only the time of 1 day, a coherent one is able to become aware of the time $t = 1$ day/bit times 10 bit = 10 days.

APPENDIX II: THE ENTROPY CONCEPT

There is no doubt that the arrow of time is determined by the maximum principle of entropy, i.e. the Second Law of Thermodynamics. This principle is the driving force of temporal dynamics, responsible, for instance, for the diffusion of particles, the direction of chemical reactions, the heat flow or the pressure in all systems. Increasing entropy determines the course of time flow, since the entropy of a (closed) system can only increase but never decrease, if all influenceable parts of the system under consideration are taken into account. It is worthwhile to note that entropy is also based on a judgement of the *informational* content of a system rather than of its energetic contents. Roughly speaking, entropy is a measure of the *potential* information of a system which has to be strictly distinguished from the *actual* information. As an example, take the toss of a coin which results either in „heads" or "tails", corresponding to one bit of information. As long as the toss is not finished, the potential information is one bit, the actual information is zero. At the moment when the result is established, the potential information is zero, but the actual information is one bit. "Potential" information accounts for the possibilities, "actual" for the resulting and unavoidable events. It is not wrong to describe the maximum entropy principle in terms of the natural tendency of all systems to approach a state of highest potential information by destroying its actual information. This is often accompanied by transitions of "ordered" states into less ordered states, appearing as the transition to randomness. However, this judgement is less reliable than the tendency of increasing *potential* information. The most reliable definition of entropy and its application originates from the phase space. The "volume" elements of the phase space are for particles of three degrees of freedom six-dimensional "phase space cells" of the volume elements ($\Delta x \Delta y \Delta z \Delta p_x \Delta p_y \Delta p_z$), where Δx, Δy, Δz and Δp_x, Δp_y, and Δp_z are the unavoidable uncertainties in the actual positions x, y, z, and the actual momenta p_x, p_y, and p_z, respectively. Since these uncertainties are subjects of the uncertainty relations $\Delta x \, \Delta p_x = h$, $\Delta y \Delta p_y = h$, and $\Delta z \, \Delta p_z = h$, where h is Planck`s constant, the „volume" of such a "phase space cell" is just h^3. The "thermodynamical probability W" is the origin of the entropy concept and defined as the number of available phase-space cells in the whole available phase space of always identical particles of the system under study. The phase space of a particle is consequently a six-dimensional hyper-sphere of the „volume" ($xyzp_xp_yp_z$), where x,y,z and p_x,p_y,p_z are the actual mean values of the positions and the momenta of the particle under consideration. The thermodynamical probability is then

$$W = V/(h)^{3N} \qquad (AII.1),$$

where $V \approx (<x><y><z><p_x><p_y><p_z>)^N$, and the $<...>$ denoting mean values over the ensemble of particles. In the case of particles with f degrees of freedom, we have to take fN instead of $3N$.

Let us note that W usually has a gigantic value for macroscopic systems, and there is in general no decisive difference between W and dW for small changes dE of the energy content E of the system, provided $N > E/dE$.

The decisive law of physics is now the experience that W of an isolated or closed system never decreases but can always only increase over the course of time. This means that the particle movement is always directed to a higher availability of the phase space under observation. The "potential" information of a closed system can always only increase by reduction of "actual" information.

It turned out that it is much more elegant to formulate this law in terms of entropy

$$S \equiv k \ln W \qquad (AII.2)$$

where k is the Boltzmann constant, instead of W. Changes in S refer to relative changes in W and describe the real physical situation more in agreement with practical experience and the imagination. In addition to lower and "realistic" values of S, the entropies of independent systems are additive and not multiplicative, as is the case for W.

Also important is the fact that S (like W) is a state function. A closed system which can be characterized by its energy content E, the number N of particles, and its volume V, displays for all E, N, and V definite values of S, e.g.

$$S = S(E,N,V) \qquad (AII.3)$$

or, in mathematical terms

$$\oint dS = 0 \qquad (AII.4)$$

In order to calculate S for a system of given E, N, and V, it is necessary to know the number of possible quantum states C_j that can be occupied by a particle of energy ε_j within the system under consideration. This degeneracy parameter is the most important term for "translating" the given physical situation into the entropy concept. For instance, if one has a unidirectional particle traveling with a definite energy, it can change the momentum without variation of its energy simply by occupying all the available phase space cells on the hyper-sphere of radius ε_j in the phase space. If this particle were bound by boundary conditions to only one direction, C_j would be 1, while it would take the value

$$C_j = V(r)\,(4\pi\, p^2\, dp)/h^3, \qquad (AII.5)$$

where p is the absolute value of the momentum and $V(r)$ the spatial volume.

A higher C_j would increase the entropy of the system, for instance by random elastic rescattering within the system.

As soon as C_j characterizes the physical boundary conditions of the system and is known, one can calculate the entropy by asking for the number of different ways w_j assigns N_j indistinguishable particles to C_j, the different quantum states.

Elementary statistics delivers the following result:

$$w_j(B\text{-}E) = (C_j + N_j - 1)!/((C_j - 1)!\,N_j!) \qquad (AII.6)$$

B-E denotes „Bose-Einstein" statistics.

It is worthwhile to note that (AII, 6) holds for bosons (which have an even spin number, i.e. photons), which are able to occupy the same phase space cell, while for fermions (which have an uneven spin number, i.e. electrons), each two fermions can occupy different quantum states only. The evaluation yields

$$w_j(F\text{-}D) = C_j!/((C_j\text{-}N_j)!N_j!) \tag{AII.7}$$

F-D denotes "Fermi-Dirac" statistics.

As a considered example that may play a fundamental role in biology and in understanding biophoton emission, take the case of a photon that has $C = 2$, corresponding to the two polarization directions. It shall be absorbed by an electron state of $C = N$ before absorption and $C = N+1$ after absorption, corresponding to induced creation of phase space volume by absorption of light. Formula (AII.6) tells us that the entropy of the photon is removed from $k \ln 2$ to 0, and from (AII.7) we see that the entropy of the electron state increases from 0 to $k \ln (N+1)$. The Second Law of Thermodynamics is satisfied in this process, since total entropy does not decrease. For a reversible process, the electron could jump down to the ground state. This is forbidden if C of the electron state is higher than 1, but it is allowed for $C = 1$. Consequently, the Second Law of Thermodynamics enables the construction of a mechanism, based on a photon trap, that absorbs photons in agreement with the maximum entropy principle and allows for a decrease in the entropy of the system by emission of stored photons. ΔN emitted photons would remove entropy $\Delta S \leq k \ln 2 \, \Delta N$ from the system as soon as they are released from the trap. On the other hand, it is possible that a photon with $C = 1$ might hit a fermion, inducing a transition into the ground state and lowering the entropy of the fermion system. The two photons carry together a higher entropy, corresponding at least to the difference of entropy loss of the fermion system, but each of them has a lower entropy than the light stimulating single photon had before. This is the basis of laser radiation. It is evident that this mechanism can be generalized to other values of C and N in the interactions of bosons and fermions. Important is the induced creation or annihilation of available phase space cells by photons. Photons can create or destroy „order" depending on whether they remove entropy in the fermi system or release it with lower entropy than they carry at the instant of absorption. For a living system it is decisive that the photon emission displays a high entropy in the case that the living state is surrounded by a heat bath, and that it decreases its entropy in response to non-random environmental influences.

For maximum entropy, the subsystems of different energy ε_j and different particle number N_j can be looked upon as independent ones and are therefore additive in the entropies. Consequently,

$$S = \sum k \ln w_j \qquad (AII.8)$$

By making use of the Stirling formula, one approximates $\ln x! \approx x \ln x - x$ (which becomes more and more accurate with increasing x), and one then gets after straightforward calculations

$$S = k \ln W = k \sum_j \left\{ \pm C_j \ln\left(1 \pm \frac{N_j}{C_j}\right) + N_j \ln\left(\frac{C_j}{N_j} \pm 1\right) \right\}$$
$$= k \sum_j \left\{ \pm C_j \left[\left(1 \pm \frac{N_j}{C_j}\right) \ln\left(1 \pm \frac{N_j}{C_j}\right) \mp \left(\frac{N_j}{C_j}\right) \ln\left(\frac{N_j}{C_j}\right) \right] \right\} \qquad (AII.9)$$

where the upper signs in $\pm$ or $\mp$ refer to (BE) and the lower to (FD) statistics.

Max Planck was the first to assign to a light beam entropy in accordance with (17), e.g.

$$S(\nu)d\nu = k/\tau \,((1+f)\ln(1+f) - f \ln f) \qquad (AII.10)$$

where $f = N/C = \dot{n}\tau$ and $C \propto 1/\tau$

However, because of the approximation of Stirling's formula, this is an approach for high enough numbers N and C, and both should exceed at least the number 10. Otherwise we have to go back to the original equation (AII.6). Equation (AII.10) is useful to estimate the entropy of light in cases that f and τ are known. By comparison with (AI.2) we can see the connection between the resonator value Q and the entropy of light. Actually, for constant f, it is easy to see that

$$\partial S/\partial \tau = -S/\tau \qquad (AII.11)$$

which means that the entropy of light decreases with increasing "classical" coherence time.

However, in general the maximum entropy depends on the constraints that the system is subjected to. The most important case is the closed system where at every instant the influx of heat energy is compensated by the output, such that the energy of the whole system is constant. The maximum entropy under this condition requires the validity of the maximum principle:

$$dS - (\partial S/\partial E)_{V,N}\, dE = 0 \qquad (AII.12)$$

Since $dS = k\, d\,(\ln w) = k\, \ln(C/N + 1)\, dN$ for constant C from (14) and $dE = \varepsilon\, dN$, where $\varepsilon = h\nu$ is the quantum energy of the photon, we arrive at:

$$k\, \ln(C/N + 1) = \varepsilon\, (\partial S/\partial E)_{V,N} \qquad (AII.13)$$

For closed systems, it turns out that $(\partial S/\partial E)_{V,N}$ is a constant. This is the reason to *define* a physical parameter that is called (absolute) temperature T by

$$T \equiv 1/\, (\partial S/\partial E)_{V,N} \qquad (AII.14)$$

Consequently, we obtain after simple straightforward calculations

$$N = C/\,(\exp(\varepsilon/kT) - 1) = Cf \qquad (AII.15)$$

This is the famous "Bose-Einstein" (or, neglecting -1 in the denominator, the "Boltzmann") distribution for photons in a closed system ("black-body radiation"). It tells us how many photons of energy $\varepsilon = h\nu$ are found in a system of volume V at temperature T. Equation (AII.15) consists of the product of available phase space cells C and the factor $f = 1/(\exp(h\nu/kT) - 1)$ that describes the probability of occupying the phase space cells.

By taking account of (AII.9) and $p = h\nu/c$ for photons, where c is the velocity of light, and considering that photons display two different polarization directions (which doubles the value C), one can calculate the value C and arrive finally at

$$(\partial^2 N/(\partial V \partial \nu)) = 8\pi\nu^2/(c^3\,(\exp(h\nu/kT) - 1)) \qquad (AII.16)$$

which accounts for the photon number density per unit of frequency ν.

Equation (AII.16) is of basic importance for understanding photon emission from all systems, even if it is confined only to closed systems. It provides a useful tool for understanding other systems by comparison of the measurable spectral photon intensities with the theoretical results of Equation (AII.16). As soon as one finds significant differences between the results of (AII.16) and the experimental results, one is justified in concluding that the system is not a closed one and „away" from thermal equilibrium.

It is worthwhile to note here that analogically to the temperature T, two other very basic terms are defined by the gradient of entropy. These are the chemical potential μ and the pressure P, e.g.

$$\mu \equiv - T\, (\partial S/\partial N)_{E,V} \qquad (AII.17)$$

$$P \equiv T\, (\partial S/\partial V)_{E,N} \qquad (AII.18)$$

The closed system is a paradigm of a fully chaotic photon field. The chemical potential μ is zero, since addition or subtraction of photons by keeping the energy and the volume constant does not change the entropy of the equilibrium system. However, this approach is not necessarily valid for a coherent photon field that couples in a non-linear way with matter such that influx or emission of photons influences the "order" of the system under study. Consequently, one has to change (AII.16) by taking account $\mu \neq 0$. Instead of photon energy ε, the reduced energy $\varepsilon^* = \varepsilon - \mu = h\nu - \mu$ has to be inserted. Note that for $\mu \to h\nu$, f may increase to considerably higher values.

REFERENCES

1. Gurwitsch, A. (1922) Über Ursachen der Zellteilung. *W.Roux' Arch.*, **52**.

2. A.Gurwitsch, (1932) Die Mitogenetische Strahlung. *Monographien aus dem Gesamtgebiet der Physiologie der Pflanzen und der Tiere.* J.Springer, Bd. 25, Berlin.

3. Reiter, T und Gabor, D (1928) Zellteilung und Strahlung. *Sonderheft der Wissenschaftlichen Veröffentlichungen aus dem Siemens-Konzern.* J.Springer, Berlin.

4. Ruth, B (1977) Experimenteller Nachweis Ultraschwacher Photonenemission aus biologischen Systemen. *Dissertation*, Universität Marburg.

5. Slawinski, J. (1988) Biophoton Emission (Multi-author Review). *Experientia*, **44**, 559-571.

6. Inaba, H. (1988) Biophoton Emission (Multi-author Review). *Experientia*, **44**, 550-559.

7. Boveris, A., Varsavsky, A.I., Da Silva, S.G. and Sanchez, R.A. (1983) *Photochem.Photobiol*, **38**, 99-104.

8. Quickenden T.I. and Tilbury, R.N. (1983) *Photochem.Photobiol*, **38**, 337-344.

9. Zhuravlev, A.I. (ed.) (1983) *Biochemiluminescence*. USSR Academy of Sci. and Moscow soc. Nature, vol.58, Nauka Publ. House, Moscow, 210-222.

10. Seliger, H.H. (1973) *Chemiluminescence and Bioluminescence*. Cormier, M.J., Hercules, D.M. and Lee, J. (eds.), Plenum Press, New York, 461-478.

11. Popp, F.A. (1976) *Molecular Aspects of Carcinogenesis*. E.Deutsch, K.Moser, H.Rainer and A. Stacher, (eds.), Thieme Verlag, Stuttgart, 47-55.

12. Popp, F.A. (1974) *Archiv für Geschwulstforschung*, **44**, 295-306.

13. Ruth, B. und Popp, F.A. (1976) *Z. Naturforsch*, **31c**, 741-745.

14. Ruth, B. (1979) *Electromagnetic Bio-Information*. Popp, F.A., Becker, G., König, H.L., Peschka, W. (eds.), Urbau & Schwarzenberg, München, 107-122.

15. Popp, F.A., Ruth, B., Bahr, W., Böhm, J., Grass, P., Grolig, G., Rattemeyer, M., Schmidt, H.G., Wulle, P. (1981) *Collective Phenomena*, **3**, 187-214.

16. Popp, F.A., Gu, Q. and Li, K.H. (1994) *Modern Physics Letters B*, vol. 8., 1269-1296.

17. Popp, F.A., Li, K.H. and Gu, Q. (1992) *Recent Advances in Biophoton Research and its Applications*. World Scientific, Singapore.

18. Chang, J.J. and Popp, F.A. (1998) *Biophotons*. Chang, J.J., Fisch, J. and Popp, F.A. (eds.), Kluwer Academic Publishers, Dordrecht, 217-237.

19. Popp, F.A., Nagl, W., Li, K.H., Scholz, W., Weingärtner, O., Wolf, R. (1984) *Cell Biophysics*, **6**, 33-51.

20. Slawinski, J. and Popp, F.A. (1987) *J.Pl.Physiol*, **130**, 111-123.

21. Rattemeyer, M., Popp, F.A. and Nagl, W. (1981) *Naturwissenschaften*, **11**, 572-573.

22. Chwirot, B. (1986) *J. Pl. Physiol.*, **122**, 81-86.

23. Popp, F.A. and Li, K.H. (1993) *Int.J.Theor.Phys.*, **32**, 1573-1583.

24. Perina, J. (1985) *Coherence of Light*. D. Reidel, Dordrecht.

25. Bajpai, R.P. (1999) *J. theor.Biol.* **198**, 287-299.

26. Popp, F.A. (1979) *Electromagnetic Bio-Information*. Popp, F.A., Becker, G., König, H.L. and Peschka, W. (eds.), Urban & Schwarzenberg, München, 123-149.

27. Popp, F.A. (1988) Biophoton Emission. Popp, F.A., Gurwitsch, A.A. Inaba, H., Slawinski, J.,Cilento, G., Li, K.H., van Wijk, R., Chwirot, W.B. and Nagl, W. (eds.), *Experientia (Multi-author Review)*, **44**, 576-585.

28. Popp, F.A. and Nagl, W. (1986) *Polymer Bull*, **15**, 89-91.

29. Popp, F.A. and Chang, J.J. (2000) *Science in China (C)*, **43**, 507-518.

30. Popp, F.A. (2000) *Biophotonics and Coherent Systems*. Beloussov, L., Popp, F.A., Voeikov V. and van Wijk, R. (eds.), Moscow University Press, 117-133.

31. Dicke, R.H. (1954) *Phys.Rev.* **93**, 99-110.

32. Galle, M. (1993) *Dissertation*. Universität Saarbrücken.

33. Galle, M., Neurohr, R., Altmann, G., Popp, F.A. and Nagl, W. (1991) *Experientia*, **47**, 457-460.

34. Schamhart, D.H.J. and van Wijk, R. (1987) *Photon emission from biological systems*. Jezowska-Trzebiatowska, B., Kochel, B. and Slawinski, J. (eds.), World Scientific, Singapore, 137-152.

35. Scholz, W., Staszkiewicz, U., Popp, F.A. and Nagl, W. (1988) *Cell Biophysics*, **13** 55-63.

36. Popp, F.A., Chang, J.J., Gu, Q. and Ho, M.W. (1994) *Bioelectrodynamics and Biocommunication*. Ho, M.W., Popp, F.A. and Warnke, U (eds.), World Scientific, Singapore, 293-317.

37. Vogel, R. and Süßmuth, R. (1998) *Bioelectrochemistry and Bioenergetics*, **45**, 93-101.

38. Etienne, J.J., Popp, F.A., Papaconstantin, E. and Niggli, H. (1992) Low Level Luminescence of Acetabularia Acetabulum as a tool for Evaluating the Quality of Cosmetic Ingredients. *Proceedings 17th IFSCC International Congress, Yokohama (Japan)*, October 13-16.

39. Yan, Y (2000) Biophoton Emission and Germination Capacity of Barley Seeds. In *Biophotonics and Coherent Systems*, Beloussov, L., Popp, F.A., Voeikov, V. and van Wijk, R. (eds.), Moscow University Press, 431-438.

40. Köhler, B., Lambing, K., Neurohr, R., Nagl, W., Popp F.A. and Wahler, J. (1991) *Deutsche Lebensmittel-Rundschau*, **82**, 78-83.

41. Cohen, S. and Popp, F.A. (1997) *J.Photochem.Photobiol. B: Biology*, **40**, 187-189.

42. Schrödinger, E. (1987) *Was ist Leben?* Piper Verlag, München.

43. Prigogine, I. (1976) *Order through fluctuation: Self-organization and social systems. Evolution and Consciousness*. Jantsch, E. and Waddington, CH., (eds.), Addison + Wesley P.C., Reading.

44. Fröhlich, H. (1968) Long-range coherence and energy storage in biological systems. *Int.J.Quantum Chem*, **2**, 641-649.

45. Slawinski, J. and Popp, F.A. (1987) Temperature hysteresis of low level luminescence from plants and its thermodynamical analysis. *J.Pl.Physiol*, **130**, 111-123.

46. Perina, J. (1971) *Coherence of Light*. D.Reidel P.C., Dordrecht.

47. Arrechi, F.T. (1969) Photocount Distributions and Field Statistics. *Ottica quantistica (Quantum Optics)*, Glauber, R.J. (ed.), Academic Press, New York.

48. Popp, F.A. and Li, K.H. (1993) Hyperbolic relaxation as a sufficient condition of a fully coherent ergodic field. *Int.J.Theor.Phy*, **32**, 1573.

49. Popp, F.A. and Yan, Y. (2002) Delayed luminescence of biological systems in terms of coherent states. *Phys.Lett.A*, **293**, 93-97.

50. Popp, F.A. et al. (2002) Evidence of non-classical (squeezed) light in biological systems. *Phys.Lett.A*, **293**, 98-102.

51. Popp, F.A. and Chang, J.J. (2000) Mechanism of interaction between electromagnetic fields and living systems. *Science in China (C)*, **43**, 507-518.

52. Popp, F.A.(1976) *Biophotonen. Ein neuer Weg zur Lösung des Krebsproblems*. Schriftenreihe Krebsgeschehen Bd.6, Verlag für Medizin Dr. E. Fischer, Heidelberg.

53. Popp, F.A. (2001) *Energieverteilung und Bewußtsein*. Vortrag auf der Biosynthese-Tagung am 5./6.10.2001, Basel, Schweiz.

54. Popp, F.A. (1979) Photon Storage in Biological *Systems*. In *Electromagnetic Bio-Information*, Popp, F.A., Becker, G., Koenig, H.L. and Peschka, W. (eds.), Urban & Schwarzenberg, München, 123-149.

55. Dürr, H.P., Popp, F.A. and Schommers, W. (2002) *What is Life?* World Scientific, Singapore-London.

Chapter 13

THE PHYSICAL BASIS OF LIFE

R.P. Bajpai
Institute of Self Organising systems and Biophysics,
North Eastern Hill University, Shillong 793022, India.
and
International Institute of Biophysics, IIB e.V.
Raketenstationen, ehem. Kapellener Strausse, Neuss, D41472 Germany.

Abstract: The properties of living systems are divisible into three classes: microscopic, macroscopic, and consciousness. The comprehensibility of a property in the semi–classical framework is the basis of division; a property in the microscopic class is comprehensible, in the macroscopic class has a few gaps in its comprehension, and in the consciousness class is incomprehensible. The properties of quantum objects are also of three types: classical, non-classical but local, and non-classical and non-local. Non-classical properties are incomprehensible in the classical framework and are of holistic character. The two classifications are suggestive of a conjecture that a living system is a macroscopic quantum object and its properties in the macroscopic and consciousness classes are manifestations of its holistic quantum features. The conjecture gets support from the behaviour of photon signals needed for fulfilling thermodynamic requirements of energy conservation and information degradation in biological processes. Photon signals associated with the quantum aspects of living systems will be in pure quantum states. It appears that the photons identified as biophotons are in pure quantum states and are emitted in the fundamental biological processes at the genetic level. Living system emerges as a collage of intermittent coherent quantum patches of moderate size from the behaviour of biophoton signals. The patches (or domains) act as quantum objects during selections and detections at genetic level. The information generated in selections is stored in an erasable memory. The contents of the memory associated with a patch are accessible to other patches. The erasure of the memory of different patches is need based and gives rise to a burst of photons.

Keywords: Living systems. Co-operative functioning. Holistic behaviour. Biophoton signal. Non-exponential decay. Probability of subsequent photon emission. Quantum search. Genetic Code. Information. Maxwell demon. De-coherence. Quantum patches. Intermittence. Photon bursts. Consciousness.

1. THE PHYSICAL PARADIGM

One of the important items in the agenda of current scientific investigations is the demystification of "life" - an attribute of many naturally occurring objects. These objects are referred as living systems and objects lacking this attribute are referred as non-living systems. A living system differs from a non-living system in many ways; the difference is so pronounced that different branches of science have evolved for their studies- Biology for living systems and Physical Science for non-living systems. These branches employ different paradigms and are quite different in approach. The paradigm employed in Biology is called biological paradigm and is based on a phenomenological and descriptive framework, while the paradigm employed in Physical Science is called physical paradigm and is based on an analytic and axiomatic framework. The scope of investigations in Physical Science is wide but is restricted in Biology to the investigations of intra-object interactions and relations. The inter-object interactions and relations among living systems are investigated in social and environmental sciences. The domain of investigations of the biological framework is a small region in space and time; it is given by the size and lifetime of living objects. The inclusion of inter-object interactions in Biology can extend this region but the extended region remains within the domain of applicability of the physical framework. A common feature of both paradigms is the identification of the chain of hierarchical structures with well-defined stabilities and spatial extensions. The chain of hierarchical structures in increasing order of complexity is biomolecule -organelle-cell-tissue-object-society-species in the biological paradigm and quark-elementary particles-atoms-molecules-condensed matter structures-planets- solar systems-cosmological structure in the physical paradigm. The space-time region utilised in the descriptions of the entire chain of the biological paradigm is contained within the space-time region utilised by one element in the chain of the physical paradigm namely the condensed matter structures. This fact is the motivation behind many attempts to understand the formation of the biological chain and the nature of its various links within the framework of physical science. The success of these attempts will make the biological framework superfluous and will provide the physical basis of life. It will then be possible to describe all properties of living systems in the physical framework. These attempts have had a partial success; many properties of living systems can be described in the physical framework and the number of such properties is increasing. The two frameworks seem to have a common domain of applicability; more and more portions of the common domain are being identified. The identification of any additional common portion raises the hope that eventually the entire biological framework will be contained in the physical framework and "life" will become merely an attribute of a few physical systems. The hope has remained unfulfilled in spite of many intensive efforts; a large class of biological properties is still incomprehensible in the physical framework. The incomprehensible properties pose the question, " Why does Biology defy its integration with Science?" The question expresses our faith in the correctness of the physical paradigm and points out its inability to comprehend biological properties. Something is wrong or missing somewhere; it is proposed to find it out by examining the basic ingredients of the two paradigms.

The basic ingredients of the physical paradigm are the concepts of space, time, matter, interactions, and the framework of quantum theory. The paradigm is capable of describing all known non-living material systems from quarks (10^{-13}cm) to cosmos (billion light years) at time scales from Planck's time (10^{-43} s) to cosmic scale (billion years)[1]. The framework used in the physical paradigm can incorporate all known basic interactions (with a possible though not final exception of general relativity) in a unified manner. The framework was developed in 1925-1926 and has not changed in any essential way. The physical paradigm has an unprecedented success in describing and predicting phenomena; no experimental evidence has been found so far that would contradict its predictions. There is no reason to doubt its validity and there is no scope or reason for any change in the paradigm. The living systems do come within the ambit of the physical paradigm and need to be described in it. The physical paradigm, however, has one disadvantage – its framework is cumbersome. The framework is based on a theory of purely mathematical nature, which makes the interpretations incomprehensible and controversial. As a result, one tries to avoid to the maximum extent the use of the correct physical paradigm based on quantum theory; one instead, resorts to approximate and abridged versions of the paradigm. The abridged versions are variants of the physical paradigm that utilise the framework of classical physics for describing many aspects of a phenomenon. The variant paradigms usually depict the behaviour of many properties with sufficient accuracy and their interpretations are much simpler. The variant paradigms are widely used in Biology. It may be noted that though variant paradigms retain the concepts of space, time, matter and interactions envisaged in the physical paradigm, yet are not approximations of the physical paradigm. The variant paradigms may yield erroneous results in some situations. The problem of getting erroneous results also exists in non-living systems where the descriptions based on classical physics of a few properties turn out to be incorrect and paradoxical. Many paradoxical properties have been identified in non-living systems and methods have been devised to circumvent the situations where these properties play significant roles. Similar situations may also occur in living systems, where variant paradigms will yield incorrect and paradoxical results for some properties. It is our contention that the problems encountered in the integration of Biology with the web of knowledge and in the demystification of life emanate from the use of variant paradigms in these situations. These situations have not been identified in living systems. Comparing the predictions of quantum theory with classical physics may give some clue to identify these situations.

Quantum theory makes predictions of the various attributes of a system. Some attributes have classical counterparts; the classical values of these attributes are obtainable in the classical framework. The classical value of an attribute may differ with the value predicted in quantum theory. The classical framework can take cognisance of this difference by an arbitrary prescription of replacing the classical value of an attribute by its predicted quantum value. The arbitrary prescription is successful in incorporating many quantum effects, e.g. uncertainties in the measurements of canonical conjugate quantities i.e. position q and momentum p, quantized values of energy and angular momentum, discreteness of quantum states, capability of a quantum system to tunnel through a potential barrier, resonating states, dependence of spin and statistics etc. The classical framework incorporates the influence of these quantum effects in biological properties by arbitrary phenomenological prescriptions. The incorporation converts the classical framework into a semi-classical framework that retains the advantage of classical interpretations and provides a better description of biological properties. The arbitrary prescription does not succeed in incorporating the influence of quantum effects without classical counterparts e.g. holistic behaviour, superposition of all possible paths, phase of wave functions, non-locality, entanglement, and quantum selection [2]. These are called subtle quantum effects. A property arising from subtle quantum effects in either living or non-living systems will appear counter intuitive and paradoxical in both classical and semi-classical frameworks. One can reverse the argument and claim that a property problematic in the semi-classical framework is a manifestation of some subtle quantum effect. One can even conjecture that problematic biological properties are manifestations of quantum effects. The holistic nature of these properties gives additional support to the conjecture.

A holistic property implies that the system while displaying a holistic property acts as a single unit. The individual identity of the spatially separated parts or constituents of the system is obliterated and these parts appear to function in a coordinated and cooperative manner. All living systems exhibit some holistic properties; the characteristic time scales of phenomena connected to a holistic property range from microseconds to years. The current paradigm used in the description of biological phenomena is based on the semi-classical framework of molecular biology. This framework is very successful in describing non-holistic biological

The physical basis of life

phenomena. It correctly identifies biomolecules, the elementary constituents of living systems, but assumes that biomolecules interact only via local chemical reactions. The framework provides a pictorial language of description, whose alphabets are biomolecules. All biological phenomena manifest only through biomolecules. The manifestation of a phenomenon may involve many stages; each stage depicts a chemical reaction. All stages have been identified only in a few phenomena; these phenomena appear as chains of regulated reactions. A chain is molecular pathway and it provides the molecular basis of a phenomenon. The molecular basis of a phenomenon does not contain information about the mechanism regulating different reactions in a pathway and is not equivalent to its physical basis. It is hoped and presumed that the regulating mechanism of a pathway will eventually be discovered. Consequently, the molecular basis of a property is usually considered at par with its physical basis.

The set of properties with known or knowable molecular basis constitutes an important class of properties in living systems. The class is named microscopic to emphasize the possibility of description of this class of properties using microscopic constituents. The other remaining properties of living systems, whose molecular basis is not knowable, constitute the class of holistic properties. This class needs holistic description in the framework of molecular biology. A holistic description requires postulating the existence of a mechanism for co-operative functioning of biomolecules located at distances many times larger than their sizes and inter-molecular distances. The descriptions of different holistic properties require different types of mechanisms; the nature of the mechanism depends upon the characteristic time of the holistic property. If the characteristic time is larger than 1ms, the typical time of bio-molecular pathways, then the mechanism will be required to integrate the effects of many molecular reactions of about 1ms duration into an effect observable at larger times. If the characteristic time is smaller than 1ms, then the mechanism will be required to be operative only during smaller time intervals say, of about a few microseconds. The mechanism should also wash out its effect at larger time intervals of the order of 1ms to retain the individualistic behaviour of molecules in chemical reactions. The requirements on the nature of mechanism for cooperative functioning divide the holistic class of properties into two sub classes; macroscopic class with characteristic time larger than 1ms and consciousness class with characteristic time smaller than 1ms. The semi-classical framework can meet in principle the requirements of the integrating mechanism, so that a macroscopic property appears comprehensible with a gap in its understanding. There are too many macroscopic properties and hence too many gaps in understanding. Many gaps have remained unfilled for too long. Perhaps, the gaps are unbridgeable and the integrating mechanism does not exist in reality. The requirements of the other mechanism cannot be met even in principle in the semi-classical framework, so that the properties in the consciousness class will remain incomprehensible. The consciousness class has always been an enigma to the semi-classical framework; many attempts have been made to deny the existence of this class by attributing it to some benign illusion.

The existing framework of molecular biology fails to describe a holistic property of macroscopic or consciousness class. We attribute the failure to the use of the semi-classical framework. The semi-classical framework needs to be replaced by the framework of quantum theory. The replacement can be done in a manner that retains the success of molecular biology. The replaced framework will recognise the existence of quantum states with classical counterparts characterised by biomolecules and will provide the correct physical paradigm. The correct physical paradigm retains the pictorial language of biomolecules and molecular pathways for describing many permissible quantum states. These states are responsible for various properties belonging to the microscopic class. In addition to these states, there are m quantum states of a living system not specifiable by biomolecules; these additional states are responsible for holistic properties. A living system in these states will function as a single unit and will behave like a macroscopic quantum object. It is not necessary for a living system to remain a macroscopic quantum object throughout its life; it can be a collage of objects, which attain quantum nature for small and non-synchronous durations. The holistic properties of a quantum object are observable as correlations among its constituents in the semi-classical framework. Some of these correlations have classical counterparts, so that the properties depending on them can be described in the semi- classical framework. These are a few additional properties of the living systems that can be put in the microscopic class. The majority of correlations are without classical counterparts; these give rise to holistic properties. The framework of quantum theory does not ascribe a special significance to the characteristic time of a property and hence does not provide a reason to divide holistic properties into two classes based on characteristic time. The origin of characteristic time lies somewhere else. It is suspected that the time enters in the paradigm through the transfer and processing of information. The speed of information processing has to be much higher in the consciousness class of properties than in the macroscopic class of properties. Information is a new ingredient of the paradigm. It also has different features in classical and quantum frameworks. The difference manifests in consciousness properties involving a fast processing and transfer of information. The consciousness properties in the quantum framework of two types; one is associated with the classically visualized aspects of information, while the other is associated with purely quantum and non- classical aspects of information. The former type will appear logical and the latter will appear counter intuitive.

The experimental support for the quantum nature of living systems comes from the phenomenon of biophoton emission [3]. Biophoton emission is the incessant emission of photons of unknown origin by living systems. The emitted photons are called biophotons and the signal is called biophoton signal. Biophoton signals have characteristic features and are easy to identify. Almost all features of biophoton signal are incomprehensible in the semi-classical framework of molecular biology. Popp suggested that the reason for incomprehensibility lies in the origin of the signal; a biophoton signal is emitted in a holistic process and it is futile to seek the source of the signal in a biomolecule. Popp visualized a holistic process as coherent in the semi-classical framework and concluded that the signal emitted in the coherent process should also be coherent. The speculations about a living system in the physical paradigm concur with the suggestion of Popp and visualize the nature of a holistic process to be quantum. Many features of biophoton signals reflect the quantum nature. The quantum nature makes these features incomprehensible in the semi-classical framework. The behaviour of a quantum selection process provides another support to our speculations in the physical paradigm. Quantum selection is the most efficient selection process [4]. If a living system resorts to quantum selection in matching during replication, transcription and protein synthesis in the optimal way, then the number of steps required in selection mode gives for the first time a reason for the occurrence of 4 types of nucleic acid bases and 20 types of amino acids and for the coding of amino acids by codons or triplets of nucleic acids [5].

The experimental observations mentioned in the previous paragraph have put the physical paradigm on the centre stage of biological thoughts. The paradigm needs to be strengthened by incorporating other quantum aspects and exploring their implications. We therefore, give a brief exposition of some quantum aspects relevant in the study of living systems in section 2. These aspects will form the background for the discussions of subsequent sections. The experimental evidence in support of the physical paradigm provided by the phenomenon of biophoton emission is critically examined in section 3. This section enumerates the critical features of biophoton signals along with the nature of difficulties encountered by the semi-classical framework in explaining these features. The sensitivity of a biophoton signal to various extrinsic and intrinsic parameters is the basis of connection between biophoton signals and holistic properties. A biophoton signal is thought to portray the dynamical behaviour of a living system and to carry digitalised signatures of ongoing holistic processes. Even though the signatures of individual processes have not been deciphered, yet the signal has been put to many uses. The deciphering of signatures is essential for proper understanding of the physical basis of life. It will require joining the various pieces of evidence in a model. The rudiments of an emerging model are presented in section 4. The model gives a few glimpses of the quantum vision of life.

2. SOME SUBTLE ASPECTS OF QUANTUM THEORY

The physical framework can describe incompletely and completely defined systems. Thermodynamics and statistical physics give descriptions of incompletely defined systems, while mechanics and electrodynamics provide descriptions of completely defined systems. The framework of descriptions can be classical or quantum for both types of systems. The observed properties in both types of systems reveal the supremacy of the quantum framework. Let us first consider the incompletely defined systems; these are macroscopic systems encountered in daily life and their properties are deeply ingrained in our psyche as routine experience. These systems obey the two laws of thermodynamics; both classical and quantum descriptions must incorporate these laws. The reason for the validity of these laws goes much deeper and the origin of these laws can be traced back to the basic assumption about the space-time structure of the physical paradigm. The First Law states that heat is a measurable attribute of a physical system; it is a form of energy that can be exchanged with the other forms of energy of the system or its environment in such a way that the total energy is conserved. The validity of the Law is not statistical; it is a consequence of the homogeneity of time and it is valid in each energy transfer. The subtlety of the First law is worth clarifying in connection with the uncertainty relation $\Delta E \Delta t \geq \hbar/2$ between energy E and time t. The origin of this uncertainty relation and its interpretation are different from similar looking uncertainty relations between canonical conjugate variables. This uncertainty relation is not derivable from the usual consideration and is on a different footing. Time is merely a parameter in the quantum theory. It is not represented by a hermitian operator and is not an observable in the strict sense. The meaning of Δt in the uncertainty relation is the time required for the completion of a process that changes the energy of a system by an amount ΔE. The process must fulfil all requirements including the conservation of energy. Thus if a system decays spontaneously by radiating energy ΔE, then the time of decay should be greater than $\hbar/2\Delta E$. The uncertainty relation can also be interpreted to give an estimate of the time required in the decay of a broadband state. If the energy spread or width of a broadband state is ΔE,

The physical basis of life

then the lifetime Δt of the state will be of the order $\hbar/2\Delta E$. This relation at no stage implies non-conservation of energy by an amount ΔE during a time interval Δt. The relation does not sanction the occurrence of non-energetic transitions. The relation does provide an estimate of the minimum amount of energy needed in a transition, e.g. if the death of an organism is modelled as a transition from one state to another occurring in say, about 1μs, then the minimum energy released in the transition should be of the order of 10^{-21} erg. If it is further assumed that the death is a manifestation of the extrication of the ghost (or some other variant) from a biological machine, then the above estimate gives the allowed minimum energy of the ghost. The estimate of energy will be much smaller if the ghost takes a larger time to extricate.

The Second Law of thermodynamics is again a universal law valid in both classical and quantum contexts. The law gives the direction of energy flow among systems. An important consequence of the law is that the transfer of energy from one form to other is not always bi-directional so that different forms of energy can be graded. The direction of energy flow is determined by a measurable attribute called entropy [6]. The change in the entropy of a thermodynamic system is measured by the flow of heat energy. The Second Law encounters problems in the classical framework. The two well-known problems are entropy of mixing in ideal gases and existence of Maxwell demon. Both problems remain insurmountable in the classical framework but are easily resolved in the quantum framework. The resolution brings out two aspects of entropy, which are relevant for understanding the physical basis of life.

The entropy of mixing is the increase in entropy because of mixing of two ideal gases at the same temperature and pressure. The entropy of mixing depends on the identity of the two gases. If the two gases are of different species however close they may be, the entropy of mixing is always positive. However, if the gases are identical, then it is always zero. The behaviour is problematic in the classical framework because identity of gases has no role in energy transactions. The dependence of entropy on the identity of molecules is a new aspect of entropy. The new aspect is easily understandable in the quantum framework where entropy of an incompletely defined system is given by the number of possible quantum states of the system. The number of quantum states remains unaltered when two identical gases are mixed. The number of quantum states increases when two non-identical gases are mixed. The increase gives a contribution to the entropy that precisely matches with the entropy of mixing. The resolution of the problem links the flow of energy from or to a system to the changes in the number of its possible quantum states. It is noted that the amount of energy flow is a measurable quantity while the number of possible states of the system is a calculable attribute in a mathematical construction. In some processes energy flow is easily measurable while in others the changes in the number of quantum states is easily calculable. The examples of the processes of the later kind are selection and measurement processes, which decrease the number of possible quantum states. These processes must be accompanied by a flow of energy from the system to the environment. These are essential processes for any act of recognition.

The possibility of Maxwell's demon has implications that are more bizarre in the classical framework. Maxwell conceived a demonic device that continuously watches the movement of molecules of a gas contained in two partitions separated by a frictionless sliding door. The demon has a capacity to operate the door at will. The demon uses his capacity in a cunning manner. He watches the molecules at both partitions to collect information about their velocities and then uses this information while operating the door. He opens the door only for a small duration in which either fast moving molecules of the first partition move into the second partition or slow moving molecules of the second partition move into the first partition or both. The operation of the door regulates the movement of molecules, which cools the first partition and heats the second. The regulated movement gives the demon a capability to generate temperature difference between two portions of a gas. The capability can be utilised for doing mechanical work or for generating new structures. The capability assumes a demonic character in classical physics, where the expenditure of energy incurred in gathering information about the velocities of molecules and in operating a frictionless door can be made negligible. The demon can perpetually generate structures without any wastage of energy, which is a violation of the Second Law. Since the existence of such a device is not ruled out by any physical considerations, Maxwell's demon has been a serious problem for the validity of classical physical theories.

It is interesting to note that the Maxwell demon is not problematic in the quantum framework because the energy required in the functioning of the demon is neither negligible nor reducible to zero. Let us analyse the energy requirement in each step. The demon receives information of the state of a molecule from a light wave reflected from it. The light wave carries energy and its production involves an expenditure of energy. The minimum energy carried by a light wave of frequency ω is $\hbar\omega$, where $\hbar$ is the usual Planck's constant. It is the energy of a single quantum or photon. The frequency of the photon has to be greater than the background noise created by the black body radiation for meaningful transfer of information i.e. $\hbar\omega > \frac{1}{2} kT$, where k is the Boltzmann constant and T is the temperature. The demon does not use this information instantaneously but stores it for future use in its memory. The information is always stored as a physical attribute. So that a memory has to be a physical device and its storage capacity will necessarily be limited. The memory will get filled after a while: it will make the demon non-functional. The limited storage capacity does not allow the demon to act perpetually. The demon can again become functional by erasing the contents of its memory. Erasing of the memory reverts the demon back to its original state but it requires an expenditure of energy. It has been estimated that the energy required in erasing 1 bit of memory is $\frac{1}{2}kT$. The amount of energy required in erasing the entire memory ensures the validity of the Second Law. The energy spent in erasing the contents of the memory is thought to be negligible in classical physics and was not even mentioned in the original formulation of the problem. The resolution of the problem has two important consequences: Maxwell demon has become a desirable device and the use of the language of information theory [7]. A physical device acting like a Maxwell's demon can create structures in a system and perform regulatory functions by extracting energy from a source other than the system later. The device merely acts as a catalyst and can continue to function perpetually. Many protein molecules form structures and perform different regulatory functions by acting like Maxwell demons.

The language based on information theory can describe a reaction or a process in both completely and incompletely defined systems, e.g. a protein molecule stores information in its conformational state, uses this information to make a choice while performing a regulatory function, and reverts back to its original state after receiving energy from the biochemical machinery. Information is a technical word and a calculable attribute of a state either of a molecule or of a thermodynamic system. Any physical attributes of a state of a system can code information and the amount of information coded in the state is calculated by the number of values allowed for the physical attribute, e.g. if an attribute of a system has two possible values with equal probability and the system is in a state with one definite value of the attribute, then the information of the state is 1 bit. The negative of the entropy in any state of a thermodynamic system gives the amount of information in the system.

The relation between information and entropy in a thermodynamic system implies many properties of information. These properties are valid even in well-defined systems where the concept of entropy is not applicable. Thus, information or a part of it can be transferred from one system to another; all information transfers tend to decrease or degrade the total information; an information transfer is usually accompanied by a flow of energy; and any structure formation in a system increases the information. These properties are valid in individual reactions or processes of molecules. The energy need not flow as heat energy in such reactions but can flow in a pure quantum state. Heat energy is described as a mixed state with random phases in the quantum framework while the energy in a pure state has a well-defined phase. Energy flowing in pure quantum state carries information about the process causing information transfer. The electromagnetic nature of reactions involving biomolecules causes the energy to flow as a photon or electromagnetic signal.

The properties of an electromagnetic signal in a pure state are different from the properties of a signal in a mixed state. The difference is observable in the effects of interference, superposition and entanglement. These effects are detected by measuring various space-time correlations. Pure quantum signals interfere and do not require any additional machinery to make them coherent. Quantum signals superpose and any number of quantum signals can be described by a single signal. A photon signal in a pure quantum state may be composite of many signals emanating from different domains or patches. There is another aspect of superposition- if a photon is presented with more than one path it will not choose one path among them; it will enter into a state of superposition of all possible paths. The situation appears as if the photon is partly following one path and partly following other paths. This will occur so long as no observation is made to determine which path was taken. The superposition affects the values of various conditional probabilities. An easily measurable conditional probability is $P_o(n, \Delta)$ the probability of no photon detection in a small time interval Δ after the detection of a photon. This probability and its dependence on signal strength n are used in identifying the presence of quantum signals [8]. Entanglement is yet another purely quantum effect of composite quantum systems. Information in entangled systems mainly resides in the correlations than in the properties of individuals; one can learn rather precisely little from the study of individuals. If a living system is in an entangled state of quantum patches, the study of the quantum patches will not give much information. A biophoton signal can also be entangled with its emitting system. The entangled signals can be used in transmitting uncorrupted and encrypted information to other systems.

3. FEATURES OF BIOPHOTON SIGNALS

A biophoton signal is determined by the number of photons detected in the successive bins of duration Δt by a photo multiplier from an in vivo sample placed inside a dark chamber. Most of the measuring systems have usually a provision to stimulate the sample by visible or ultraviolet radiation. The controllable parameter of stimulating radiation are intensity I, time of stimulation τ, wavelength λ_{st}, and polarisation ε_{st}. One can determine similar parameters of the emitted radiation, which will be the time t at the middle of a bin, wavelength λ, and polarisation ε. The dependence of the number of photons n detected in a bin on various factors is explicitly shown by writing n (I, τ, λ_{st}, ε_{st}; t, λ, ε, Δt). The dependence of n on I and τ is complex and shows a saturation effect. The dependence becomes very weak in the saturation regime and the signal becomes stable. The measurements are done in this regime. The choice of I and τ is specific to a measuring set-up. τ is kept from 5s to 10s. The sample is usually stimulated with visible light, though stimulation by UV light has been found to be more useful in cell cultures. The determination of the wavelength of stimulation requires the use of a mono-chromator. No experiment has been performed so for in which a polarised light excites the sample. The signal emitted by a sample is very strong just after stimulation but most of the contribution to the signal comes from fluorescence processes, which becomes negligible in a few milliseconds. As a result, a biophoton signal is measurable 5 to 10ms after the stimulation. A delay of 5ms is usually built in most measuring systems so as to protect the scintillation counters of the system from intense fluorescence signals. The choice of bin interval Δt depends on the system but has to be larger than 10µs. The properties of the scintillation crystal and the electronics (TTL logic) set this lower limit. The contribution of different wavelengths in the emitted signal is determined by the use of band pass or interference filters. The signal is very weak after passing through interference filters. Any measurement of the polarisation of the emitted signal has not been made. The behaviour of a typical biophoton signal is depicted in Fig.1. The figure is not drawn to any scale and is indicative only.

Different linear time scales depict pre, during and post stimulation regions of the signal. The measurements are made only for a few minutes before the light stimulation. The signal in this region is essentially constant and is only a little more intense than the background noise. The measurements are difficult during the light stimulation of a sample; one has not been able to separate the signal from the stimulating light. Consequently, the signal in the region 1 has not been measured so far. The sharp rise shown in the figure is based on expectations from observing the signals with stimulations of short durations. The region 2 of the signal is the most widely observed region. It has been observed from bacteria to human tissues. The duration of this region is a bit longer (of $\cong$200s) only in systems containing chlorophyll. The signal's shape and strength in this region is very sensitive to environmental conditions and physiological parameters. Finally, the region 3 is observable for hours. The average value of the signal is structure less but the signal has fluctuations, which are different from white noise in living systems.

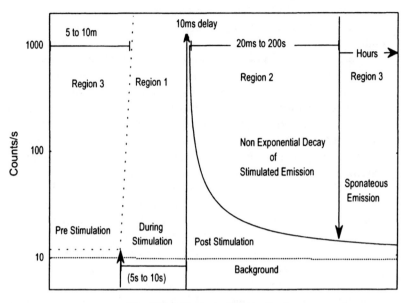

Figure 1: Different regions of a typical biophoton signal: Various regions of a biophoton signal are depicted. The portion of the signal during stimulation is not observable.

The amount of available biophoton data is very vast and is growing rapidly. The objectives of the different types of measurements have been to identify the features of biophoton signals that are unexplainable in the semi-classical framework and to determine the links between biophoton signals and the parameters of a living system. These measurements establish the holistic nature of biophoton emission. The web site www.lifescientists.de is a good source of biophoton data. A brief description of the repeatedly observed unexplained features relevant for understanding the physical basis of life is given below.

3.1 Ultra Weak Strength Mainly in the Visible Region [3]

Biophoton signals are usually detected by wide band scintillation counters sensitive in the range (300-800) nm. The choice of the band indicates that the biophotons lie mostly in the visible range. The spectral distribution of a signal is determined by the use of interference and band pass filters. The spectrum of a biophoton signal is continuous with a substantially large portion in the red region. The spectral distribution varies from system to system and does not resemble the spectral distribution of black body radiation. The strength of the signal is also many orders of magnitude higher than that of a black body radiation. Emission of a photon in the visible region needs (3-5) eV of energy, which is much higher than the energy available in the (ATP-ADP) biochemical machinery. The up-conversion of low energy photons of biochemical machinery into visible range photons needs a mechanism that has not been discovered so far. The ultra weak strength of the signal is another problematic feature; the strength is 6 to 7 orders of magnitude weaker than expected in a forbidden transition from the number of atoms or molecules present in a living system. The behaviour of a biophoton signal does not allow attributing its origin to some rare random events.

3.2 Dependence of Biophoton Signal on Stimulating Frequency

The entire range of visible light can stimulate a living system to emit a biophoton signal. Non-damaging ultraviolet radiation also stimulates living systems to emit biophoton signals. The stimulation by infrared radiations is usually negligible. The wavelength of the stimulating radiation affects the strength of the biophoton signal but not its shape. If the strength is measured by counts registered in the first measuring bin, then the curve $n(\lambda_{st})$ gives the dependence of the strength on the wavelength of stimulating radiation. The curve has a few broad maxima; the properties of these maxima namely, position, width and height, appear to characterize a living system as well. The spectral distribution of a biophoton signal does not depend much on the wavelength of the stimulating radiation. A noteworthy feature is the occurrence of a blue component in the biophoton signal emitted after the stimulation by red light.

3.3 Non-Exponential Nature of Decay of a Biophoton Signal

The nature of decay of both spontaneous and stimulated biophoton signals is not exponential. The signal of spontaneous biophoton emission does not show any appreciable decay for hours. A non-decaying signal cannot arise from any number of exponentially decaying terms. The signal of stimulated biophoton emission decays; its decay curve is a characteristic feature of the emitting system. The decay curve can be fitted by exponentially decaying terms but such fittings do not yield consistent and stable values of the life times and strengths of different terms. It is attributed to the non-exponential nature of the decaying signal. The use of red, blue, or green cut-off filters, reduces the strength of a biophoton signal but does not alter its nature of decay; the decay curves obtained with filters are similar in shape with the decay curve obtained without a filter. The ratio of counts with a filter to the counts without filter in a bin around a definite time after stimulation does not vary much in the entire duration of a decay. A much better fitting of a biophoton signal is obtained from the functional form $n = n_o(t_o+t)^{-m}$, where n_o, t_o, m are variable parameters. This functional dependence is referred to as hyperbolic decay in the biophoton literature. This type of functional dependence does not provide a procedure to compare the strengths of signals; n_o can be used as a strength parameter only if the value of m is kept constant. The difficulty is circumvented in the following successful form proposed for a biophoton signal

$$n(t) = B_0 + \frac{B_1}{(t_o + t)} + \frac{B_2}{(t_o + t)^2} \tag{1}$$

where B_o, B_1, B_2, and t_o are variable parameters. The functional form is based on the assumptions that a biophoton signal is in a pure quantum state; the state is a squeezed state generated by a frequency stable damped harmonic oscillator with time dependent damping and mass terms [9]. Quantum evolution of the squeezed state confers time dependence to the expectation value of the photon number operator that is observed as a time dependent signal. The shape of the signal contains information about the dynamic of the system. The parameter t_o is related to the strength of the damping and is system specific. All other parameters are situation specific. B_o gives the contribution observable as spontaneous emission, while B_1 and B_2 give the decaying contribution observable only in stimulated emission. Either B_1 or B_2 can be used as a strength parameter of the signal and their ratio B_1/B_2 as a shape of the signal.

3.4 Sensitivity of the Biophoton Signal to Physiological and Environmental Factors

Biophoton signals are sensitive to many factors; the sensitivity is much more pronounced in stimulated emission. The decay curve can sense a minute change in the physiological condition of a system or in some of its environmental factors [10]. A change in a biophotonic decay curve is easily detectable but difficult to quantify. Changes occur in both strength and shape of a signal. It is possible to calibrate a two dimensional scale consisting of B_1 and $\log(B_1/B_2)$ to detect the changes in various factors affecting a living system. A simple approach is quite often sufficient in which the decay signal is characterised by the number of counts detected in a predefined time interval. The starting point of the time interval and its duration depend upon the living system and measuring apparatus. It is a single bin measurement of duration ranging from 50ms to 3m. It is also a single point characterisation of the signal or its decay curve. We shall call it biophotonic activity of a sample. The behaviour of the biophotonic activity of a few selected systems is given below to demonstrate its linkage with the metabolic processes.

a. Lichen is a stable symbiotic association of an algae and a fungus. The metabolic activity of a sample of lichen depends strongly on its water content; so is the case with its biophotonic activity, which increases by nearly 3 orders of magnitude in a few minutes when a small amount of water is added to a dry sample of lichen [11]. The added water slowly evaporates and the wet sample becomes dry in 15 to 20 hours. The biophoton activity of the sample decreases and reverts to its earlier value. The decrease in the activity is gradual. The dry-wet cycle of biophotonic activity can be repeatedly observed because a sample of lichen is a quasi-stable system whose growth in a few weeks is hardly noticeable.

b. A fungicide can kill lichens in a few minutes. The death does not cause any visible change and one has to wait for months to ascertain the death of a lichen sample [12]. The biophotonic activity drops immediately and provides a quick and reliable method of ascertaining the death of a lichen sample. This is true for any other living sample and has been a basis of many applications involving dying material.

c. The biophotonic activity of a sample of *Daphnia Magna* changes in an oscillatory manner with the change in the cell density [3]. Similarly the biophotonic activity of cancerous and normal cells changes differently with increase in the density.

d. Chlorella is a single cell alga. A sample of cell suspension of chlorella suspends its many metabolic activities temporarily if put in the dark for a prolonged period say, 10 to 15 hours; the activities are restored to the normal rate only after 5 to 10s exposure to light. The biophotonic activity of the sample also follows the behaviour of metabolic activities. The sample if put in the dark for a prolonged period does not immediately respond to light stimulation. The level of biophotonic activity of the sample is reduced to a few percents of its normal value. However, the biophotonic activity is restored to its normal value after 5 to 10s exposure to light. The biophotonic activity of decaying leaves also exhibits similar behaviour. We monitored the biophotonic activity of many plucked leaves for 15 days. There was no delay in the response to light stimulation for the first 8 days. Subsequently, some delay was observable, which grew up to 20s in few samples.

e. Biophoton activity is considerably enhanced during mitosis and cell death [3].

3.5 Photo Count Statistics of the Signal of Spontaneous Biophoton Emission

The signal of spontaneous biophoton emission lacks any gross structure; the average number of counts per second is unchanging for hours. These signals are suitable for determining the statistics of photo counts. Many measurements have been made to determine the photo count statistics; the size of a bin (i.e. dwell time) ranges from 100ms to 3m. The mean and standard deviation of the photo count in a bin are nearly equal for different bin sizes and numbers of bins in the measurement. It is suggestive of Poisson statistics; so is the moment analysis of photo counts. The word suggestive is used because the signal to noise ratio was around 0.1 in these measurements.

3.6 Dependence of $P_o(n, \Delta)$ on Signal Strength

The conditional probability of no photon detection in an interval Δ after the detection of a photon and its dependence on signal strength has been accurately measured in the decaying portion of stimulated biophoton signals. The value of Δ was varied in the range (10µs –10ms) and it corresponded to the variation of signal strength in the range (0.05-10) counts/Δ. The measured values were around the curve

$$P_o(n, \Delta) = Exp(-n(\Delta, I)), \qquad (2)$$

where $n(\Delta, I)$ is the signal strength given by the average number of photon detected in the interval Δ in a signal of intensity I. These measurements provide sustainable evidence for the quantum nature of biophotons [13].

3.7 The Nature of Fluctuations of Spontaneous Biophoton Signals

The fluctuations observed in time series data of a spontaneous biophoton signal are not due to white noise alone. Fourier series analysis of data indicates the nature of the signal to be aperiodic. In our analysis of the large time series data of many signals, we divided the entire time series into many pieces of sufficiently long duration and performed Fourier series analysis in each piece separately. The Fourier analysis gave a few large period periodograms of higher strength in each piece; the periods and strengths of these periodograms varied from piece to piece. The variations in the periods of large strength periodograms in the pieces of a signal emitted by the cells in a medium were more than in a similar signal emitted by the medium alone. Another intriguing feature of a biophoton signal is an occasional detection of a large number of counts in a bin. It appears as if photons suddenly burst out from a living system. The number of counts in a bursting bin is 3-10 times the number of counts in its neighbouring bins. The time of emission of a photon burst is probabilistic. We identified the emission of bursts in a signal by the criterion that the photo counts in bursting bins were greater than N_{min}. We then determined the intervals between successive bursts. The distribution of the interval was usually geometrical, which implies the existence of an intrinsic measurable property of spontaneous biophoton signals. The property is conditional probability p of subsequent photon burst emission in the next bin. The probability of subsequent photon burst emission in the nth bin is related to p and is given by $p(1-p)^{n-1}$. The value of p varied with N_{min}. The variation of p with N_{min} is plotted in Figure2 for spontaneous biophoton signals emitted at 37^0C and at 4^0C by the same sample of cells. The figure also contains similar determinations of p for the signal emitted by the control sample without any cells. These measurements were simultaneously made at two different channels, so that the figure contains two sets of determinations. Both sets of measurements indicate that the value of p in the sample of cells is higher than the control sample for different choices of N_{min} and further its value in the same sample of cells at 37^0C is higher than its value at 4^0C.

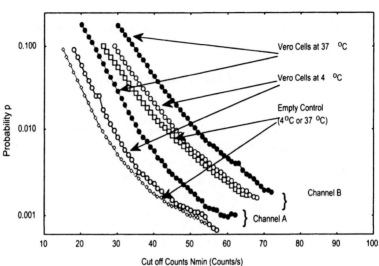

Figure 2: Probability of subsequent photon burst emission in spontaneous biophoton signals: The probability p is plotted as a function of N_{min} for empty control, Vero cells at 4^0C and 37^0C. The bin size was 1s and the probability at both temperatures in the control sample was nearly the same. The data used in the analysis is of M.Lipkind and is unpublished so far.

4. EMERGING SCENARIO

Holistic behaviour is a common feature of living systems, quantum framework and biophoton signals. It is our contention that the three are connected; the connection is slowly unfolding and endowing life with a physical basis. A living system turns out to be a collage of intermittent quantum patches mainly localised in the nucleus or mitochondria. The quantum patches are responsible for cooperative functioning and emission of biophotons. The fundamental biological processes at the genetic level are quantum in nature and can be considered as quantum selections. A selection manifests as a reaction in which some state of a system (or molecule) is picked up from many available states (or molecules) to form a new state (or compound). The selection increases the information of the state and has to be accompanied either by a decrease in the entropy of the selected system or an increase of the entropy of the surrounding environment. The system can decrease its entropy by emitting energy, so that energy-emitting selections (or exothermic reactions) can occur spontaneously. The energy is emitted as electromagnetic radiation of a definite frequency in a selection; the frequency is determined by the difference between the energies of reactants and products. It is possible to visualise energy-emitting selections in the semi-classical framework but these visualisations do not capture all features of quantum selections [14]; some important features are missed. The significant missing features are: energy emission in a pure quantum state, emission of photons in high frequency mode that appears as up-conversion of low energy photons in semi- classical framework into, and entanglement of the quantum state of emitted photons with the state of a living system. Energy emission in a pure quantum state implies that the emitted photons are coherent and ought to exhibit non-classical features. Energy emitted in the semi-classical framework is in a mixed quantum state of photon; the state is incoherent and lacks non-classical features. Another aspect of energy emission in a selection is the frequency; the frequency has a definite value both quantum and semi-classical frameworks. It is a characteristic of the selection. The emission in a selection is local process in the semi-classical framework; it can be local or non-local in the quantum framework. The non-local character can give rise to a giant selection that may appear as many simultaneous selections in the semi-classical framework. The frequency of the emitted photons can be used to ascertain the correct scenario. The frequency will be higher for giant selection; the intensity will be higher for many selections. It will appear as if many photons emanating from semi-classical selections get them converted into a high-energy photon in a giant selection. The up conversion of the photon frequency implies that the many semi- classical selections should appear to occur simultaneously and the part of the system involved in these selections behaves as a cohering unit. Only coherent quantum systems can produce a high-energy biophoton in place of many low energy photons expected in the semi-classical framework. The up-conversion of electromagnetic energy may be viewed as an indicator of the coherence and quantum nature of living systems. Entanglement is again a purely quantum phenomenon without classical counterparts. The photons emitted in the quantum state can remain entangled with the system and do not acquire separate identity. The system can up-convert entangled low energy photons into a high-energy photon. Further, all low energy photons need not be emitted at the same time in the semi-classical sense. It calls for attributing a wider meaning to simultaneous selections; all selections whose energies remain entangled should be considered as simultaneous selections. Perhaps, a more appropriate word for selections capable of up-converting energy is coherent selections. Both quantum coherence of a system and entanglement of energy state do not last for a long time because of the presence of de-cohering interactions with the environment. Even the detection of an emitted photon spoils the entanglement of energy and makes uncontrollable changes in the system, which may make the system describable in the semi-classical framework.

The classical and quantum selections are not equally efficacious in selecting an object. An elementary selection step is capable of picking a desired object from a collection of two objects in a classical selection and from a collection of four objects in a quantum selection. Similarly, three elementary selection steps are capable of picking the desired object from a collection of eight objects in a classical selection and from a collection of 20 objects in a quantum selection. The selection made in three elementary steps is slightly erroneous in the quantum case but is exact in the classical case. The occurrence of (1 step, 4 objects) and (3 steps, 20 objects) in quantum selections appears to have a deeper relevance for living systems. The numbers in the two pairs coincide with the numbers occurring in the two fundamental selection processes in molecular biology, namely; one nucleic acid molecule selects its base pair from 4 different types of nucleic acids and a sequence of 3 nucleic acid molecules, called codon, codes for one of the 20 amino acids. The coincidence is too glaring to be dismissed as incidental. Perhaps, it is indicative of the quantum nature of fundamental biological processes occurring at the genetic level such as transcription, signalling, and protein synthesis etc. The base pairing is a one step selection process and each nucleic acid makes its own selection. A single step pairing is only possible with four types of nucleic acids; the pairing is error free and of high fidelity. The occurrence of 5 or more types of nucleic acids would require more elaborate machinery for base pairing. The selection of an amino acid is a three steps process; the steps are counted by the three nucleic acids constituting a codon. The occurrence of more than 20 types of amino acids would require more than 3 steps in quantum selections, while less than 20 amino acids would not be an optimal utilisation of the capability of a codon. One can visualise that evolution has endowed living systems with a capability to use the most efficient and optimally efficacious method of selection. It is an intriguing scenario; it requires supporting evidence that can come from the study of the properties of photons emitted in these selections. As pointed out earlier, the photons in a quantum selection should be in a pure quantum state and of frequency determined by the energy balance in the selections. The energy required to be transferred in the selection of a single base pair or an amino acid is very small and photons with such small energy have not been detected so far. This rules out the selection of a single base pair or amino acid. It appears that thousands of these semi-classical selections occur coherently and these selections give rise to photons mostly in the visible range. We do not know what makes these selections coherent and why coherence is not extended to millions or billions of semi-classical selections. Since millions or billions of coherent selections will produce ultraviolet or more energetic photons that are capable of destroying most living systems, it is possible to attribute some role to evolution in restricting the number of units (of nucleic acids or codons) participating in coherent selections to a range that produces visible range photons. Alternatively, it can be taken as a phenomenological input; the reason for its validity in all living systems needs to be uncovered. The restriction to the number of units involved in a coherent selection will probably arise from de-cohering interactions, which restrict the size and duration of quantum patch in non-living objects. The de-cohering interactions do permit the development of many finite size quantum patches for short durations. It is conceivable that the interaction between quantum genetic material and its in vivo environment is such that many finite size quantum patches exist in a living system. It is possible to conjecture a scenario in which a living system is a collage of many coherent quantum patches or domains, the size of a quantum patch does not vary much in different living systems, the coherence of a patch remains intact only for a small duration, and quantum patches are continuously created and destroyed. It is a genetic level scenario for its motivation is the assumed capability of nucleic acids to make quantum selections and observation of biophoton emission. Since nucleic acids are localised in a small region inside a nucleus or mitochondria, the quantum patches should also be localised mainly in the nucleus or mitochondria. The biophotons emitted by these quantum patches should appear to emanate mostly from the nucleus and mitochondria.

Only a bare skeleton of a toy model has emerged so far, that too with many missing links. The model needs dressing up; it needs a device to store information and instructions, and a mechanism to supply energy. Information and instructions need to be stored for directing and coordinating actions of different patches; energy is needed to perform these actions. Protein molecules can undertake both types of responsibility. One type of protein molecules can store information while another type can act as transducer to convert chemical energy into metabolic energy. The protein molecules can perform these functions perpetually only in the quantum framework. These functions require the association of both types of protein molecules with a coherent quantum patch. The association will increase the size of a quantum patch and make it more susceptible to de-cohering interactions. Similarly, the loss of coherence of a patch will loosen the association of protein molecules. The molecules supplying energy seem to recoup the lost energy individually and incoherently by the normal biochemical machinery. Such a molecule acts like Maxwell's demon and uses the incoherent biochemical energy for generating coherent structures and for producing photons in pure quantum states. The molecules storing information and instructions probably keep information and instructions intact for some time even after the loss of coherence of the patch. The other quantum patches can access this information and coordinate their actions accordingly. The retaining of information provides continuity and is essential for a smooth functioning of a living system. The efficient management of the memory resources requires the erasure of superfluous information. The erasure has to be need based and should be quickly accomplished. Erasure utilises energy and produces photons. The erasure of the memory associated with a large number of patches in an interval will produce a large number of photons in the interval. These photons will appear as a burst of photons if the interval is small. The need-based erasure of memory will depend upon the ongoing dynamics of the system and is likely to be probabilistic in the quantum framework. The probability of emission of a subsequent photon burst is expected to be sensitive to physiological parameters and environmental conditions. This probability is a measurable in spontaneous biophoton signals and it provides a new attribute of living systems, whose significance is similar to the strength of a stimulated biophoton signal.

Biophoton signal plays a passive role in fulfilling the thermodynamic requirements in fundamental biological processes; this role is instrumental in the emergence of a quantum vision of life. A biophoton signal can also play an active role, e.g. it may trigger some processes or coordinate the holistic functioning of the living system. The signal can play an active role with its different attributes, which can carry information and can transmit it at the speed of light. The information can be put to use in many ways. The information may contain instructions to coordinate the functioning of different quantum patches; it requires continuous association of the signal with the living system. The information may contain messages for other systems; it requires in other systems a capability to detect a biophoton signal and decipher its messages. The information may also be used for exploring the environment; it requires radar-like capability in the living system to infer from the differences in the emitted and returned signals. These are some of the possible active roles of biophoton signals. These roles appear speculative at present because of the inability of a non-living detector to decipher the message contained in low intensity quantum signals. Our inability should not be used to rule out the possibility of a superior detecting capability of living systems. Evolution may have conferred a living system with capability to detect the quantum state of a biophoton signal and decipher the message contained in it. If it is so, then living systems are equipped with a hitherto undiscovered optical mode of communication, in which information is transmitted via quantum channels that are more efficient, specific, and secure than classical channels. The existence of such a mode of communication has been indicated by experiments with onion roots, yeast, bacteria, fish eggs etc. These experiments observe the influence of optical connectivity between two living systems and demonstrate the transfer of information in an optical mode [15]. The information transferred shows specificity. A biophoton signal is expected to contain information specific to the emitting system but this information has a component that elicits observable response from some living systems. It appears as if a living system can detect only a specific type of biophoton signals. The specificity in biophoton signals is built in at both emission and detection level. Both types of specificity can be attributed to the quantum nature of metabolic processes. Specificity is a hallmark of living systems; the semi-classical framework attempts to incorporate specificity by postulating the existence of some hidden fields, like morphogenetic fields. These fields have never been discovered and quite a few properties of these fields are unphysical. The quantum vision of life explains specificity of living systems without recourse to any new field. Specificity is merely a manifestation of the involvement of pure quantum states in the active role of biophoton signals. The information transfer is almost loss less in squeezed and coherent states and these signals are detectable even at intensities below the noise level. The information transfer through a quantum channel offers a possibility of sending a completely secure message that is decipherable only by the intended recipient.

The ability of some persons to perform arithmetical computations faster than a digital computer [16] and to acquire knowledge of events at unconnected locations [17] (telepathy) provide supportive evidence of the involvement of electromagnetic signals in living systems. The former ability implies that some persons possess a better technique to process information contained in electromagnetic signals in their brains; they perhaps, employ parallel or quantum processing techniques. The latter ability implies that the information about events does not decay completely but persists for a long time and some persons can decipher this information. The information stored in a stable state of electromagnetic radiation does remain intact for a long time. If the stable states happen to be low energy pure quantum states of a photon, then only biological detectors will be able to read it. Biophoton signals are suitable for storing information and coordinating its processing can be a biophoton signal or its variants.

REFERENCES

1. Weinberg, S. (1977) *The First Three Minutes: A Modern View of the origin of the Universe*. Basic Books, New York.

2. Aspect, A., Grangier, P. and Roger, G. (1982) Experimental realization of Einsein-Popdlsky-Rosen-Bohm-Gedanken experiment:a new violation of Bell's inequalities. *Phy. Rev. Lett.*, **49**, 91-94.

3. Popp, F.A. (1992) Some Essential Questions of Biophoton Research and Probable Answers. *Recent Advances in Biophoton Research and its Applications*, Popp F.A., Li, K.H. and Gu, Q. (eds.), World Scientific, Singapore, 1-46.

4. Grover L. (2001) Searching with Quantum Computers *Dr. Dobb's Journal.* **April 2001**. (Also http://xxx.lanl.gov/abs/quant-ph/ **95122032** and Grover, L. K. "From Schrodinger's Equation to the Quantum Search Algorithm, http:// www.bell-labs.com/user/lkgrover/papers.html).

5. Patel A. (2000) Quantum Algorithms and the Genetic Code. http://xxx.lanl.gov/abs/quant-ph/ /0002037.

6. Denbigh, K.G., and Denbigh, J.S. (1985) *Entropy in Relation to Incomplete Knowledge.* Cambridge University Press.

7. Loewenstein, W.R. (1999) *The Touchstone of Life.* Oxford University Press.

8. Walls, D.F. and Milburn, G.J. (1994) *Quantum Optics.* Springer-Verlag, Berlin.

9. Bajpai, R.P., Kumar S. and Sivadasan, V.A (1998) Biophoton emission in the evolution of a squeezed state of frequency stable damped oscillator. *Applied Mathematics and Computation*, **93**, 277-288.

10. Bajpai, P.K. and Bajpai, R.P. (1996) Biphasic decay of biophoton emission from plant tissues. *Asian Jour. Phys.*, **5**,139-150.

11. Bajpai, R.P. and Bajpai, P.K. (1992) Light-induced Biophotonic Emission from Plant tissues. *Jour. Bio. and Chemi-luminescence* ,**7**, 177-184.

12. Bajpai, P.K., Bajpai, R.P. and Sinha, G.P. (1992) A Fast Technique for Investigating the effect of Biocides on Lichens. *Int. Jour. Bio-deterioration and Bio-degradation*, **30**, 1-8.

13. Bajpai, R.P. (1999) Coherent Nature of the radiation emitted in delayed luminescence of leaves. *Jour. Theo. Bio.*, **198**, 287-299.

14. There are many web sites giving introductory account and references on various quantum aspects. A few of them are: http://www.imaph.tu-bs.de/qi/lecture/qinfl0_html; http://www.research.ibm.com/quantuminfo/teleportation/; http://www.qubit.org/QuantumComputationFAQ.html.

15. Beloussov, L., Popp, F.A., Voeikov, V., and von Wijk R. (2000) *Biophotonics and Coherent Systems*, Moscow University Press.

16. Shakutala Devi is considered to be a human computer. She holds an undisputed place in Guinness Book of Records for multiplying two 13 digit numbers in 1980 in 28s. She extracted the 23rd root of a 201 digits number in 50s.

17. Narby J. (1998) *The Cosmic Serpent: DNA and the Origins of Knowledge* Tarcher/Putnam.

Chapter 14

DEFINITION OF CONSCIOUSNESS
IMPOSSIBLE AND UNNECESSARY?

Michael Lipkind
Unit of Molecular Virology,
Kimron Veterinary Institute,
Beit Dagan, PO Box 12,
50250 ISRAEL;
International Institute of Biophysics,
Neuss-Hombroich, D-41472 GERMANY

PREFACE

Wow! What a title! An author daring to write something under such a title immediately makes himself look ridiculous. He must either be totally ignorant of what is going on in the field, or have up his sleeve something like a panacea or 'philosophical stone' - a recipe for the solution of a problem on which the greatest minds of mankind have broken their brains. The author's arrogant impudence is even more emphasized by his double affiliation, which appears rather suspicious and not suited to an attack on such an impregnable bulwark that has excited and tried human intellectual capacity since the very outset of civilization. Moreover, if this is just another essay or review, it is doubtful that something new in principle can be found: everything has already been masticated many times over. Therefore, it may seen that only a lazy curiosity could make a reader go on reading the paper. The author can only declare that the title's intention is quite serious and this presentation is not a review.

1. INTRODUCTION

Numerous attempts to define consciousness have resulted in an accumulation of metaphoric expressions, jokes, and other aphoristic utterances. Such abundant "scientific folklore" instead of an unequivocal, scientifically valid, definition may be indicative of the epistemological limit for human cognition. The peculiarity of the situation is that, on one hand, the meaning of consciousness is so self-evident and intuitively clear that it needs no any explanation while, on the other hand, any practical attempt at "understanding", or "explanation", or, especially, "definition" of consciousness meets insuperable obstacles[1]. As a rule, attempts to explain of the "puzzle" of consciousness result in comprehensive highly erudite essays describing and reviewing the problem, while the attempted definitions of the consciousness are either tautological, or reduced merely to counting its features and attributes. Commonly, any such analysis inevitably comes to the dead end that is the **psycho-somatic gap**. The stunning effect of the latter has cast doubt on the general capacity of the scientific approach to be applied for analyzing the consciousness problem. This is perfectly illustrated by the famous utterance: "How it is that anything so remarkable as a state of consciousness comes about as a result of irritating nervous tissue, is just as unaccountable as the appearance of Djin when Aladdin rubbed his lamp" (Huxley, 1866). Such an agonizing situation is in contrast to the general conviction on the omnipotence of the scientific approach in general. That is why, according to the tacit consent of experts at the beginning of this century, it is not worth hurrying with the consciousness definition since that "search for the most adequate definition would be paralyzing or lead to the evocative, but hardly satisfactory suggestion, that consciousness can only be defined in terms of itself" (Angell, 1904). As a result, the definition of consciousness has been acquiescently declared as being both **impossible and unnecessary**. In this respect, the following utterance is instructive: "No matter how it is ultimately defined, it seems we do not at present have any intuitively compelling explanations of how to bridge the psychophysical gap" (Van Gulick, 1995). This situation has led either to a totally agnostic attitude, perfectly expressed by the famous "*Ignoramus et ignorabimus*" by Du Bois-Reymond (1872) and quite recently followed by McGinn (1989, 1995) and Bieri (1995), or to a desperate appeal for the establishment of a totally new paradigm based on a counter-intuitive approach (Shear, 1995). However, since, as Metzinger clearly has claimed, "it is not at all clear ... what we would accept as a convincing solution" (Metzinger, 1995), the appeal for a new paradigm has been followed by no practical suggestion: the hope for that looks like expectation for a *"Deus ex machina"* of the medieval theater, or the advent of Messiah.

Such a situation is in discrepancy with the recently proclaimed movement "Toward a Science of Consciousness" (1996, 1998, 1999, 2000, 2001,

---

[1] "What is meant by consciousness we need not discuss; it is beyond all doubt" (Freud, 1964); the same intuitive utterance is about **time:** "What then is time? If no one asks me, I know; if I wish to explain to one who asks, I know not" (St. Augustine "Confessions", cited from [Metzinger, 1995]).

2002), since in science, a strict and unequivocal definition of any notion or concept is a fundamental and obligatory operation. In the ideal case, the definition in natural sciences is realized via mathematical expression by which it becomes sharply specific and unequivocal. Therefore, if the definition of consciousness is considered deliberately as being **impossible in principle**, the scientific approach to the problem of consciousness becomes doubtful. Therefore, the trend "Toward a Science of Consciousness" becomes an obligatory demand **to define** this notion.

Thus, in the light of the above reasons, the analysis suggested below, leading to a non-tautological definition of consciousness, would look like an ambitious pretension immediately provoking a skeptical attitude and suspicion of adventurous amateurship. Therefore, against such a suspicious background, it is worth starting the whole analysis from a closer consideration of the meaning of the word "definition" itself. Namely, the question is what this mental operation which is called definition means, i.e. what it must include.

2. WHAT IS THE MEANING OF A "DEFINITION"?

Apart from the common demand for conceptual clarity and absence of contradictions, a valid definition must satisfy the following conditions: It must not be tautological or synonymous, must not be reduced to mere listing of attributes, properties, and features of an entity to be defined, must be valid syntactically, i.e. contain a subject (the Defined) and a predicate (the Defining), and should be constructed in accordance with the "minimization principle", i.e. contain only necessary and sufficient condition(s).

In this respect, two kinds of definitions are accounted: **formal (taxonomic)**, and **naturalistic**. The former determines, first, which sort (amongst the already established ones) of things the defined something relates to, and, secondly, what differences are possessed by this something as compared to the other things of the same sort. Essentially, such taxonomic definition means the determination of the exact place of the defined entity within the currently existing hierarchic system of the considered world. The naturalistic definition is connected with the intrinsic "nature" of the defined entity, or its "mechanism" if the defined entity is, for example, a process or a dynamic state. Usually, the naturalistic definition is a reduction of a certain yet unknown "whole" to its parts (portions, ingredients, constituents, fractions, elements etc.) up to something that is based on unquestionable fundamentals or axioms. Hence, it is closely connected with scientific exploration of the phenomenon to be defined, and directly depends on the theories or hypotheses currently associated with that phenomenon[2].

As far as the consciousness is concerned, it is a realm with no taxonomic subjection to any other realms. Consequently, in view of the present state of the art, it seems that the formal taxonomic definition of consciousness is not possible in principle, i.e. the consciousness, as cited above, can only be "defined in terms of itself" (Angell, 1904). Therefore, any scientific exploration of consciousness should be associated only with its naturalistic definition that is closely connected with the development of an adequate theory of consciousness.

3. THE EXISTING THEORIES OF CONSCIOUSNESS: DO THEY PROVIDE THE NATURALISTIC DEFINITION?

In accordance with the line of the analysis, all the published theories of consciousness may be distributed into two uneven groups: those of biological origin and all the others[3]. In the comprehensive review summarizing all the modern theories of consciousness (Burns, 1990, 1991), almost all of them have a physical and mathematical origin and background. The essence of the above distribution is that the 2nd group consists of **modeling** theories that are constructed by **analogical** principle. Namely, from the whole spectrum of the consciousness manifestations, a certain particular property(ies) is selected as a factor to determine the essence of consciousness, and this property is modeled using abstract mathematical, cybernetic or physical laws and regularities in an analogical way. The modeling approach is a powerful tool in the exact sciences, predicting possible experimental confirmation of the analyzed theory on the basis of which an adequate formal or naturalistic definition can be developed. However, as far as consciousness is concerned, the above modeling theories cannot provide any naturalistic definition since they are not based on the biological reality.

However, the situation in neuroscience gives an impression that the main interest is concentrated on the fact that the brain is "the most complicated of all material objects" (Edelman, 1993). It seems that namely this incredible complication, *i.e.* a puzzling combinatorial network of neuronal interconnections, attracts the most attention. What has been called the

---

[2] For example, the definition of light is closely connected with the studies on its nature, thus depending on the respective theories, e.g. electromagnetic theory or quantum theory. A similar situation concerns the definition of temperature, gravitation, magnetism, etc.

[3] The latter group, however, includes only the theories based on the concept of the "embodied mind" meaning that the consciousness is anyhow associated with the functioning brain.

"meat machine" is analyzed as a **"machine"** - a functioning gadget - and as comprehensive as possible **imitation** of its function is believed to uncover the appearance of mind. As to its "meat", this is considered as just stuff that can be modeled successfully and, thus, "function" even better than the "native" original. Consequently, it is not taken into account that this "meat" is alive, originates from an initial zygote, overcomes massive apoptosis, metabolically incredibly active, highly sensitive to any physiological disturbances and, at the same time, demonstrates tremendous regulation capacity for functional recovery after drastic physical injury. In such a situation, the theories developed on the basis of the neuronal network are similar to the above-described modeling theories that cannot provide a naturalistic definition. The latter can be derived only from a biological theory of consciousness that must be based on genuine living manifestations, like metabolism, heredity, morphogenesis, reproduction, growth, differentiation, restitution, maintenance of the internal milieu, cell cycle, i.e. all the **phenomenology of Life**. Unfortunately, the life sciences are the least popular amongst the theorists dealing with the consciousness explanation: the fact that the human consciousness belongs to subjects who are also biological objects, does not "work" in theoretical considerations. The common conviction is that a future theory of consciousness should not be based on biological concepts since they are "dim and not strictly defined", as opposed to physical fundamentals (Chalmers, 1995). Consequently, if to consider the most elaborated biological theories by Sperry (1980), Eccles-Popper (Popper & Eccles, 1977; Eccles, 1986, 1990, 1994), Crick-Koch (Crick & Koch, 1990; Crick, 1994a, 1994b), and Edelman (1989, 1993), it becomes clear that although each of the above theories is based on different grounds (mentalist-non-dualist, dendron-psychon, neuron firing network, and re-entrant neuronal selection, respectively) it is finally reduced to the process of **conduction** of the nerve impulse. The latter, being expressed by physical parameters, inevitably leads to the same familiar psycho-physical gap.

The conclusion is that any theory based on the reduction of consciousness to the physical fundamentals inevitably rests on the dead-end. Therefore, any further progress cannot be made in the "ordinary" way: a totally new mode of analysis is needed (Metzinger, 1995, Shear, 1995).

Quite recently, the psycho-physical gap has been expressed by formulation of the "Hard Problem" of Consciousness.

4. THE "HARD PROBLEM" OF CONSCIOUSNESS AND GÖDEL'S THEOREM

The "Hard Problem" of consciousness proclaimed and further developed by Chalmers (1995, 1996) has exposed with clear acuity the question how and why performance of any form of neural activity can give rise to subjective experience. The problem is formulated in a more general abstract form, namely: "How can a physical system of any degree of complexity be aware of itself?" By critical consideration of a number of theoretical works on consciousness (Jackendoff, 1987; Allport, 1988; Wilkes, 1988; Baars, 1988; Crick & Koch, 1990; Crick, 1994a; Dennet, 1991; Edelman, 1993; Flohr, 1992; Humphrey, 1992; Clark, 1992; Hardin, 1992), Chalmers demonstrates their insolvency concerning the "hard problem" and claims that "a full theory of consciousness must build an explanatory bridge" between neurophysiology and consciousness. In order to achieve this, i.e. "to account for conscious experience, we need an **extra ingredient** in the explanation" because "no more account of the physical process will tell us why experience arises", i.e. "the emergence of experience goes beyond what can be derived from the physical theory" (Chalmers, 1995, the boldface is mine). On the basis of these clearly dualistic utterances, Chalmers dwells on possible candidates for such extra ingredients, namely, non-algorithmic processing, nonlinear and chaotic dynamics (Penrose, 1989, 1994), quantum mechanics (Hameroff, 1994), and even "future discoveries" in neurophysiology. He convincingly proves that there is "nothing extra" in those ingredients as far as their explaining power is concerned – still, the same question remained unanswered: why should each of those processes give rise to experience?

Thus, the "hardness" of the problem lies in the, in principle, impossibility to ascribing to a system characterized by purely physical parameters the property of conscious experience. From the point of view of mathematical formalism, the "hard problem" may be considered as an example of the limitations to the decidability or verifiability pointed out by Gödel (Gödel, 1931; Nagel & Newman, 1958). According to Gödel's theorem, no formal system (within itself) can fulfill the following three conditions altogether, namely, it cannot be (a) finitely describable, (b) consistent (free of contradictions), and (c) complete (be proven to be true or false). The Chalmers' "hard problem" falls into the trap of the "undecidable Gödel sentences", thus, bringing the problem to "Gödel's Gate" (Antoniou, 1992) from which there are several paths depending on which of the above three conditions is given up. The path based on giving up the second condition (consistency, absence of contradiction) should be beyond consideration as incompatible with scientific method in general. The path based on giving up the third condition (completeness) means choosing the consistency and finiteness and, thus, dealing only with constructive proofs (Beeson, 1985); then whenever the one choosing this path meets unavoidable incompleteness, she/he stops studying the specific problem and transfers to another task of the same logical sort (consistency and finiteness without completeness). The tasks decidable by the way of this path belong, using Chalmers' terminology, to the sort of "easy" problems. The path based on giving up the first condition (finite description) means choosing the consistency and completeness and, hence, losing the finite description, so that the one choosing this path gets no constructive proofs and, thus, obtains the notorious "hard" problem which, according to Gödel's theorem, is non-solvable within the frame of the considered formal system. The golden middle path between the finite but incomplete ("easy" problems) and complete but infinite ("hard" problem) is to try to go beyond the theory (beyond a given formal system) by constructing a larger one. "Any apparent disharmony can be removed only by appropriately widening of the conceptual framework" (Bohr, 1958).

5. SEARCH FOR THE NATURALISTIC "EXTRA INGREDIENT"

Thus, a formulation of the "extra ingredient" can be a basis for a naturalistic definition of consciousness. Since the latter, in the present context, must be associated with elaboration of the biological theory of consciousness, the task is to search for the naturalistic extra ingredient in the realm of Life.

This search can be done in two ways. The first one is based on the analysis of such biological phenomena that can be reduced to physical fundamentals, while the second one is connected with those biological phenomena which are not immediately clear and have not yet been explained (and probably **cannot** be explained in principle) using the physicalistic glossary.

5.1. Analysis of the Reducible Biological Phenomena: Neuronal Firing

The concept of neuronal firing is one of the main concepts of modern neuroscience, serving as the basis for a number of theoretical explanations concerning brain functioning. Most of the explanations are based on abstract (cybernetic, combinatorial) considerations of the incredibly complicated networks of the enormous (practically endless) number of possible neuronal interconnections. However, the physiological expression of neuronal firing is based on (can be reduced to) processes connected with the **conduction** of a nerve impulse along a nerve fiber. In fact, this is conduction of a signal starting from the receptor stimulation and ending by the psychic experience associated with the signal perception, i.e. it covers the enigmatic way: **external (physical) stimulus – physiological conduction – mental experience.** In the frame of this triad, there is no logical "jump" between the external stimulus and physiological conduction: both can be described, in principle, by physical fundamentals. However, analyzing the second and the third components, one inevitably reveals that a chain of consecutive processes occurring within the somatic sphere and proceeding to the psychic sphere reaches what can be designated as **"the break of continuity"** or **"the gap in entirety"**[4]. It means that the chain of processes occurring within the somatic part can, in principle, be fully described using all the glossary of physico-chemical, biochemical, biophysical, and physiological notions. After the break of continuity, the description can be done only in psychological language with no Rosetta stone at one's disposal for its translation into the life sciences' glossary.

The main approach to the above-formulated problem from the classic neuronal conception of brain functioning is based on views about supercomplexity of the neuronal interconnections of different orders of hierarchy and can be described in two different ways - more radical (formal, definite) and more moderate (rather intuitive), namely:

a. **Radical conclusion.** All the events within the somatic part of the chain starting from the initial one just after the excitation of receptor and including the last one which is that just before the "break of continuity" **do not differ in principle from each other.** In accordance with this, the task is to establish an unequivocal relation between the content (expression) of the **last** event (process) in the somatic part of the chain and the essence (expression) of the corresponding experience at the psychic end of the chain. Then, the somatic part of the chain can be reduced to the **conduction** of the stimulus from the receptor to the "end" of the somatic part (just "before" the "break of continuity"). In such a case, the "break of continuity" is so drastic that there is no hope to establish any "unequivocal connections" between the two spheres.

b. **Moderate conclusion.** The processes in the somatic part of the chain are getting more and more complicated so that the last one before the "break

---

[4] This is the expression by A. Gurwitsch uttered first in his manuscript written in 1954 which was published in 1991. The Gurwitsch's "break of continuity" has a certain nuance as compared to the more recently used "explanatory gap" (Levine, 1983, Chalmers, 1995). The latter is an ontological notion meaning the dualistic break between matter and mind, the impossibility of bridging them. Gurwitsch's gap concerns a particular problem related to a **discrete somato-psychic act** starting from the excitation of a receptor by an external (physical) stimulus, reaching the "physiological end", "jumping" through the "break of continuity" and keeping on its further development at the psychic level including its binding to other such discrete acts, simultaneously occurring or previously occurred and memorized, as well as to the whole memory background and to the general personal background however identified ("I", "self", "psyche" etc.).

of continuity" must be **fundamentally different** from the earlier events. In this case, the somatic part of the chain **is not a mere conduction** but a certain **qualitative** processing of the moving stimulus, resulting in the latter exhibiting an extraordinary quality of being able to "jump" through the brain-mind gap[5]. However, neither in the arsenal of physico-chemical notions, nor in the neurophysiological data are there means for an adequate description of the development of this new fundamental quality.

Thus, the analysis of neuronal firing demonstrates that the search for the naturalistic extra ingredient amongst the concepts that are reducible to physical fundamentals results in a negative outcome: such concepts are not efficient in overcoming the explanatory gap. Then, the search may be attempted amongst those biological phenomena that are not immediately clear and cannot be explained using the contemporary physicalist glossary.

5.2 Analysis of "Unexplainable" Biological Phenomena: Two Opposite Processes, Segregation and Integration, in Relation to the "Binding Problem"

The binding problem in neurobiology and cognitive science appears from the established evidence on the segregated nature of brain functioning. This means that particular features of any perceived object are processed separately (in different brain areas) while the object appears unitary, i.e. the whole dynamically changing world is perceived as integral and coherent. Therefore, the synthesis of all these disjuncted, dispersed, and separately processed elements of complex signals from the continually changing picture of the external world must be realized by binding together neurological states occurring in different brain areas.

Historically, the binding problem was connected first with psychology, being dependent on general philosophical views on spatial-temporal contiguity of mental representation of the external world (Hume, 1735/1958; 1777/1975). However, the modern version of the binding problem is expressed on neurological level, being based on well established evidence on the disjunctive way of processing of visible percepts. Namely, the visual data are processed separately within about fifty functionally segregated specialized cortical areas (Hubel &Livingstone, 1987, Livingstone & Hubel, 1987), each one being responsible for a specific feature, like movement, colour, texture, size, curvature, some topological properties like height/width ratio, stereoscopic depth, orientation of lines and edges, and so on

---

[5] This is similar to the views by F. Crick about the "transfer from implicit to explicit" (Crick, 1994b).

(Treisman, 1986; Ramachandran & Anstis, 1986; Ramachandran, 1990; Zeki, 1992). At the same time, multi-modal association areas in the cortex in which single perceptual features could be unified into a final perceptual image have not been found, so that the question is how the disjuncted features of any perceived object are linked together. The common conclusion is that there is **no** something which could be called either **central cortical "information exchange"** (Hardcastle, 1994), or **"master map"** (Treisman, 1986), or **multi-modal association areas** (Damasio, 1989) in the cortex. Nevertheless, still there is a certain (rather emotional) hope for future finding of "grandmother" neurons and convergence zones (Hardcastle, 1994).

However, together with the segregation phenomenon, there is simultaneously proceeding an integration phenomenon associated with the **same** visual perception, the factual scheme being as follows. An object being seen is projected upon the retina and its visible image irritates the retinal receptors. Then, the immutable histological fact is that about 300 retinal rods (the first neuron of the visual network) are structurally (histologically) connected via bipolar cells (the second neuron) with one ganglionar cell (the third neuron). Consequently, these 300 adjacent rods form a microarea on the retina which during the vision process may contain heterogeneous pictures as projected from a (micro)part of the visible object (even if the latter is not moving). Thus, the axon of the ganglionar cell must conduct forward an impulse carrying such complex (integrated) visual information. This already is incompatible with the classic neuronal theory according to which neuron firing means only the **conduction** of a signal that is triggered according to the "all-or-none" principle.

Thus, two opposite processes associated with the same current visual signal go on simultaneously. One of them consists of the above-described **splitting** of the perceived object's image into dozens of quite different "components" of the vision (features of the perceived object) like shape, movement, colour etc. which are processed separately in distinct cortical areas, while another one consists of the anatomically determined **confluence** of the distinct signals carrying different spatial parts of the perceived object. The possibility itself of the coexistence and simultaneous realization of these two antidromic processes within the same anatomic unit (system) is totally incomprehensible: confluence (merging, junction, maybe synthesis) of a number (hundreds!) of axonal impulses **before** they come to the brain centers, on one hand, and in parallel occurring disjunction (splitting, breaking, maybe analysis) of the object as a whole - **not into parts** (portions, pieces) - but into so drastically different causally disconnected (somewhat categorial) features, like colour, form, movement etc., on the other hand.

This insuperable obstacle for the neuronal theory[6] is to be considered in the framework of fundamental computational mechanism of brain functioning. According to modern computational language, the problem is formulated as claiming that the objects or their different aspects have to be represented (to co-exist) within the same physical "hardware" (brain) that result in the **"superposition catastrophe"** (Von der Malsburg, 1987). Therefore, if to define the main postulate of the neuronal theory in way that each mental representation ("symbol") of the external objects is represented by the corresponding subsets of coactive neurons within the same brain structure, then if more than one of such "symbols" become active at a moment (which must occur within the real functioning pattern), they become superimposed by coactivation (structural "overlapping" of the activated subsets). In such a case, any information carried by the "overlapping" subsets must be lost, that being the mortal verdict for the whole foundation of the classic neuronal theory.

Consequently, the binding problem, being not a purely theoretical construction but arisen from the very heart of the neurobiological and psychological reality, is incomprehensible within the limits of the anatomical-physiological regularities, and, hence, cannot be employed as a source which could provide a theoretical basis for building a bridge over the explanatory gap. Therefore, the binding problem not only cannot help to overcome the explanatory gap, but itself appears to be an enigmatic phenomenon which needs an extra ingredient for its own explanation.

6. THE VITALISTIC PRINCIPLE

Thus, the general conclusion is that the sought for extra ingredient must be associated with the development of an entirely new paradigm based on a **"non-physical** feature". The latter, however, is not accepted in present day science as a "working tool" due to the epistemological premises deeply ingrained into the principles of scientific exploration since the time of Galileo and Newton[7]. However, whenever the concept of the extra ingredient is associated with living systems, the whole idea is inscribed into the frame

---

[6] Described first in the manuscript by A. Gurwitsch written in 1954 and published in 1991 (Gurwitsch, 1991).

[7] Noam Chomsky noticed (in private conversation with philosopher John Searle) that as soon as we come to understand anything, we call it "physical" (Searle, 1992). Consequently, anything in the world is either physical, or unintelligible and, hence, the "understanding" of any natural phenomenon, including both life and consciousness, means ultimate reduction of the observed phenomenon to physical fundamentals.

of the vitalistic trend that inevitably induces immediate dislike before any specific consideration.

6.1. Vitalism under Anathema: Peculiarities of the Situation

Nowadays, the term "vitalism" has become a synonym of something which, in the best sense, is mystical but in general is merely "anti-scientific". However, for mysticism *per se,* e.g. parapsychology, there is quite a decent place in the modern "interdisciplinary" disputes on "scientific basis of consciousness". Such notions as "self", "soul", "I", "psyche", "psychon", "suspended animation", "conscious mental field", "psi"-phenomena, "ghost in the machine", "unembodied mind" – in general, all the spectrum of a dualistic and spiritualistic glossary – have a loyal privilege to be discussed in the most scientifically oriented meetings on consciousness while there is a strong taboo against any mention of the word "vitalism" – not even usage of the heretic word but just a medieval suspicion in being defiled by this evil. A Physicist theorizing on the most general and "hot" problems of the Universe, Life, and Consciousness has a full right and freedom to include into her/his theoretical considerations any notions which can be beyond all the known physical laws, ike Free Will, Universal Mind, Act of Creation, without being labeled as Mystic. A Biologist has no such right and freedom: if her/his theoretical considerations are beyond the dogmas of Molecular Genetics – a Bible and physicalistic bulwark of the Modern Biology – she/he may be very easily blamed and ostracized as Vitalist.

The essence of the vitalistic principle as applied for its naturalistic extra ingredient is that its "extra-ness" (irreducibility to the physical fundamentals) **is related not to the conscious (human) mind but to a species-specific living entity**. The constructive character of the extra ingredient must be connected with its non-tautological definition. Unfortunately, according to the commonplace (or, better to say, vulgar) idea of the vitalistic principle, the latter is a kind of an abstract metaphysical **omnipotent** "vital spirit" ("*vis vitalis*", "elan vital"), "organizing" ("animating") material components of a living system (organism), thus determining its structural appearance and coherent functioning. Such evidently tautological character of the vitalistic concept has neither explanatory power, nor scientific value.

6.2. Classic Vitalism versus Classic Dualism

While the reductionist philosophy is based on pure physical laws, the dualistic philosophy jumps from the physical sphere directly to the sphere of consciousness, skipping off the phenomenon of life. Classical dualism establishes a principal gap between **consciousness and non-consciousness** realms. The latter realm includes both physical and biological phenomena with conviction that the living state can be reduced to (comprehensively explained by) the same physical laws. Vitalism establishes a principal gap between **living and non-living** phenomena (i.e. all forms of Life versus the "dead" Physical World) with conviction that Consciousness can be reduced to (comprehensively described by) the autonomous principles of Life. In present day science, the dualistic view is considered as legitimate while the vitalistic view is totally intolerable with a determined inclination to exorcise this evil from science.

A rather simple but crucial shift in the division between the realms (location of the "gap") is the most important step of the suggested approach. It means, that the primary question is "What is Life?" and the question "What is Consciousness?" becomes the secondary one. In other words, this shift is hoped to provide a construction of a larger formal system (Life versus Consciousness) which, by breaking the limits of Gödel's ring (1931), would make the notorious "hard problem" solvable.

However, before dwelling specifically on a "working" vitalistic theory, both Life and Consciousness are to be analyzed, first, formally, the task being to deduce the latter from the former using an abstract formal approach based on the original **morphic** principle.

7. THE DEFINITION OF LIFE

By the majority of authors dealing with the Consciousness problem, including "New Mysterians" (Chalmers, 1995, 1996), "Liberal Naturalists" (Rosenberg, 1996), and devoted materialists (Dennett, 1991, Hardcastle, 1996), Life is not strictly defined but is described by a set of various life manifestations. There is a simple enumeration of arbitrarily chosen life manifestations, the choice depending on the authors' educational background, taste and emotional inclination, for example: "DNA, adaptation, reproduction, and so on" (Chalmers, 1995), in one case, and "reproduction, development, growth, self-repair, immunological self-defense and the like" (Dennett, 1996), in another one. Such a situation is intolerable in Physics whose basic definitions are grounded in consideration of mathematically expressed "necessary" and "sufficient" conditions. Therefore, besides the semantic meaning, the syntax - the hierarchy between and amongst the connected notions and their attributes - determines the necessity and sufficiency of the definition conditions.

As opposed to the above invalid "set-like" kind of life definition, the following one is suggested:

Life (any living system) is evolving and aging species-specific non-equilibrial (i.e. needing incessant energy influx) self-reproducing and self-preserving GEOMETRICAL FORM which is continuously refilled by specific substances.

From the formal point of view, the advantage of this definition, as compared to the above-mentioned "set-like" ones, is its syntax. Namely, instead of a dull enumeration of manifestations of Life as the defined subject, the definition has one predicate (*"FORM"*) around which there are necessary attributes (including a subordinate clause) hierarchically subordinated to the predicate. Together with this, the suggested definition is neither tautological, nor synonymic. In this definition, however, there is a certain inevitable logical "imbalance" amongst the attributes to the predicate "Form". Namely, some of the attributes are related to the notion of Life as related to a living individual (e.g. "evolving and aging") while the others are related to the general notion of Life as a cosmic phenomenon (e.g. "non-equilibrial"). Evidently, any of the genuine life manifestations like metabolism, reproduction, heredity, growth, differentiation, morphogenesis, etc. are **covered** by this definition.

Apart from the syntactic estimation, the crucial point of the above definition - the **morphic** principle - is not trivial. Usually, the other Life manifestations, like metabolism, heredity, reproduction, internal milieu constancy, evolution, and, accordingly, different principles, like biochemical, biophysical, genetic, thermodynamic, synergetic (self-organization) have been regarded as being of paramount importance (Loeb, 1906, 1912; Lotka, 1925; Szent-Gyorgi, 1957; Monod, 1971; Dawkins, 1976; Cairns-Smith, 1982; Caplan & Essig, 1983; Dennett, 1995; Prigogine & Stengers, 1984; Haken, 1977; Babloyantz, 1986; Waechterhauser, 1988; Eigen, 1992, 1993; Elitsur, 1994; Lifson, 1987; Kauffman, 1996). These principles are clearly associated with a reductionist trend of explanation contrary to the morphic principle: the **form** in the above definition is that biological entity which cannot be derived from the physical fundamentals.

If the essence of Life is based on the morphic principle, the question is how the evolving species-specific **geometrical form** of a living system is made up if *a priori* it is not determined by (not supervenient on) the material stuff constituting (filling) it. This **"how"**-question becomes entangled with the **"why"**-question, namely, why the molecules entering the living species-specific system behave differently as compared with their "*in vitro* behavior" determined by their canonical (physico-chemical) properties. In such a case, the explanatory gap is between the molecules' physico-chemical properties determining their chaotic movement in solution, on one hand, and the molecules' "in vivo behavior" characterized by **vectorized** mode of their movement within the body of living cell, on the other hand. Accordingly, the molecules entering the living system display an additional quality of moving along certain preferable directions which are non-linear and dynamically changing. This quality is not determined by the intrinsic canonical properties of the molecules themselves: the molecules' vectorized movement is forced upon them by the living system as a whole. The trajectories of the coordinated movements of the totality of the involved molecules will provide somatic (morphological) species-specific appearance of the living system as well as its coherent functioning. Thus, the **molecules' vectorized movement *in vivo* versus their chaotic movement *in vitro* is the expression of the explanatory gap on the molecular level of living systems.**

The above morphic definition of Life is a basis for formulation of the non-tautological definition of consciousness to be dealt with further.

8. THE ANALOGOUS "PRE-NATURALISTIC" DEFINITION OF CONSCIOUSNESS BASED ON THE MORPHIC PRINCIPLE

By analogy with the Life definition, the syntactically valid formal definition of consciousness based on the same morphic principle is asserted to be possible. Accordingly, further analysis concerns the consecutive consideration and definitions of the suggested concepts of **protoconsciousness, primordial consciousness** and, finally, **consciousness *per se*,** the former being related to the fundamental essence of consciousness as an attribute of life while the latter being related to the consciousness of a normal adult human individual. The mode of the consideration will be based on the "developmental" approach including embryogenesis (ontogenesis) of consciousness as well as its phylogenetic appearance.

Before specific consideration, the inescapable problem of the emergence of consciousness is to be considered first.

8.1. The Generation Problem or Panpsychism: The Fatal Choice

Any attempt to understand the emergence of consciousness inevitably rests at the crossroads of two extreme paths: generation problem and panpsychism. The first alternative immediately leads to the "hard problem" with the explanatory gap (Chalmers, 1995), while, according to the second one, the phenomenal quality of consciousness is reduced indefinitely (infinitely?) to certain properties of (certain? any? every?) elementary physical entities (molecules? atoms? elementary particles? photons? quarks?) which, hence, are not "purely physical" but also "mental". At first glance, these two assertions seem to be really unequivocal alternatives, i.e. those excluding each other: the fatal choice is either to dive into the bottomless panpsychist well, or stop before the unbridgeable fathomless abyss of the "explanatory gap". As opposed to the generation problem whose logical incomprehensibility has been evidenced generally (Tyndall, 1879; Clifford, 1879, James, 1890), the panpsychist doctrine seems to be logically more consistent. However, the comprehensive analysis of the panpsychist concept reveals such an impression to be delusive, so that both extremes are not so contrasting. Namely, a number of inconsistencies can be found (Seager, 1995), the **combination problem** being the main one. In accordance with it, "the panpsychist must proclaim that it is consciousness itself that divides down to the elemental units; otherwise the generation problem returns with its full force" (Seager, 1995). Then, the question arises how - by what regularity - these **mental units** are combined together into hierarchically highly complicated associations corresponding to different experiential phenomenology. "It is far from clear that it even makes sense to speak of the combination of basic mental elements, even granting they are in some sense conscious, into distinct and more complex conscious experiences" (Seager, 1995). If these combinations bearing experiential phenomenology are formed by causal physical laws, the mental quality becomes explanatorily epiphenomenal, and, hence, superfluous. Then again the generation problem "returns with its full force" (Seager, 1995) leading to the same unavoidable explanatory gap.

The vitalistic theory of consciousness to be outlined below is an attempt to avoid the fatality of the choice between the generation problem and panpsychism.

8.2. The Psychic Ontogenesis

The basic assertion is that the ontogenetic development of the somatic and mental spheres goes in inalienable association. Then, the question is how this inalienability is realized, the different possibilities being possible, for example, both somatic and mental developments may go in parallel or the mental component appears with a certain delay, and so on.

The analytical approach consists in imaginable backward (retrospective) evolution. The starting point is the stage of a mature individual's consciousness associated with functioning brain that can be expressed as a continual stream of variegated experiences (only partly depending on environment) overlaying a certain personal *"psychic background"* which is slowly changing and advancing along with age and which can be identified with the individual "I"[8]. The current "experience stream" contributes to the "I's" changing and aging. The psychic ontogenesis relates to this psychic background.

Consistent analysis of the imaginable backward evolution (retrospective involution) of psychic phenomenology reveals the impossibility of indicating a moment of the first appearance (emergence) of the initial psychic phenomenology during the brain development. Moreover, the conceivable retrospective involution cannot be limited to the developing brain but must continue deep into earlier stages of the developing embryo (before cell differentiation). Consistently performed retrospective analysis inevitably arrives at the only initial point of the individual psychic development, that is the individual's **zygote**.

The above involution evidently must be followed by successive "simplification" of the psychic phenomenology. It is not *a priori* clear up to what level such conceivable involution is realizable, especially during the early – "pre-cerebral" (before cell differentiation) period of the development, *i.e.* till what level the corresponding "simplified" ("rudimentary") psychic phenomenology would still have its specific attributes ("psychic gleams"). Success in this way depends on formulation of such **rudimentary psychic phenomen** which would be taken for a further analysis of the notion of **protoconsciousness** to be postulated and defined. Such a rudimentary psychic phenomenon has been suggested (Gurwitsch, 1991): this is **knowledge**. The choice of this notion was conditioned by two evident advantages: (a) knowledge can be considered as the basis of any expression of psychic phenomenology; (b) the notion of knowledge permits quantitative estimation like "less - more" and "simpler - more complicated" throughout the evolution process, so that no "break of continuity" could occur along this way

Thus, at this point of the analysis, two conclusions are taken together: in the human individual psychic development, the initial point is the zygote where the psychic activity is reduced to **knowledge**[9]. But **what** is this

---

[8] The "I" here is considered, as opposed to Deikman (1996), not as pure awareness but as its content.
[9] There is a need to emphasize the difference between the notion of **knowledge** in this context and the notion of **information** suggested by Chalmers (1995). The latter has a certain abstract mathematical cybernetical meaning and can be applied to the "dead" world as well. In the present context, **knowledge**

knowledge about? This is a crucial point determining whether the proposed formulation of the extra ingredient is indeed non-tautological. In the Drieschian style, it would appear as follows: "An embryo (in our case, a zygote) **knows its own state**" (Driesch, 1908). The tautology here is clearly evident, being connected with the word "state" which is too general. Therefore, if "knowledge" is accepted as a rudiment of consciousness, the main task, in order to escape the tautology, is to indicate that limited quality of the notion of "state" which "knowledge" must be related to. Such a quality is suggested as being expressed by the notion of **form** taken in a strictly defined manner. Accordingly, the utterance "the zygote knows its own **state**" is changed to the utterance "the zygote knows its own **form**". Such expression of rudimentary consciousness is not tautological and can serve as a basis for the definition of consciousness we seek.

8.3. The Definition of Protoconsciousness

On the basis of the above-formulated definition of life and the concept of knowledge, the notion of **protoconsciousness** as a basic abstract fundamental related to **any** living entity of **any** origin is suggested. A preliminary (auxiliary) formal definition of protoconsciousness is as follows:

Protoconsciousness is an embodied immanent capacity of any living entity to FEEL its own evolving dynamically fluctuating species-specific FORM (synonym for this capacity: *"geometrical feeling"*)

First of all, the above definition is not tautological: the subject and the predicate in the syntactic hierarchy of the definition are not at all identical[10]. The expression "any living entity" means any species-specific cell as an elementary living system. In accordance with this definition, protoconsciousness is an inalienable attribute of life, i.e. any cell whenever it is alive feels its own geometry. The latter is conceived not only as an external shape characterized by morphological contours: the geometrical feeling transpierces through the whole system, i.e. each geometrical (stereometrical) dot within the three-dimensional living whole is felt, sensed, being aware of by that whole.

---

means (is inalienable of) its immediate experience, *i.e.* the knowledge of a momentary "state" means a kind of its "possession" and the possession is already a feeling of belonging to a certain possessor which is a kind of "**Self**".

[10] A tautological version of this definition would be as follows: "The protoconsciousness is the immanent capacity of any living entity to feel its own **state**" (instead of **form**).

The concept of protoconsciousness includes another attribute – the cell's capacity to preserve its species-specific morphology by reacting morphogenically to any disturbing factor. This means that if any physical factor acts upon the material substrate "filling" (constituting) the species-specific form of any living cell, this causes a "tension" (upsets the balance) between the "ideal" species-specific geometry (stereometry) of the cell, on one hand, and the real spatial distribution of the material stuff within this "ideal" form, on the other. This tension is "felt" by the cell which "reacts" morphogenically "to smooth" the disharmony between the geometry (morphology) of this ideal form and the real distribution of the physical substrate which "fills" (constitutes) this morphic (geometrical) "receptacle". This "physical substrate" includes, essentially, the whole totality of all the intracellular molecular substances involved into processes that are under "usual" biochemical and genetic regularities. Consequently, it is not passive stuffing inside the "ideal form" but a biochemically highly active medley of different substances including numerous enzymes involved in a dynamic network of metabolic pathways. Besides, this physical substrate is partly included in different subcellular structures (e.g. cytoskeleton) that are not rigid but dynamically flexible, being under continual regulational and restitutional changes. Therefore, the material substrate filling (occupying) the geometrically ideal morphic receptacle is in the state of incredible dynamic heterogeneity and turbulence, thus "gushing over" the external and internal geometrical "borders" of the ideal form. Just this dynamic **non-congruence** between the ideal geometrical form and the real spatial distributions of the material substances filling this form is currently felt (experienced) by the cell. Therefore, the above auxiliary definition of the protoconsciousness is exchanged for the corrected version that is as follows:

Protoconsciousness is the embodied immanent capacity of any living cell to FEEL any spatial non-congruence between the cell's evolving species-specific "ideal" GEOMETRIC FORM, on one hand, and the real distribution of the material stuff "filling" this form, on the other.

This is the basic abstract definition from which further definitions including consciousness *per se* are to be inferred. In accordance with this definition, the "geometrical feeling" is not the feeling of the geometrical form but the feeling of the non-congruence between this ideal geometrical form and its physical (material) realization. The living cell "feels" the three-dimensional shape of its own body as soon as the actively fluctuating (dynamically "boiling") material stuff constituting this body "violates" the geometrical abstract borders of this shape. Consequently, such "non-congruence" is, essentially, a formalized expression of the psycho-somatic gap which, allegorically, is analogous to the discrepancy between the ideal geometry and its physical realization (Abstract Mathematics versus Solid Physics), e.g. geographical meridian upon the globe versus the corresponding particular relief on the Earth.

Another postulated attribute of protoconsciousness is the capacity of a living cell to **preserve** its species-specific morphology by means of the morphogenic reaction smoothing the non-congruence between the "ideal" geometrical form and the material stuff constituting it. Consequently, the combination: **"geometrical feeling"** – **"morphogenic reaction"** can be considered as a **rudimentary psychic act ("morphological mind")**.

Although the above morphic definition of protoconsciousness is formally consummate, there is a feeling of a certain deficiency, immediately provoking urges for further elucidation. Namely, there still remains a gap between the postulated geometrical feeling as the expression of protoconsciousness and the full phenomenological bouquet of consciousness *per se*. However, now it is not that drastic unbridgeable psycho-somatic gap between matter and mind: both the defined protoconsciousness and consciousness *per se* to be defined belong to the same ontological level (with no "break of continuity"), being described by the same **psychic** vocabulary. Namely, both have at least one common attribute – feeling: either the postulated "geometrical feeling" as an "elementary" fundamental for psychic phenomenology (protoconsciousness), or the "ordinary" feeling as a basic constituent of the full spectrum of manifestations of consciousness *per se*. However, it is not at all self-evident that the abstract morphic principle together with the postulated geometrical feeling would be sufficient for description of consciousness *per se*. Accordingly, the subsequent step is to develop a definition of the consciousness *per se* based on the concept of geometrical feeling. The main task in this way is to elucidate the central syntactic point of the protoconsciousness definition: the predicate "FORM" with its main attribute "species-specific" ("species-specific 'ideal' GEOMETRICAL FORM").

In this respect, the postulated "geometrical feeling" concerns each geometrical dot within the living stereometric morphological whole. Accordingly, the main effort is to formulate the naturalistic factor determining the species-specific morphology. In the context of the present consideration, such a factor is that "extra ingredient" based on the vitalistic principle whose necessity was proclaimed above. Consequently, the morphic principle that could be considered as an epistemological tool must be expressed in ontological meaning. Therefore, the postulated concepts of the "geometrical feeling" and the "morphogenic reaction", which are rather complicated and highly specific to serve as axiomatic concepts, must be inferred from a more basic fundamental expressing the extra ingredient. Such a fundamental can be provided by the theory of the biological field by Alexander Gurwitsch.

9. A NON-TAUTOLOGICAL VERSION OF THE VITALISTIC PRINCIPLE: THE THEORY OF THE BIOLOGICAL FIELD BY A. GURWITSCH

The theory of the biological field by Alexander Gurwitsch is an unprecedented example of "working vitalism" (the expression of Gurwitsch himself). This is a unique combination of the logical structure of the vitalistic principle expressed in its full philosophical integrity with specific postulations based on the whole Life phenomenology manifested at all three levels of the biological organization - molecular, cellular and morphological. The format of this article does not permit detailed description of Gurwitsch's theory of biological field, which can be found elsewhere[11].

The vitalistic essence of the Gurwitschian field is that it is irreducible to any known physical fields. The mode of the field action expressed on the morphological level is defined as subjection of equipotential elements (cells) to integral morphogenic field causing the cells' **spatial** orientation or/and movement. According to Gurwitsch's field theory, any living cell is a source of the field that is described by strictly defined postulates. The latter concern vectorial repulsive character of the field, its anisotropy, field sources, nature of elementary field "flash", formation of integral microfields and macrofields (by geometric composition), dynamics of field tension and its association with metabolic rate (Gurwitsch, 1991, Lipkind, 1998a, 1998b). The explanatory capacity of the field theory was tested by Gurwitsch using different levels of the biological organization: molecular (metabolism), cellular (mitosis, differentiation and histogenesis) and organismic (morphogenesis, neuro-muscular system, brain cortex structure and functioning), on one hand (Gurwitsch, 1991; Lipkind, 1998b) and different manifestations of the consciousness (feeling, stream of incoherent chaotic

---

[11] Gurwitsch was the first who introduced the notion of field, taken from the physics glossary, into biology (1912), this fact having been acknowledged in contemporary reviews (Von Bertalanffy, 1933; McDougall, 1938, Weiss, 1939) as well as in the recent works (Haraway, 1976, Sheldrake, 1986; Goodwin, 1986; Welch, 1992; Laszlo, 1993). Since then, the non-tautological versions of the field principle were developed by Gurwitsch during all his life until his efforts were crowned by the final version - vectorial cellular field described as a mathematical model based on twelve postulates deeply rooted in biological reality and describing different biological phenomena manifested at molecular, cellular and morphological levels. However, the theory was developed in the Stalinesque Soviet Union in the absence of any immediate contact of its author with Western scientists: the majority of his publications were done in Russian (Gurwitsch, 1944, 1947, 1991), this being the reason for poor (if any) familiarity of Western readers with the last version of his theory. The first comprehensive review of Gurwitsch's field theory written in a European language (German) was published in 1987 (Lipkind, 1987a, 1987b). The first English review on the theory was published in 1992 (Lipkind, 1992). Some excerpts from Gurwitsch's book written in 1954 and published only in 1991 in Russian (Gurwitsch, 1991) were translated into English in 1994 (Beloussov, 1994). The most detailed English reviews have been published just recently (Lipkind, 1996, 1998a, 1998b).

thoughts, memory and recollection), on the other hand (Gurwitsch, 1991; Lipkind, 1996, 1998b).

Application of Gurwitsch's field theory to the psychic sphere was based on the analysis of the brain cortex cytoarchitectonics and consideration of the brain as a geometrical continuum of the integral field.

The unprecedented advantage of Gurwitsch's field theory is its wide descriptive capacity: by the **same** postulates, the theory is efficient when applied to somatic as well as psychic spheres. This makes it possible to understand and comprehend consciousness as a **biological** reality described by the **common** language with other biological phenomena.

10. DEFINITION OF PROTOCONSCIOUSNESS IN THE LIGHT OF THE THEORY OF THE BIOLOGICAL FIELD BY A. GURWITSCH

In accordance with such purpose the above-postulated "geometrical feeling" and "morphogenic mental reaction" assumed as bases for the above-described concept of protoconsciousness are integrated with the postulates of Gurwitsch's field theory. On the basis of such amalgamation, a naturalistic definition of the protoconsciousness is suggested being as follows:

Protoconsciousness is the immanent capacity of any living system to FEEL any spatial non-congruence between the configuration of the system's integral Gurwitschian FIELD, on one hand, and the material stuff being within the field-influenced space, on the other.

The vitalistic essence of this definition is that the Gurwitschian field is not reducible to any known physical fields and, hence, can be considered as the sought for extra ingredient.

11. DEFINITION OF THE "PRIMORDIAL CONSCIOUSNESS" IN THE LIGHT OF GURWITSCH'S THEORY OF BIOLOGICAL FIELD

Thus, it was accepted by the introspective involution analysis, that the initial point of the individual psychic development relates to the zygote. The starting point of the further analysis is connected with the moment of the first cleavage when the field sources of the first two blastomeres interact geometrically, the integral field is composed, and the system of two blastomeres, that since this moment **is** the **embryo,** feels (experiences) configuration of its integral field. This is the beginning of the great game of the individual somatic development immanently associated with (led by) the mutually developing integral field. This game is due to the current knowledge experienced by the developing embryo that "**acts morphogenically**" towards smoothing geometrical disharmonies (non-congruence) arising from either "normal" environmental fluctuations, or extravagant interventions. Consequently, the primordial consciousness concerning the developing embryo at early ("pre-cerebral") stages (before the cell differentiation) is defined as follows:

Primordial consciousness is the capacity of an early non-differentiated ("pre-cerebral") embryo to FEEL any non-congruence between the geometrical configuration of the evolving INTEGRAL Gurwitschian FIELD, on one hand, and the real spatial distribution of the material stuff within the field-influenced space, on the other hand, and to REACT morphogenically to smooth the non-congruence.

This rudimentary psychic phenomenology during "pre-cerebral" stages of embryo development includes the embryo's **knowledge** about (**experience of**) the continually changing momentary configuration of the embryo's integral field as well as the morphogenic **action** as a reaction to that knowledge (experience)[12]. Consequently, the early embryogenesis can be imagined as the chain of the embryo's actions, each one being associated with the **act of choice** by the embryo between different possibilities. Such embryonic "free will" becomes evident in the case of the experimental interference that has been clearly demonstrated by the classic experiments by H. Driesch ("harmonic regulations") having led to the notion of **equifinality** based on the fact that the same final species-specific form may be achieved by quite different ways of development (Driesch, 1891, 1908). In the glossary of the modern theory of supervenience (Horgan, 1982, Kim, 1984), the equifinality can be expressed as the **absence of isomorphic identity** during morphological development: the **same** final species-specific morphology can be realized via quite **different** processes occurring on cellular and molecular levels.

---

[12] This "knowledge" (experience) is realized via the feeling of the same non-congruence – now between the embryo's integral field configuration ("ideal" geometry) and its material realization.

12. TRANSFORMATION OF THE MORPHOLOGICAL MENTALITY INTO REFLEXIVE NEUROPHYSIOLOGY: CELL DIFFERENTIATION AND THE DEVELOPMENT OF NEURAL SYSTEM, SENSE ORGANS, AND BRAIN

Cell differentiation as a specification of the cells' roles in the whole organism is a turning point of the individual development, being associated with drastic and irreversible (normally) changes which are as follows:

a) The decrease and then the loss of the capacity for the Drieschian "harmonic regulations".

b) Morphological "parcellation" of the embryo leading to its geometrical complication. As a result of it, the embryo loses its "compact" form (morphological "wholeness") and its different parts (*e.g.* limbs) become morphologically separated. Accordingly, just because of the purely geometrical premises, the integral field of the whole embryo ceases, turning into separated field areas of separated organs geometrically not interacting with each other.

c) As the morphological development comes nearer to the end, the morphogenically (geometrically) coordinated cells' movements give place to physical (muscular) movements of spatially separated parts of the shaped body.

d) All the above morphological processes go in parallel to the specific neural cell differentiation and rapid development of the nervous system (neural network) with its central part - the brain, that becomes the main integrating factor for the coherent functioning of the organism as a whole.

e) According to this scheme, the early **"conscious" whole embryo** that perceives (is aware of) the configuration of its integral field associated with the whole body turns into a **highly sophisticated automaton** integrating via a neural network all the multiple parts of the still developing, morphologically complicated embryo body into a coherently functioning whole. Thus, the ancient **mental** ("simple", slowly reacting, morphological) **activity** of the early **non-differentiated** embryo turns into the **neural** (highly sophisticated and quickly reacting instinctive/reflexive) **behavior** of the swiftly complicating (via further differentiation) maturing organism. Then, the embryo's "behavior" is a result of the neurophysiological activity that includes input and processing of all the signals coming from the developing receptors. The chain of the embryo's actions is now incredibly more effective but such instinctive-reflective activity, to a certain respect, resembles a "zombie-like" behavior.

13. THE DEVELOPMENT OF BRAIN AS GEOMETRICAL CONTINUUM: CONVERSION OF PRIMORDIAL CONSCIOUSNESS INTO CONSCIOUSNESS *PER SE*

The development of the brain includes two remarkable features: together with the development of highly complicated cytoarchitectonics, the brain develops as a **morphologically compact geometrical whole**. In view of Gurwitsch's field theory, this leads to the formation of the brain integral field – a kind of the field geometrical continuum reminding, in this respect, the morphological integrity of the whole undifferentiated embryo. As a result, a remarkable combination is established: unsurpassed complexity of the integral field geometrical configuration is combined with incredible sophistication of the neural cytological infrastructure. There are all the anatomic premises for manifestations of the mental phenomenology in its highly developed form. This stage of the development means "resurrection" of the disappeared initial embryonic consciousness but now this is not that "ancient" rudimentary embryo's "knowledge" (experience) of its integral field with a slow "morphological action". Now, the whole embryo integral field is substituted by the integral field of the brain that is exposed to the current stream of diverse signals (impulses) coming from external and internal somatic receptors via highly sophisticated neural network. All these impulses are conducted neurophysiological by, i.e. the physical energy causing specific excitation of the respective receptors is transformed into the neural impulses described by biophysical and biochemical terms that transfer the initial signals from the receptors towards the corresponding areas of the brain. The current waves of such stream flow into (engulfed by) the Gurwitschian integral field of the brain, that causing dynamic "disharmonies" in the brain field configuration. These disharmonies are due to the same **non-congruence** – this time between the ideal geometry (stereometry) of the whole brain integral field continuum (internal "lines of force"), on one hand, and the real distribution of the structured physical stuff filling this geometric frame, on the other hand. This distribution deviates from the spatial limits of this ideal geometric form due to and in accordance with the physical characteristics of the coming impulses. Just as in the case of the primordial consciousness, this non-congruence is experienced by the individual, but now the non-congruence caused by the signals **conducted physiologically** is **experienced phenomenologically**. Since the non-congruence is related to the whole brain integral field, while the comprehensive current stream of the signals includes the whole total of all the stimuli from all the body receptors, the experience of this non-congruence will reflect the whole external[13] world as perceived in all its integral totality.

Incidentally, the notion of "information" usually employed in connection with perception (Velmans, 1995) is devoid of any biological (naturalistic) meaning.

Thus, the definition of consciousness *per se*, is as follows:

Consciousness per se is the capacity of a mature human brain (cortex) to FEEL any non-congruence between the geometrical configuration of the integral Gurwitschian FIELD of the brain, on one hand, and the current fluctuations of the spatial distribution of the physical stuff within the field-influenced space, on the other hand.

The important elucidation to be added to this definition relates to the above-uttered expression **"current fluctuations of the spatial distribution of the physical stuff"**. The detailed version appears as follows: **"current fluctuations of the spatial distribution of the physical stuff"** *that are caused by the current stream of afferent neural impulses proceeding from the sense organs (including the totality of the external, internal, and proprioceptive receptors) and conducted into the brain.* This addition could be syntactically inserted into the complete definition but that would make it too cumbersome. Therefore, it seems to be better arranged in the above form of a supplementary (auxiliary) elucidation.

The above-stated non-congruence means the same discrepancy between the geometrical ideal form and its material realization, i.e. "filling" ("stuffing") the geometrical frame with the physical "meat". Evidently, the absolute congruence cannot be achieved in principle, like the absolute zero temperature. Then, the maximal conformity meaning practical (not absolute) congruence between the geometrical ideal form and its physical realization is accomplished through minimal fluctuations, e.g. weak inevitable agitation of the molecules filling the field-determined geometrical framework. Since according to the suggested postulation, just the non-congruence is felt and experienced, the maximal congruence would mean on the psychic level

---

[13] This includes the "internal" stimuli, which are "external" to the brain.

something like *nirvana*. On the background of such quasi-congruence, any physical perturbations within the "molecular substrate" disturb the physical-versus-geometrical conformity and cause the non-congruence. The current dynamic stream of the afferent neuronal firing originates from various receptors which are excited by different physical stimuli causing specific sensations, e.g. mechanic (tactile, equilibrial, proprioceptive), chemical (olfactorial, gustatory), photonic (visual), and acoustic (auditory). Specificity of the conducted impulses is determined by the corresponding physical characteristics of the initial stimuli. Since the current dynamic stream of the impulses causing the non-congruence proceeds from all the totality of the receptors bombarded ceaselessly by all the possible stimuli coming from the external world and the own body, the experience of this non-congruence unequivocally reflects current state of the world as-perceived.

Thus, the causal chain is as follows: 1) Irritation of the receptors by the physical stimuli leading to their excited state; 2) Conduction of this excitement via neuronal network into the brain; 3) Disturbance by the neuronal firing of the conformity between the field-determined geometrical frame and the molecular substrate filling this frame (the non-congruence); 4) Feeling (experience) of the dynamically changing non-congruence which reflects in every detail a coherent and comprehensive picture of the external world (including the own body) as perceived by the individual.

14. PHYLOGENETIC ANALYSIS

The phylogenetic analysis, i.e. the evolution of the consciousness as a species-specific property, is an especial topic that will not dwelt on in detail here. However, the idea that the ontogenetic development repeats somehow the phylogenetic history (Haeckel, 1892, 1899) forces us to search for the carriers of the suggested consecutively developed models of consciousness amongst the appropriate representatives of all the Life kingdoms. In this respect, it can be mentioned only that all the unicellular species remained to possess only the protoconsciousness; all the plants and some representatives of the most primitive ("unneural" or having diffuse poorly differentiated neural system) animal *Metazoa* types, e.g. *Sponga, Coelenterata*, and possibly some primitive *Plathelmintes*, can be considered as having primordial consciousness ("morphological mentality"). The reflective-instinctive kind of the "unconscious" behavior has reached its culmination in the case of *Insecta*. The remarkable advancement of primordial consciousness towards consciousness *per se* amongst the invertebrates occurred in the case of *Cephalopoda* (e.g. octopus, squids etc.): these animals, the only representatives amongst the invertebrates, demonstrate memory and the capacity to learn the symbolic significance of events (Young, 1965, 1977). This correlates with a rather large compact brain (as opposed to separate ganglia of other *Mollusca*), highly developed eyes with lens accommodation and the other sense organs, some of them being unique (Young, 1973, 1977). But the most amazing is their brain cytoarchitectonics demonstrating layered structure with geometrical orientation of the neuron axes in some brain lobes resembling the mammalian cortex while the texture of parallel fibers is strikingly like those of the mammalian cerebellum (Young, 1977). The development of the compact brain in the case of such so taxonomically remote types as *Vertebrata* and *Mollusca* is a remarkable example of the evolutionary convergence.

15. CODA

The definitions of protoconsciousness, primordial consciousness, and consciousness *per se* have been formulated in consistent succession, being derived from the suggested definition of life based on the morphic principle. The crucial point concerns the concept of the "geometrical feeling" as a non-reducible axiomatic fundamental on the basis of which the concept of protoconsciousness was uttered as an **inalienable** attribute of any living system. Accordingly, consciousness in general, being an immanent property of Life, can be realized in a wide spectrum of its expressions, from the **"geometrical feeling" - "morphological response"** as an **elementary psychic act** (**"morphological mentality"**) – through the **instinctive/reflective behavior** – up to the highly developed manifestations of the mature human individual's psychic phenomenology that is **consciousness** *per se*. Such a path of consciousness development was supported by ontogenic and phylogenic analysis.

The whole analysis could be done, in principle, on the level of the pure abstract morphic principle without considering its "nature". However, the use of the Gurwitschian field principle accounted in its holistic interpretation (i.e. its irreducibility to the known physical fields) has added to the abstract morphic principle a more specific expression that permitted a more concise and elegant description of the problem, as well as more evident relationships of the psychic phenomenology with other life manifestations.

16. EPILOGUE

The general impression from the above analysis may seem disappointing. Namely, the puzzling constellation of the most basic problems of the mind-matter relationships, namely: (a) the psycho-somatic abyss and explanatory gap; (b) the hard problem of consciousness, Gödel's theorem and the necessity for the extra ingredient; (c) the fatal choice between the generation problem and panpsychism; (d) the appeal for a new paradigm based on counter-intuitive thinking – all that was reduced to the obsolete dusty vitalistic philosophy. Such a result would not be considered as achievement.

However, if to overcome emotional dislike towards the concept of vitalism, a dryly rational account shows that the naked kernel of the vitalistic idea is the evident analytic and empirical conclusion that the known physical fundamentals are not sufficient to describe the life phenomenology, so that an additional axiomatic fundamental(s) is needed. This is quite an "innocent" statement, which is no more mystical than the widely discussed concept of the extra ingredient (Chalmers, 1995). The simplest - just formal - truth is that if at all the concept of the extra ingredient is accepted as valid, its further association with anything from reality is the necessary condition to make this concept "work". Then, any attempt to search for the extra ingredient within the Life realm means in fact recognition of the vitalistic principle. Hence, formulation of the hard problem with an appeal for the extra ingredient and at the same time rejection of the vitalistic principle as obsolete (Chalmers, 1995) is a conceptual mistake.

However, an abstract vitalistic principle (*vis vitalis*) is a toothless declaration unless it has a non-tautological expression, that having been the main task of the whole consideration. Such non-tautological formulation has been attempted and analyzed here being based on the **morphic** principle, which is accounted as a genuinely intrinsic feature of the living systems non-reducible to the physical fundamentals. The latter conclusion can be disputed: it is possible to claim that the morphological shape can be inferred finally from the canonical properties of the whole totality of all the molecules involved into the system. However, such a case can be in principle realizable only on the basis of isomorphic identity (Horgan, 1982; Kim, 1984, 1987) between the living state in its macro-expression and the respective processes occurring on the molecular level, while the absence of the isomorphic identity is evident at all the levels of biological organization (Lipkind, 1998c). In particular, a number of possible states on the molecular level would correspond to the same living macro-phenomenon observed at a higher level (cellular and morphological). The important point is that such diversity is not a result of statistical distribution, which fluctuates around a principal rigid pivot-like regularity but is a result of the top-down causation typical for the living systems (Lipkind, 1999, 2000).

The morphic principle has been more concretized (and at the same time generalized) by Gurwitsch's theory of biological field that permitted to express the morphic principle not only on morphological and cytological levels but also on the molecular one. This means, essentially, the reductionist operation within the holistic conceptual framework (field). The main achievement of such reductionist operation is the postulate about the "molecular substrate" of the field action that was expressed by the field-directed **vectorization** of the molecules' movements (or orientation) *in vivo* as opposed to their chaotic movements *in vitro*.

Thus, Protoconsciousness can be imagined as current awareness by a living cell of the gap, that being expressed as non-congruence, divergence, incompatibility, collision, conflict – between the ideal geometrical form and its physical realization. This means that such discrepancy between the **Ideal Geometry** and **Robust Physics is "felt"** by the cell. The ideal geometry is species-specific, initially pre-existing, and pre-determined, while the robust physics is actually occurring and constantly fluctuating to adapt, to adjust, to fit, to accommodate, to approximate to the ideal geometry. Accordingly, the living process can be expressed as continuous dynamic approximation of the "real" physical form to its geometrical "ideal". Since this approximation is felt until there is the non-congruence between physical and geometrical (which can diminish only asymptotically, i.e. the physical will never coincide with geometrical), then the geometrical feeling (protoconsciousness) is an inalienable part of any living entity. Hence, the Geometrical Feeling is suggested for the role of the Protophenomenal Fundamental alongside physical fundamentals (Mass, Charge, Time/Space).

As to the deep ontological meaning of the concept of "Geometrical Feeling", it could be analogized with the "Universal Grammar" by N. Chomsky (1988), which may be considered as "intrinsic part of the structure of matter ever since the Big Bang, or a necessary part of the eternal Platonic world of logic and mathematics" (Harnad, 2001).

17. REFERENCES

Allport, A. (1988) What concept of consciousness? In *Consciousness in Contemporary Science*, A. Marcel and Bisiach, E. (eds.), Oxford University Press, Oxford, pp. 161-182.

Angell, J.R. (1904) *Psychology: an Introductory Study of the Structure and Functions of Human Consciousness*. Holt, New York.

Antoniou, I. (1992) Is consciousness decidable? In *Science and Consciousness*, Proceedings of the 2nd International Symposium, January 3-7, 1992, Athens, pp. 69-91.

Babloyantz, A. (1986) *Molecules, Dynamics, and Life*. John Wiley & Sons, New York.

Baars, B.J. (1988) *A Cognitive Theory of Consciousness*. Cambridge University Press, Cambridge.

Beeson, M. (1985) *Foundations of Constructive Mathematics*. Springer, E. Bishop, Berlin.

Beloussov, L.V. (1994) Alexander Gavrilovitsch Gurwitsch (1874-1954). R*ivista di Biologia – Biology Forum,* **87**, 119-126.

Bertalanffy, L. von (1933) *Modern Theories of Development: An Introduction to Theoretical Biology.* Oxford University Press, Oxford.

Bieri, P. (1995) Why is consciousness puzzling? In *Conscious Experience,* Metzinger, T. (ed), Schoening, Imprint Academic, Paderborn, pp. 45-60.

Bohr, N. (1958) *Atomic Physics and Human Knowledge.* Wiley & Sons, New York.

Burns, J. (1990) Contemporary models of consciousness: Part I. *The Journal of Mind and Behavior,* **11,** 153-172.

Burns, J. (1991) Contemporary models of consciousness: Part II. *The Journal of Mind and Behavior,* **12,** 407-420.

Cairns-Smith, A.G. (1982) *Genetic Takeover and the Mineral Origin of Life.* Cambridge University Press, Cambridge.

Caplan, S.R. and Essig, A. (1983) *Bioenergetics and Linear Nonequilibrium Thermodynamics.* Harvard University Press, Cambridge, Massachusetts.

Chalmers, D.J. (1995) Facing up to the problem of consciousness. *Journal of Consciousness Studies,* **2**, 200-219.

Chalmers, D.J. (1996*) The Conscious Mind: In Search of a Fundamental Theory.* Oxford University Press, New York - Oxford.

Chomsky, N. (1988) *Language and Problem of Knowledge.* MIT Press, Cambridge, Massachu-setts.

Clark, A. (1992) *Sensory Qualities.* Oxford University Press, Oxford.

Clifford W.K. (1874) In *Lectures and Essays,* Stephen, L. and Pollock (eds), Macmillan, 1879, London, pp. 31-70.

Crick, F.H.C. (1994a) *The Astonishing Hypothesis - The Scientific Search for the Soul.* Simon and Schuster, London.

Crick, F.H.C. (1994b) Interview with J. Clark at the launch of the new book 'The Astonishing Hypothesis'. *Journal of Consciousness Studies,* **1**, 10-17.

Crick, F.H.C. & Koch, C. (1990) Toward a neurobiological theory of consciousness. *Seminars in the Neurosciences,* **2**, 263-75.

Damasio, A.R. (1989) The brain binds entities end events by multiregional activation from con-vergence zones. *Neural Computation,* **1**, 123-132.

Dawkins, R. (1976) *The Selfish Gene.* Oxford University Press, Oxford; New edition (1989), Oxford University Press, Oxford.

Deikman, A.J. (1996) 'I' = Awareness. *Journal of Consciousness Studies,* **3**, 350-356.

Dennett, D.C. (1991) *Consciousness Explained.* Little-Brown & Co.; 2nd edition (1993), Penguin Books, Harmondsworth.

Dennett, D.C. (1995) *Darwin's Dangerous Idea.* Simon and Schuster, New York.

Dennett, D.C. (1996) Facing backwards on the problem of consciousness. *Journal of Consciousness Studies,* **3**, 4-6.

Driesch, H. (1891) Entwicklungsmechanische Studien. 1. Der Werth der beiden ersten Furchungszellen in der Echinodermenentwicklung Experimentelle Erzeugung von Theil- und Doppelbildungen. *Zeitschrift für Zoology,* **53**, 160-178.

Driesch, H. (1908) *Science and Philosophy of the Organism.* Adam & Charles Black, London.

Driesch, H. (1915) *Vitalism: Its History and System.* Russian Edition, A.G.Gurwitsch (ed), the authorized translation from German into Russian by A.G. Gurwitsch (In Russion).

Du Bois-Reymond, E. (1872) *Über die Grenzen des Naturerkennens.* Quoted from "Vorträge über Philosophie und Gesellschaft" (1974), Meiner, Hamburg.

Eccles, J. C. (1986) Do mental events cause neural events analogously to the probability fields of quantum mechanics? *Proceedings of the Royal Society. London B,* **227**, 411-428.

Eccles, J. C. (1990) A unitary hypothesis of mind-brain interaction in the cerebral cortex. *Proceedings of the Royal Society, London B,* **240**, 433-451.

Eccles, J. C. (1994) *How the SELF Controls its BRAIN.* Springer-Verlag, Berlin & Heidelberg.

Edelman, G. (1989) *The Remembered Present: A Biological Theory of Consciousness.* Basic Books, New York.

Edelman, G. M. (1993) Morphology and mind: Is it possible to construct a perception machine? *Frontier Perspectives,* **3**, 7-12.

Eigen, M. (1992) *Steps Towards Life - A Perspective on Evolution,* Oxford University Press, Oxford.

Eigen, M. (1993) The origin of genetic information: viruses as models. *Gene,* **135**, 37-47.

Elitsur, A.C. (1994) Let there be life: Thermodynamic reflections on biogenesis and evolution. *Journal of Theoretical Biology,* **164**, 429-459.

Flohr, H. (1992) Qualia and brain processes. In *Emergence or Reduction?: Prospects for Nonreductive Physicalism.* Beckermann, A. Flohr, H. and & Kim, J. (eds), De Gruyter, Berlin.

Freud, S. (1964) *New Introductory Lectures in Psychoanalysis,* Starchey, J. and Freud, A. (eds.), W. W. Norton, New York.

Gödel, K. (1931) Über formal unentscheidbare Sätze der Principia Mathematica und verwandter System. *Monatshefte für Mathematik und Physik,* **38**, 173-198.

Goodwin, B.C. (1986) Is biology a historical science? In *Science and Beyond,* Rose, S. and Appignanese, L. (eds.), Blackwell, Oxford, pp. 47-60.

Gurwitsch, A.G. (1912) Die Vererbung als Verwirklichungsvorgang. *Biologische Zentralblatt,* **32**, 458-486.

Gurwitsch, A.G. (1930) *Die histologischen Grundlagen der Biologie.* G. Fischer Verlag, Jena.

Gurwitsch, A.G. (1937) Degradational radiation from central nerval system. *Arkhiv Biologiche-skikh Nauk (Archives of Biological Sciences)* **45**, 53-57 (In Russian).

Gurwitsch, A.G. (1944) *The Theory of the Biological Field.* "Sovetskaya Nauka" Publishing House, Moscow (In Russian).

Gurwitsch, A.G. (1947) Une theorie du champ biologique cellulaire. *Bibliotheca Biotheoretica,* (Leiden), ser. D, v. **11**, pp. 1-149 (in French).

Gurwitsch, A.G. (1954) *Principles of Analytical Biology and the Theory of Cellular Fields.* Manuscript, Moscow (In Russian).

Gurwitsch, A.G. (1991) *Principles of Analytical Biology and the Theory of Cellular Fields.*
Nauka Publishers, Moscow (In Russian).

Gurwitsch, A.G. and Gurwitsch, L.D. (1948) *Introduction into the doctrine of mitogenetic radi-ation.* USSR Academy of Medical Sciences Publishing House, Moscow (In Russian).

Gurwitsch, A.G. and Gurwitsch, L.D. (1959) *Die mitogenetische Strahlung.* G. Fischer Verlag, Jena.

Haeckel, E. (1892) *Monism.* A. Block, (ed), English Translation published in 1894, London.

Haeckel, E. (1899) *The Riddle of the Universe.* Watts (ed), English Translation published in 1929, London.

Haken, N. (1977) S*ynergetics: An Introduction.* Springer, Heidelberg.

Hameroff, S. (1994) Quantum coherence in microtubules: A neural basis for an emergent consciousness? *Journal of Consciousness Studies* **1**, 91-118.

Hameroff, S., Kaszniak, A. & Scott, A. (eds), (1996) *Toward a Science of Consciousness.* MIT Press, Cambridge, MA, pp 1-751.

Hameroff, S., Kaszniak, A. & Scott, A. (eds), (1998) *Toward a Science of Consciousness II.* MIT Press, Cambridge, MA, pp. 1-738.

Hameroff, S., Kaszniak, A. & Chalmers,D. (eds), (1999) *Toward a Science of Consciousness III.* MIT Press, Cambridge, MA, pp. 1-504.

Haraway, D.J. (1976) *Cristals, Fabrics, and Fields: Metaphors of Organicism in Twentieth-Century Developmental Biology.* Yale University Press, New Haven.

Hardcastle, V.G. (1994) Psychology's binding problem and possible neurobiological solutions. *Journal of Consciousness Studies,* **1**, 66-90.

Hardcastle, V.G. (1996) The why of consciousness: a non-issue for materialists. *Journal of Consciousness Studies,* **3**, 7-13.

Hardin, C.L. (1992) Physiology, phenomenology, and Spinoza's true colors. In *Emergence or Reduction?: Prospects for Nonreductive Physicalism,* Beckermann, A., Flohr, H. and Kim, J. (eds.), De Gruyter, Berlin.

Harnad, S. (2001) No easy way out. *The Sciences,* **41**, 36-42.

Horgan, T. (1982) Supervenience and microphysics. *Pacific Philosophical Quaterly,* **63**, 29-43.

Hubel, D.H. and Livingstone, M.S. (1987) Segregation of form, color, and stereopsis in primate area 18. *The Journal of Neuroscience,* **7**, 3378-3415.

Hume, D. (1735/1958) *A Treatise on Human Nature.* Oxford University Press, Oxford - New
York.

Hume, D. (1777/1975) *Enquiries Concerning Human Understanding and Concerning the Prin-ciples of Morals.* Oxford University Press, Oxford - New York.

Humphrey, N. (1992) *A History of the Mind.* Simon & Schuster, New York.

Huxley, T.H. (1866) *Lessons in Elementary Physiology.* Quoted by N. Humphrey (1992).

Jackendoff, R. (1987) *Consciousness and the Computational Mind.* MIT Press, Cambridge, Massacusetts.

James, W. (1890) *The Principles of Psychology.* Vol. 1, Henry Holt & Co., New York; reprinted by Dover Books, 1950.

Kauffman, S. (1996*) At Home in the Universe: the Search for Laws of Self-Organization and Complexity.* Viking, New York.

Kim, J. (1984) Concepts of supervenience. *Philosophy and Phenomenological Research,* **65**, 153-176.

Kim, J. (1987) 'Strong' and 'global' supervenience revisited. *Philosophy and Phenomenological Research,* **68**, 315-326.

Laszlo, E. (1993) *The Creative Cosmos: A Unified Science of Matter, Life and Mind.* Floris Books, Edinburgh.

Levine, J. (1983) Materialism and qualia: the explanatory gap. *Pacific Philosophical Quaterly,* **64**, 354-361.

Lifson, S. (1987) Chemical selection, diversity, teleonomy and the second law of thermodynamics: Reflections of Eigen's theory of self-organization of matter. *Biophysical Chemistry*, **26**, 303-311.

Lipkind, M. (1987a) A.G. Gurwitschs Theorie vom biologischen Feld. I. Voraussetzungen, Ursprung, logische Structur und konceptionelle Entwicklung. *Fusion* (German edition), **8**, 30-49 (In German).

Lipkind, M. (1987b) A.G. Gurwitschs Theorie vom biologischen Feld. II. Anwendung der Theorie des vektoriellen biologischen Feldes zur Analyse der verschiedener Phaenomene lebender Systeme. *Fusion* (German edition), **8**, 53-65 (In German).

Lipkind, M. (1992) Can the vitalistic entelechia principle be a working instrument? (The theory of the biological field of Alexander G. Gurwitsch). In *Recent Advances in Biophoton Research and its Applications*, Popp, F.-A., Li, K. H. and Gu, Q. (eds), World Scientific, Singapore – New Jersey - London - Hong Kong, pp. 469-494.

Lipkind, M. (1996) Application of the theory of biological field by A. Gurwitsch to the problem of consciousness. In *Current Development of Biophysics*, Zhang, C., Popp, F.-A. and Bischof, M (eds.), Hangzhou University Press, Hangzhou, pp. 223-251.

Lipkind, M. (1998a). Alexander Gurwitsch and the concept of the biological field. I. Major prerequisites, origin, logical structure and conceptual development. *21st Century Science & Technology*, **11/2**, 36-51.

Lipkind, M. (1998b) Alexander Gurwitsch and the concept of the biological field. II. Analysis of different phenomena in living systems by means of the theory of the vectorial biological field. *21st Century Science & Technology*, **11/3**, 34-53.

Lipkind, M. (1998c) The concepts of coherence and 'binding problem' as applied to life and consciousness realms (Critical consideration with positive alternative). In *Biophotons*, Chang, J.J. and Popp, F.-A. (eds), Kluwer Academic Publishers, Dordrecht, pp. 359-373.

Lipkind, M. (1999) The holistic quality in biology: ontology, epistemology, and causation. *Coherence – International Journal of Integrated Medicine*, No. 2, 4-18.

Lipkind, M. (2000) The holistic quality in biology in view of the Gurwitschian field principle: ontology, epistemology, and causation. In *Biophotonics and Coherent Systems*, Beloussov, L., Popp, F.-A., Voeikov, V. and Van Wijk, R. (eds.), Moscow University Press, Moscow, pp. 27-41.

Lipkind, M. and Beloussov, L. (1995) Alexander Gurwitsch: personality and scientific heritage. In *Biophotonics: Non-Equilibrium and Coherent Systems in Biology, Biophysics, and Biotech-nology*, Beloussov, L. and Popp, F.-A. (eds), Bioinform Services Co, Moscow, pp. 15-32.

Livingstone, M.S. and Hubel, D.H. (1987) Psychophysical evidence for separate channels for the perception of form, color, movement, and depth. *The Journal of Neuroscience*, **7**, 3416-3468.

Loeb, J. (1906) *The Dynamics of Living Matter*. Columbia University Press, New York.

Loeb, J. (1912) *The Mechanistic Conception of Life.* University Chicago Press, Chicago.

Lotka, A. (1925) *Elements of Physical Biology.* Williams & Wilkins, Baltimore; Republished in 1956 as *Elements of Mathematical Biology,* Dover, New York.

McDougall, W. (1938) *The Riddle of Life: A Survey of Theories.* Methuen, London.

McGinn, C. (1989) Can we solve the mind-body problem? *Mind,* **98,** 349-366.

McGinn, C. (1995) Consciousness and space. *Journal of Consciousness Studies,* **2,** 220-230.

Metzinger, T. (1995) The problem of consciousness. In *Conscious Experience,* Metzinger, T. (ed.), Schoening, Imprint Academic, Paderborn, pp. 3-43.

Monod, J. (1971) *Chance and Necessity: An Essay on the Natural Philosophy of Modern Biology.* Knopf, New York.

Nagel, E. and Newman, J. (1958) *Gödel's Proof.* Routledge & Kegan Paul, London – Boston Henley.

Penrose, R. (1994) *Shadows of the Mind.* Oxford University Press, Oxford.

Penrose, R. (1989) *The Emperor''s New Mind.* Oxford University Press, Oxford.

Popper, K. and Eccles, J.C. (1977) *The Self and Its Brain.* Springer International, Berlin.

Prigogine, I. and Stengers, I. (1984) *Order out of Chaos.* Bantam, New York.

Ramachandran, V.S. (1990) Visual perception in people and machines. In *AI and the Eye,* Balke, A. and Trosciankopp, T. (eds), John Wiley and Sons Ltd, New York, pp. 21-77.

Ramachandran, V.S. and Anstis, S.M. (1986) The perception of apparent motion. *Scientific American,* **254,** 101-109.

Rosenberg, G.H. (1996) Rethinking nature: hard problem within the hard problem. *Journal of Consciousness Studies,* **3,** 76-88.

Searle, J. (1992) *The Rediscovery of the Mind.* MIT Press, Cambridge, Massachusetts.

Seager, W. (1995) Consciousness, information and panpsychism. *Journal of Consciousness Studies,* **2,** 272-288.

Sheldrake, R. (1986), Morphogenic field. In *Dialogues with Scientists and Sages (the Search for Unity).* Weber, R. (ed.), Penguin Group, Arkana, London; first published by Routledge and Kegan Paul, 1986, pp. 71-88.

Shear, J. (1995) Explaining consciousness – the 'Hard Problem', Editor's Introduction. *Journal of Consciousness Studies,* **2,** 194-199.

Sperry, R.W. (1980) Mind-brain interaction: Mentalism, yes, dualism, no. *Neuroscience,* **5,** 195-206.

Szent-Gyorgi, A. (1957) *Bioenergetics*. Academic Press, New York.

Tyndall, J. (1879) *Fragments of Science: A Series of Detached Essays, Addresses and Reviews*. Longmans, London.

Toward a Science of Consciousness (2000), Conference Research Abstracts, (a service from the Journal of Consciousness Studies), Tucson, April 10-15, 2000, pp. 1-175.

Toward a Science of Consciousness (2002), Conference Research Abstracts, (a service from the Journal of Consciousness Studies), Tucson, April 8-12, 2002, pp. 1-162.

Toward a Science of Consciousness (Consciousness and Its Place in Nature) (2001), Conference Research Abstracts (a service provided in cooperation with the Journal of Consciousness Studies), Skïlvde, Sweden, August 7-11, 2001, pp. 1-95.

Treisman, A. (1986) Features and objects in visual processing. *Scientific American*, **254**, 114-125.

Van Gulick, R. (1995) What would count as explaining consciousness? In *Conscious Experience*, Metzinger, T. (ed.), Schoening, Imprint Academic, Paderborn, pp. 61-79.

Welch, G.R. (1992) An analogical "field" construct in cellular biophysics: history and present status. *Progress in Biophysics and Molecular Biology*, **57**, 71-128.

Velmans, M. (1995) The relation of consciousness to the material world. *Journal of Consciousness Studies*, **2**, 255-265.

Von der Malsburg, C. (1987) Synaptic plasticity as basis of brain organization. In *The Neural and Molecular Bases of Learning*, Changeux, J.-P. and Konishi, M. (eds.), John Wiley and Sons, New York, pp. 411-432.

Waechterhauser, G. (1988) Before enzymes and templates: theory of surface metabolism. *Microbiological Reviews*, **52**, 452-484.

Weiss, P. (1939) *Principles of Development*. Holt, Rinehart, & Winston, New York.

Wilkes, K.V. (1988) Yishi, Duh, Um and consciousness. In *Consciousness in Contemporary Science*, Marcel, A. and Bisiach, E. (eds.), Oxford University Press, Oxford, pp. 17-41.

Young, J.Z. (1965) The organization of a memory system. *Proceedings of the Royal Society B*, **163**, 285-320.

Young, J.Z. (1973) Memory as a selective process. *Australian Academy of Science Report: Symposium on Biological Memory*, pp. 25-45.

Young, J.Z. (1977) What squids and octopuses tell us about brains and memories. *46th James Arthur Lecture on the Evolution of the Human Brain*, The American Museum of Natural History, New York.

Zeki, S. (1992) The visual image in mind and brain. *Scientific American*, **254**, 68-77.

Lightning Source UK Ltd.
Milton Keynes UK
25 October 2010

161857UK00007B/2/A